Fundamentals of Design of Experiments for Automotive Engineering

Volume I

Fundamentals of Design of Experiments for Automotive Engineering

Volume I

YOUNG J. CHIANG, PHD

Warrendale, Pennsylvania, USA

400 Commonwealth Drive
Warrendale, PA 15096-0001 USA
E-mail: CustomerService@sae.org
Phone: 877-606-7323 (inside USA and Canada)
724-776-4970 (outside USA)
Fax: 724-776-0790

Library of Congress Catalog Number 2023948382
http://dx.doi.org/10.4271/9781468606034

ISBN-Print 978-1-4686-0602-7
ISBN-PDF   978-1-4686-0603-4
ISBN-epub   978-1-4686-0604-1

To purchase bulk quantities, please contact: SAE Customer Service

E-mail: CustomerService@sae.org
Phone: 877-606-7323 (inside USA and Canada)
724-776-4970 (outside USA)
Fax: 724-776-0790

Visit the SAE International Bookstore at books.sae.org

Publisher
Sherry Dickinson Nigam

Product Manager
Amanda Zeidan

Production and Manufacturing Associate
Michelle Silberman

From the Publisher

I am thrilled to introduce our groundbreaking series *Design of Experiments for Product Reliability Growth,* which is divided into four distinct volumes. The first installment, *Fundamentals of Design of Experiments for Automotive Engineering,* marks the beginning of our exploration of this critical subject.

We are publishing each volume separately to ensure in-depth coverage of each topic, allowing readers to focus on specific aspects of the subject matter for a deeper understanding. After careful consideration of how our readers will be using the content, we opted to continue to the page numbering system throughout the volumes for continuity.

Volume II   Product Design and Testing for Automotive Engineering
Volume III   Manufacturing Reliability Growth for Automotive Engineering
Volume IV   Business Reliability Growth for Automotive Engineering

Thank you for joining us on this journey. We look forward to your continued support and engagement with the upcoming volumes in the series.

Sincerely,

Sherry Nigam
Publisher, SAE Books

Contents

CHAPTER 2

Full Factorial Design 2^K 27

CHAPTER 3

Fractional Factorial Design 2_R^{K-P} 75

CHAPTER 5

Composite Designs   175

Preface

This is a book about product design for reliability and sustainability via the design of experiments. It is an application-oriented text with an emphasis on deploying the design of experiments for product reliability growth throughout an enterprise as a "statistical thinking" tool, instead of focusing on a specific statistical approach to designing experiments. Emphasis is laid on how to grow product reliability using the design of experiments. Product reliability growth via process improvements and system optimizations is demonstrated using real-world examples, mostly extracted from common practices in automotive engineering. Theoretical concepts that are important to the understanding of the design of experiments are rephrased in practical exercises.

Product development efforts are concurrently focused on a product configuration at the lowest feasible cost by means of a thorough understanding of its functionality, reliability and sustainability, while today's central theme for proactive improvement calls for building high reliability into products. It is imperative to avoid field failures during a product's estimated lifetime in service, which consequently leads to delighted customers and elimination of the high expense of repairing failed goods and fixing underlying causes. Henry Ford said, "Failure is simply the opportunity to begin again, this time more intelligently." Product reliability growth is in a perpetual motion, but intellectual vitality rolls with us in light of the design of experiments.

Much of the challenge and opportunity in engineering design comes from stochastic material properties and manufacturing processes in nature. "A large safety factor does not necessarily translate into a reliable product. Instead, it often leads to an overdesigned product with reliability problems," as addressed in the article "Failure Analysis Beats Murphy's Law," *Mechanical Engineering* (September 1993). On problem solving, a thinker is encouraged to poise oneself on the understanding of physics of failure that is related to exploiting distinguished product properties and characterizing specific identifiable variables and attributes. Drawing on their extensive product development experience, the authors present a comprehensive process for ensuring product performance and reliability over the entire lifecycle.

We hope this will be a comprehensive textbook of applied design of experiments for product reliability growth and a reference manual for practicing engineers in various functional areas of product-oriented business, including research, design, testing, manufacturing, operation, supplier quality assurance, marketing, and warranty analysis. While lessons hereby learned from the book can be appropriated for product development in the real world, comments and suggestions are motivated for further improvements.

Young J. Chiang, PhD
会而必议 议而必决 决而必行 行而必果

Acronyms

AAI - Appearance Approval Inspection

AALA - American Association for Laboratory Accreditation

AAR - Appearance Approval Report

ADV - Analysis/Development/Validation (GM Specific)

AIAG - Automotive Industry Action Group

AIQ - Average Incoming Quality; Average quality level before inspection

ALT - Accelerated Life Testing

ANCOVA - Analysis of Covariance

ANN - Artificial Neural Network

ANOM - Analysis of Means

ANOVA - Analysis of Variance

AOQ - Average Outgoing Quality; Average quality level leaving the inspection point after rejection and acceptance of a number of lots

AOQL - Average Outgoing Quality Limit-Maximum value of the AOQ

AOZ - Accept on Zero

AP - Advanced Product

APQP - Advanced Product Quality Planning

AQC - Attribute Quality Characteristic

AQL - Acceptable Quality Level

AQP - Advanced Quality Planning

ARIMA - Autoregressive Integrated Moving Average

ARL - Averaged Run Length

ARMA - Autoregressive Moving Average

ASIL - Automotive Safety Integrity Level

ASME - American Society of Mechanical Engineers

ASQ - American Society of Quality

ASTM - American Society for Testing and Materials

AVL - Approved Vendor List

A2LA - American Association of Laboratory Accreditation

BCWP - Budgeted Cost of Work Performed

BCWS - Budgeted Cost of Work Scheduled

BDD - Binary Decision Diagram

BEV - Battery Electric Vehicles

BIT (BIST) - Built-in Test (Built-in Self-Test), as part of Testability Analysis

BJT - Bipolar Junction Transistors

BLDC - Brushless Direct Current

BOM - Bill of Material

BPI - Business Performance Improvement

B7 (Q7) - Basic 7 QC Tools, i.e. Fishbone Diagram, Histogram, Pareto Analysis, Flowchart, Scatter Plot, Run Chart, and Control Chart

CA - Corrective Actions

CAAM - China Association of Automobile Manufacturers

CAD - Computer-aided Design

CAE - Computer-aided Engineering

CAM - Computer-aided Manufacturing

CAPA - Corrective and Preventive Action

CBP - Customer Benefits Package; Tangibles & Intangibles Make Up a Service

CC - Critical Characteristic

CCD - Central Composite Design

CCPM - Critical Chain Project Management

c-Chart - Count Chart-Attribute Quality Control Chart for Number of Defects

CDF - Cumulative Distribution Function

CF - Carbon Fiber

CFT - Cross Functional Team

CI - Confidence Interval

CIP - Continuous Improvement Process

CLT - Central Limit Theorem

CMM - Coordinate Measuring Machine

CMM - Capability Maturity Model

CONC - Cost of Non-conformance

COP - Customer-Oriented Process

COQ - Cost of Quality

COR - Cost of Reliability

CPM - Critical Path Method, either PERT or Gantt Chart

CPP - Critical Process Parameter

CPPD - Collaborative Product Process Design

CPV - Cost per Vehicle

CQI - Continuous Quality Improvement

CQM - ASQ Certified Quality Manager

CRD - Completely Random Design

CRM - Customer Relationship Management

CRS - Constant Returns to Scale

CS1 - Control Shipping 1

CS2 - Control Shipping 2

CTQ - Critical-to-Quality

CUSUM - Cumulative Sum Charting

CV - Coefficient of Variation

DCO - Document Change Order

DCP - Dimensional Control Plan

DCP - Dynamic Control Plan (Ford)

DDDM - Degradation-Data-Driven Method

DEA - Data Envelopment Analysis

DETMAX - Determinant Maximizing Algorithm, i.e. D-Optimal

DFA - Design for Assembly

DFM - Design for Manufacture

DFMA - Design for Manufacture and Assembly

DMAIC - Define, Measure, Analyze, Improve, and Control (Kaizen- 5 Steps)

DOC - Depth of Charge

DOE - Design of Experiments

DOE-E - Design of Experiments Based on Exponential Distribution

DOE-F - Design of Experiments Based on F Distribution

DOE-ln - Design of Experiments Based on Lognormal Distribution

DOE-t - Design of Experiments Based on t-Distribution

DOE-W - Design of Experiments Based on Weibull Distribution

DOE+ML - Design of Experiments and Machine Learning, Joint Application of

DRACAS - Data Reporting Analysis and Corrective Action System

DF or DOF - Degree of Freedom

DFMEA - Design Failure Modes and Effect Analysis

DMADV - Define, Measure, Analyze, Design, Verification

DMU - Design Making Unit

DRACAS - Data Reporting Analysis and Corrective Action System

DRF - Datum Reference Frame

DSC - Dynamic Stability Control

DSD - Definitive Screening Designs

DVP&R - Design Verification Plan and Report

ECFA - Event and Causal Factor Analysis

ECN - Engineering Change Notice

ECR - Engineering Change Request

EMI - Electromagnetic Interference

EMOO - Evolutionary Multi-Objective Optimization

EMS - Expected Means Squares

ERP - Enterprise Resource Planning

ES - Engineering Specification

ES - Earned Schedule

ESD - Electrochemical Storage Devices

ESS - Environmental Stress Screening

EV - Electric Vehicle

EVM - Earned Value Management

EVOP - Evolutionary Operation Process

EWMA - Exponentially Weighted Moving Average

EWMV - Exponentially Weighted Moving Variance

EWS - Early Warning System

Exp, exp, e - Exponential Function

FAI - First Article Inspection or Final Acceptance Inspection

FAST - Function Analysis System Technique Diagram

FDI - Fault Detection and Isolation

FDM - Fused Deposition Modeling

FEA - Finite Element Analysis

FEM - Finite Element Methods

FFA - Functional Failure Analysis

FFF - Fused Filament Fabrication

FHA - Functional Hazard Analysis

FIFO - First in First Out

FIM - Fisher Information Matrix

FIR - Fast Initial Response

FMEA - Failure Modes and Effect Analysis

FMECA - Failure Modes, Effect, and Criticality Analysis

FMEDA - Failure Modes, Effects, and Diagnostics Analysis

FMVT - Failure Mode Verification Test

FOZ - Feature-of-Size

FPGA - Field Programmable Gate Array

FQPR - Field Quality Problem Report

FRACAS - Failure Reporting, Analysis, and Corrective Action System

FRB - Failure Review Board

FSLT - Full System Life Test

FSR - Functional Safety Requirement

FTA - Fault Tree Analysis

FTC - First Time Capability

DOE-F - F Distribution-Based Design of Experiments

GA - Genetic Algorithm

Gage R&R - Gage Repeatability and Reproducibility

GaN - Gallium Nitride

GD&T - Geometric Dimensioning and Tolerancing

GD^3 - Good Design, Good Discussion, Good Dissection

GERT - Graphic Evaluation and Review Technique

GF - Glass Fiber

GK - General Knowledge

GPa - Gega Pascal

GPC - Gage Performance Curve

GR - Gage Repeatability

GR&R; GRR - Gage Repeatability and Reproducibility

GTD - Getting Things Done

GTO - Gate Turn-off Thyristor

GVDP - Global Vehicle Development Process, a GM Practice

HACCP - Hazard Analysis and Critical Control Points

HARA - Hazard Analysis and Risk Assessment

HALT - Highly Accelerated Life Test

HASS - Highly Accelerated Stress Screening

HAZOP - Hazard and Operability Study

HCF - High Cycle Fatigue

HEV - Hybrid Electric Vehicle

HFCV - Hydrogen Fuel Cell Vehicle

H$_o$ - Null Hypothesis

H$_1$ - Alternative Hypothesis

IATF - International Automotive Task Force

IC - Integrated Circuit

IDOV - Identify, Design, Optimize, Validate

IEEE - Institute of Electrical and Electronics Engineers

IGBT - Insulated-Gate Bipolar Transistor

IGCT - Integrated Gate Commutated Thyristor

iid - Independent and Identically Distributed

IM - Induction Motor

IMDS - International Material Data System

IMP - Integrated Master Program

IPMA - International Project Management Association

ISIR - Initial Sample Inspection Report, e.g., PPAP Warrant and Inspection Report

ISO - International Organization for Standardization

IsoPlot - 3-D Iso-surface Plot

ITPV - Incidents per Thousand Vehicles

JIT - Just in Time

KAIZEN - Change for the Better

KBF - Key Business Factors

KCC - Key Control Characteristic

KISS - Keep it Simple, Stupid

KLT - Key Life Tests

KPC - Key Product Characteristic

KPI - Key Product (Process) Indicator or Key Performance Index

LCF - Low Cycle Fatigue

LCL - Lower Control Limit

Lidars - Laser Light Detection and Ranging Systems

Li-air - Lithium-Air (Battery)

Li-ion - Lithium-Ion (Battery)

LMC - Least Material Condition

LN, Ln, or ln - Lognormal

LRP - Logistics Requirement Plan

LRU - Line Replaceable Unit

LSL - Lower Specified Limit

LSM - Linear Scheduling Method

LTBF - Longest Time between Failures

LTPD - Lot Tolerance Percentage Defective Acceptable in Production Lots

M&TE - Measurement and Test Equipment

MBD - Management by Decree

MCO - Manufacturing Change Order

MEOST - Multiple Environment Over Stress Tests

MILTFP-41 - Make it Like the Finished Print- For Once

ML - Machine Learning

MLE - Maximum Likelihood Estimator

MMC - Maximum Material Condition

MME - Method of Moments Estimator

MRR - Median Rank Regression

MOPSO - Multi-Objective Particle Swarm Optimization

MOSFET - Metal-Oxide-Semiconductor Field-Effect Transistor

MORT - Management Oversight and Risk Tree

MPa - Mega Pascal

MRS - Marginal Rate of Substitution

MSE - Mean Squared Error

MTS - Make to Stock

MTTF - Mean Time to Failure

MTBF - Mean Time between Failures

MTTR - Mean Time to Repair

MVBs - Manufacturing Validation Build Saleable (GM)

NCP - Non-Conforming Products

NCR - Non-Conformance Report

NDT - Non-Destructive Test

NIST - National Institute of Standards and Technology, Dept. of Commerce, USA

NPI - New Product Introduction

np-chart - Attribute Quality Control Chart for Number of Defective Products

NTF - No Trouble Found

NTSA - National Transportation Safety Administration, Dept. of Transportation, USA

N$_2$ - An Interface Matrix, Representing Interfaces between System Elements

N7 - New 7 QC Tools (Affinity Diagram/Brainstorming, Relation Diagram, Tree Diagram, Matrix Diagram/QFD, Arrow Diagram/PERT, Process Decision Chart, and Matrix Data Analysis/Principal Component Analysis)

OACD - Orthogonal-Array Composite Designs

O&SHA - Operational and Support Hazard Analysis

OC Curve - Operating Characteristic Curve

OCV - Open Circuit Voltage

OEE - Overall Equipment Effectiveness

OEM - Original Equipment Manufacturer

OVAT - One Variable at a Time

PAA - Product Application Agreement

PAT - Process Analytic Technology

PBS - Product Breakdown Structure

P-chart - Attribute Quality Control Chart for Proportion of Nonconforming Units

P-Diagram - Parameter Diagram

PCA - Permanent Corrective Actions

PCB - Printed Circuit Board

PCBA - Printed Circuit Board Assembly

PCI - Process Capability Index (Indices)

PCP - Production Control Plan

PDAS - Plan-Do-Study-Action

PDF - Probability Density Function

P-Diagram - Parameter Diagram

PDCA - Plan-Do-Check-Act cycle

PDF - Probability Density Function

PDF, pdf - Probability Density Function

PDM - Product Data Management

PDPC - Process Decisions Program Chart

PDPC - Process Decision Program Chart

PERT - Program Evaluation and Review Technique

PFD - Process Flow Diagram

PFMEA - Process Failure Modes and Effect Analysis

PHR - Part Handling Review

PL - Product Line

PLM - Product Life Management

PMA - Program Management Administration

PMBoK - Project Management Body of Knowledge

PMI - Project Management Institute

PMSM - Permanent Magnet Synchronized Motor

PMO - Preventive (or Planned) Maintenance Optimization

Poka Yoke - Mistake Proofing; Fool Proofing

POSEC - Prioritize by Organizing, Streamlining, Economizing, and Contributing

PPAC - Product Performance Agreement Center

PPAP - Production Part Approval Process

PPC - Production Planning and Control

PPHR - Parts per Hundred Rubber Parts

PPM - Parts per Million

PPV - Product and Process Validation

PQP - Product Quality Planning

PQRR - Program Quality Readiness Review (GM)

PRR - Problem Reporting and Resolution

PSO - Particle Swarm Optimization

PSS - Passive Safety Systems

PSW - Part Submission Warrant

PV - Part-to-Part Variation

PWM - Pulse Width Modulation

QA - Quality Assurance

QALT - Quantitative Accelerated Life Test

QbD - Quality by Design

QC - Quality Control

QC - Quality Circles

QCC - Quality Control Circle

QCCAR - Quality Control Corrective Action Request

QFD - Quality Function Deployment

QIP - Quality Improvement Program

QIT - Quality Improvement Team

QL - Quality Lever

QLS - Quality Loss Function or Quadratic Loss Function

QMD - Quality Measurement Data

QMS - Quality Management System

QSA - Quality System Assessment

QSB - Quality System Basics

QS 9000 - Quality System Requirements 9000

QSR - Quality System Requirements or Quality Service Reviews

QVR - Quality Verification Report

Q1 - Ford Q1, considered worldwide as an indication of exceptional quality

R, R_{adj}, R_{pred} - Regression, adjusted, and predictive correlations, respectively

R&C - Reliability and Confidence Level; e.g., A combination of $R \geq 95\%$ and $C = 75\%$ means that someone is 75% Confident that Reliability $\geq 95\%$

RBD - Reliability Block Diagram

RCA - Root Cause Analysis

RDT - Reliability Demonstration Test

R-chart - Chart Used to Monitor the Dispersion of a Process

RCDQ - Reactive Customer-Driven Quality

RCL - Robustness Checklist

RCM - Reliability-Centered Maintenance

RDM - Reliability Demonstration Matrix

Red X - Primary Cause

REML - Residual Maximum Likelihood

RFD - Reliability Function Deployment

RFQ - Request for Quotation

RFS - Regardless of Feature Size

RFT - Right First Time

RG - Reliability Growth

RGCA - Reliability Growth Cause Analysis

RMA - Return Material Analysis

ROCOF - Rate of Occurrence of Failures

RoHS - Restriction of Hazardous Substances Directive 2002/95/EC

ROI - Return on Investment

RPN - Risk Priority Number; $1 \leq RPN \leq 1000$

RQL - Rejectable Quality Level

RQT - Reliability Qualification Tests

RSM - Response Surface Method

RSS - Root Mean Squared

R@R - Run at Rate (GM)

RDOE - Reliability Design of Experiments

SAE - Society of Automotive Engineers

SASIG - Strategic Automotive product data Standards Integration Group

SBU - Strategic Business Unit

S Chart - Control Chart for Standard Deviations

SCM - Supply Chain Management

SEV - Smallest Extreme Value

SFDC - Shop Floor Data Collection

SFMEA - System Failure Mode and Effects Analysis

SiC - Silicon Carbide

SIF - Stress Intensity Factor

SLA - Supplier Launch Audit (GM Terminology)

SLM - Ship Logistic Management

SLS - Selective Laser Sintering

SMED - Single Minute Exchange of Die for flexible manufacturing

SoC - State of Charge (Batteries)

SOP - Start of Production

SOP - Standard Operating Procedure

SOTIF - Safety of the Intended Functionality

SPC - Statistical Process Control

SPOF - Single Point of Failure

SQA - Supplier Quality Assurance

SQE - Supplier Quality Engineer or Supplier Quality Engineering

SQIP - Supplier Quality Improvement Program

SORP - Start of Regular Production

SS (6-Sigma) - Six-Sigma

STBF - Shortest Time between Failures

Supplier CR - Supplier Customer Readiness

SVHC - Substances of Very High Concern

SWIP$_3$E - Standard, Workpiece, Instrument, Person/Procedure/Policy, and Environment
        Six Essential Elements of a Generalized Measuring System

SWOT - Strengths, Weaknesses, Opportunities, and Threats (SWOT Analysis)

TAAF - Test, Analyze, And Fix

TBD - To Be Decided or To Be Determined

TCO - Total Cost of Ownership

TE - Tooling and Equipment Certification

TAAT - Test-Analyze-And-Fix

TGR - Things Gone Right

TGW - Things Gone Wrong

TNI - Trouble Not Identified

TOP - Time Optimized Process

TOPS - Team Oriented Problem Solving (Ford Motor Company)

TPM - Total Productive Maintenance

TPMS - Triply Periodic Minimal Surfaces

TQM - Total Quality Management

TRIZ - Russian, i.e. Theory of Inventive Problem Solving (TIPS)

TV - Total Variation

TWI - Training within Industry

UCL - Upper Control Limit

UCS - Ultimate Compressive Strength

USL - Upper Specified Limit

UTS - Ultimate Tensile Strength

u-chart - Attribute Quality Control Chart for Frequency of Defects

VDA6.3 - Verband der Automobilindustrie (German); Automotive Quality Management System Process Audit

VEVA - Value Engineering and Value Analysis

VIF - Variance Inflation Factor

VMEA - Variation Mode and Effect Analysis

VOC - Voice of Customer

VRS - Variable Returns to Scale

VSM - Value Stream Mapping

WBS - Work Breakdown Structure

2_R^{k-p} - 2-level fractional factorial design having k variables with resolution R

3_R^{k-p} - 3-level fractional factorial design having k variables with resolution R

3 Spheres - Quality Management, Assurance, and Control

3T's - Task, Treatment, and Tangibles in Service Design

360° Evaluation - Evaluating Performance with inputs from supervisors, peers, & other employees

4P - People, Place, Procedure, and Policies

4S - Surroundings, Suppliers, Systems, and Skills

5M - Manpower, Machine, Method, Material, and Measurement

5R - Responsiveness, Reliability, Rhythm, Responsibility, and Relevance

5S - Select/Sort, Straighten/Set in Place, Shine, Standardize, and Sustain (Sources of Variation)

5w2h - Who, What, When, Where, Why, How, and How Much

6M - Manpower, Machine, Method, Material, Measurement, and Mother Nature

6S - Select, Set in Place, Shine, Standardize, Sustain, and Safe

7P - Proper Prior Planning Prevents Pitifully Poor Performance

7 Factors - Team, Provider, Patience, Task, Work environment, Equipment & Technology, and Organization & Management

7 Wastes - Over-production, Transport, Waiting, Inventory, Defects, Over-Processing, and Unnecessary Movement

8D - 8 Disciplines for Structured Step-by-Step Problem Solving

80/20 Rule - 80% of the Problems Resulting from 20% of the Causes

Nomenclature

A, B, C, …	Design factors
a, b, c, …	Coded variables
C	Confidence level
$\mathbf{C_b}$	Battery capacity
$\mathbf{C_p}$	Process capability index, usually for symmetric distribution; two-sided
$\mathbf{C_{pk}}$	Process capability index, usually for skewed distribution; one-sided
$\mathbf{C_{Pm}}$	Process capability index, a departure of mean μ away from process target
[C]	Conference matrix
[D]	Design matrix
$\mathbf{D_c}$	Depth of charge
E (GPa)	Modulus of elasticity or Young's modulus (Mechanics)
$\mathbf{E_{11}, E_{22}, E_{33}}$ **(GPa)**	Moduli of elasticity, namely Young's moduli, for orthotropic materials
$\mathbf{E_T}$ **(GPa)**	Tensile modulus of elasticity
E[ ]	Expected value
F	F distribution
F (N)	Force (Newton)
f( )	Function of ( )
G (GPa)	Shear modulus of elasticity
$\mathbf{G_{23}, G_{31}, G_{12}}$ **(GPa)**	Shear moduli of elasticity for orthotropic materials
H, h, hr	Hour
$\mathbf{H_V}$ **(kgf/mm$_2$, GPa)**	Vickers hardness: rectangular pyramidal indenter
$\mathbf{I_{zod}}$ **(kJ/m^2 or J/m^2)**	Izod notched impact strength at 23 °C (ISO 180/1A)
[I]	Information matrix or Fisher information matrix
L( )	Likelihood
$\mathbf{L_{ln}}$**( )**	Likelihood of natural-Log transformed data
Ln( ) or ln( )	Natural-Logarithmic transformation of ( )
Log( ) or log( )	Logarithmic transformation of ( )
LR	Likelihood ratio

P_p	Process performance
P_{pk}	Process performance index for a stable process
R	Reliability
R	Resistance, electric
R	Model correlation, e.g., regression model
R_{adj}	Adjusted model correlation
R_{pred}	Predictive correlation, based on predicted values
R_a (μm)	Surface roughness- average
R_{rms} (μm)	Surface roughness- root Mean Squared (rms)
R_z (μm)	Surface roughness- maximum
S_h	State of health
SS	Sum of squares
T (°C), Temp (°C)	Temperature in °C
T_k (°K)	Temperature in °K
T (Ton)	Mass for mm-sec-ton system
t	t-distribution (Student t-distribution)
t (sec)	Time
S	Sample standard deviation or sample error
S_c	State of charge
S_2	Sample variance
u, v, w (m; mm)	Displacements in x, y, and z directions, respectively
V	Volts
(X, Y, Z), (x, y, z)	Cartesian coordinate system
Y	Response or objective function
Y_p	Predicted value for Y
α	p-value
α (μm/m/°C)	Coefficient of linear thermal expansion; ⊥ and // to mold or casting flow
$\alpha_x, \alpha_y, \alpha_z$ (μm/m/°C)	Coefficients of linear thermal expansion in x-, y-, and z-directions
$\alpha_1, \alpha_2, \alpha_3$ (μm/m/°C)	Coefficients of linear thermal expansion in 1-, 2-, and 3-directions
β	Coefficients of linear moisture expansion
$\beta_x, \beta_y, \beta_z$	Coefficients of linear moisture expansion in x-, y-, and z-directions
$\beta_1, \beta_2, \beta_3$	Coefficients of linear moisture expansion in 1-, 2-, and 3-directions
ε	Residual
ε	Strain
ε_{Creep}	Creep rupture strain
ε^e & ε_e	Elastic strain
ε^p & ε_p	Plastic strain
ε_{eq}	Equivalent strain
$\varepsilon_{eq}{}^p$	Equivalent plastic strain

$\varepsilon_{xx}, \varepsilon_{yy}, \varepsilon_{zz}$	Normal strains
$\varepsilon_{yz}, \varepsilon_{zx}, \varepsilon_{xy}$	Shear strains
ε_{ucs}	Ultimate compressive strain
ε_{uts}	Ultimate tensile strain
$\varepsilon_1, \varepsilon_2, \varepsilon_3$	Principal strains
$\varepsilon_{11}, \varepsilon_{22}, \varepsilon_{33}$	Normal strains defined in the (1, 2, 3) coordinate system
$\varepsilon_{23}, \varepsilon_{31}, \varepsilon_{12}$	Shear strains (tensor) defined in the (1, 2, 3) coordinate system
$\varepsilon_{11c}, \varepsilon_{22c}, \varepsilon_{33c}$	Ultimate compressive strains along the primary orthotropic material axes
$\varepsilon_{11t}, \varepsilon_{22t}, \varepsilon_{33t}$	Ultimate tensile strains along the primary orthotropic material axes
$\varepsilon_{23u}, \varepsilon_{31u}, \varepsilon_{12u}$	Ultimate shear strains in primary orthotropic material coordinates (1, 2, 3)
μ	Mean
ρ, ρ_f, ρ_m **(g/cm³)**	Density (overall), density of fiber, and density of matrix, respectively
σ	Standard deviation
σ **(MPa)**	Stress
σ_{eq} **(MPa)**	Equivalent stress, e.g., von Mises stress
σ_f **(MPa)**	Fatigue limit, also called endurance limit
$\sigma_f{}'$ **(MPa)**	Fatigue strength coefficient
σ_{uts} **(MPa)**	Ultimate tensile strength
σ_{ucs} **(MPa)**	Ultimate compressive strength
$\sigma_{xx}, \sigma_{yy}, \sigma_{zz}$ **(MPa)**	Normal stresses
$\sigma_{xy}, \sigma_{yz}, \sigma_{zx}$ **(MPa)**	Shear stresses
$\sigma_y, \sigma_{0.2\%}$ **(MPa)**	Yield strength, i.e., stress at 0.2% (tensile) or -0.2% (compressive) strain σ_{YC} (MPa)
	Yield strengths in compression
σ_{YT} **(MPa)**	Yield strengths in tension
$\sigma_1, \sigma_2, \sigma_3$ **(MPa)**	Principal stresses
σ^2	Variance
$\tau, \tau_{yz}, \tau_{zx}, \tau_{xy}$ **(MPa)**	Shear stresses
ν	Degrees of freedom
χ^2	Chi-square distribution

Introduction

Design of experiment (DOE) is the most promising tool to identify influential factors (e.g., dimensions, parameters, and concepts) and consequentially implement corrective actions for product reliability growth. In the process of product development, puzzling occasions that lead to good opportunities for applying the design of experiments include:

(a) The relationship between the desired objective and design/process factors are not clear

(b) A good number of factors that need to be narrowed down to a few

(c) Product performance or product reliability is not meeting the functional requirement

(d) How to deal with an incomplete design when there is a lack of time and/or resources

(e) How to deal with multiple objective functions

The fundamental design of experiments presented in the first six chapters of the book (Volume I) is intended to meet these needs.

Chapter 1: Basic prerequisites for deploying design of experiments, such as product reliability realization process, uncertainty of measurement, unreliability and failure, concept of design for reliability, classification of design of experiments, and commercial design of experiments software are presented. The technical skills introduced in this chapter are intended to answer the following question: "How to identify the potential design/process parameters for a given objective function amid a blaze of design of experiments?"

Chapter 2: Full factorial design 2^K is often used in experimentation as a system identification tool to relate an objective function to potential factors, especially for exploring an unfamiliar phenomenon. Both main and interactive effects can be exploited using factorial design 2^K. How to conduct a design of experiments and a conceptual recognition of statistical significance based on t-distribution are introduced. The generalized procedure for conducting a study based on full factorial design 2^K was demonstrated using the real-world examples. As a screen design, full factorial design 2^K is intended to answer the following question: "What are the influential design/process factors when given a large number of potential factors that need to be narrowed down to a few?"

Chapter 3: Fractional factorial design 2_R^{K-P} is often used in experimentation as an economical way to relate an objective function to influential factors. It is frequently used as a screening design to identify significant main effects and 2-factor interactions in a reduced number of experimental tests. As a screen design with a great number of factors, fractional factorial design 2_R^{K-P} is intended to answer the question: "What is the potential relationship between the objective function and design/process parameters?"

Chapter 4: General fractional factorial design L^{K-P} based on F-distribution, is a statistical technique that systematically determines which inputs have a significant impact on the output, when multiple design levels are required to explore the nonlinear behavior. It is one of the response surface methods based on the fit of a nonlinear polynomial equation to the experimental data. The concept of analysis of variance (ANOVA) and its related formulations are deployed. Fractional factorial design L^{K-P} is intended to answer two questions: "What is the relationship between the objective function and design/process factors?" and subsequentially "What is the optimal design configuration?"

Chapter 5: Composite designs are the most widely used experimental design as a response surface method because the additional star points (e.g., center points) can clear the main effects of aliasing with two-factor interactions in a nonlinear model and unravel quadratic terms. These designs are particularly useful for design optimization that contains only a few (3–5) factors for finalizing the decision. In a concise factorial design, a composite design is intended to answer the following two questions: "What is the relationship between the objective function and design/process factors?" and "What is the optimal design configuration?"

Chapter 6: Traditional balanced orthogonal experimental designs, i.e., design matrices presented in Chapters 2–5, may not be available for every practical problem. For example, some treatments in a balanced orthogonal design are not feasible. A "short" design of experiments, which is defined as the design with less treatments required for meeting the conditions of a traditional balanced orthogonal one, may be applied instead. In such a situation, an incomplete design matrix with a preferred approximating scheme via minimizing the variance-covariance or predictive error, such as D-optimality, A-optimality, I-optimality (i.e., V-optimality), and G-optimality, can be applied. In such a handicap game, optimal designs are deployed to answer three questions: "What is the relationship between the objective function and design/process factors based on an incomplete design matrix?", "What is the optimal design configuration?", and "What would be the optimal design configuration with multiple objective functions?"

1

Reliability Deployment

Design of experiments (DOE) is a disciplined application of statistical methods based on inferential mechanisms (e.g., natural laws and polynomials) in pursuit of the desired reliability at a comfortable confidence level. Design for reliability and sustainability is intrinsic to product development while dependent on the choice of product architecture and robustness of execution in the process. The challenge moving forward in the manufacturing industry is not who first provides the ultimate technology but who best timely realizes that technology in a robust manner over time. Product reliability deployment throughout an organization shall be planned as early as possible.

1.1. Product Reliability Realization

A robust system attempts to identify and prevent design and manufacturing issues in advance, especially in the early development phase rather than have problems found in the hands of customers. Reliability deployment is a process specifically designed toward achieving high and long-term reliability. The focus is on finding and preventing problems—way more than completing forms and checklists. For example, a vehicle "produced as designed" is never finished, so should the product evolvement continue even after it is introduced in the market? To design it right, one has to establish design reliability requirements and develop a plan exhibiting how requirements are to be met. Both qualitative and quantitative analysis methods and tools would be used to verify and validate if requirements are met in the long run [Safie 2010]. A design for reliability (DFR) plan is a high-level plan that calls all the reliability tests, including design verification tests, highly accelerated life tests, reliability demonstration tests, accelerated life tests, and ongoing reliability tests.

Globalization, individualization, digitalization, and increasing competition are changing the face of product reliability. The evolution, referred to as Industry 4.0 or the Fourth Industrial Revolution, challenges product teams to react quickly and make behavioral predictions based on an enormous amount of feedback information, especially in response to warranty data. Reliability calls for a firm system of total product life-cycle management. Think on the global and act on the local. Modern development processes should be able to convert local requirements into a global product definition, which is then rolled out locally again potentially with part of the work being done in local affiliates.

1.1.1. Product Quality, Robustness, and Reliability

Product quality is a measure of the current state as delivered to the customer relative to an ideal state by design when fulfilling the functional requirements of the product. The inherent product reliability, also defined as product robustness over time, shall be primarily addressed in two aspects of quality—robustness and reliability, respectively. Product robustness means that all functional requirements meet the market needs and the operations are insensitive to environmental noises and product variability [Taguchi 1987], but product reliability means that the functional requirements are met over the specified product lifetime with no failure. A robust product is insensitive to sources of variability while a reliable product is insensitive to sources of variability over time. In other words, reliability is the ongoing index of product performance.

1.1.2. Realization Process of Product Reliability

Product performance evolves over time because there is no one-to-one correlation of a product behavior to a functional requirement. Failure calculations are based on complex models, which include factors using specific system data such as material endurance, working temperature, environmental impacts, and applied loads. All of these behave stochastically. The recognition of reliability strategy design and realization is summarized as follows:

1. Reliability is science.
2. Reliability deployment starts at the stage of early conceptual design.
3. Reliability practice is integrated into the product development cycle.
4. How to meet functional requirements and achieve the reliability goal act hand in hand.

One reliability realization process is exhibited in Figure 1.1.1. Statistics and physics of failure run concurrently because a large safety factor does not necessarily translate into a reliable product. Instead, a design with a large safety factor often leads to an overdesigned product with reliability problems. The overall plan for product reliability, including the design, materials, product specifications, manufacturing plans, procedures for manufacture and testing, quality assurance requirements, and labeling, shall bear record to the product development, often called the reliability "Master Record."

FIGURE 1.1.1 Reliability realization.

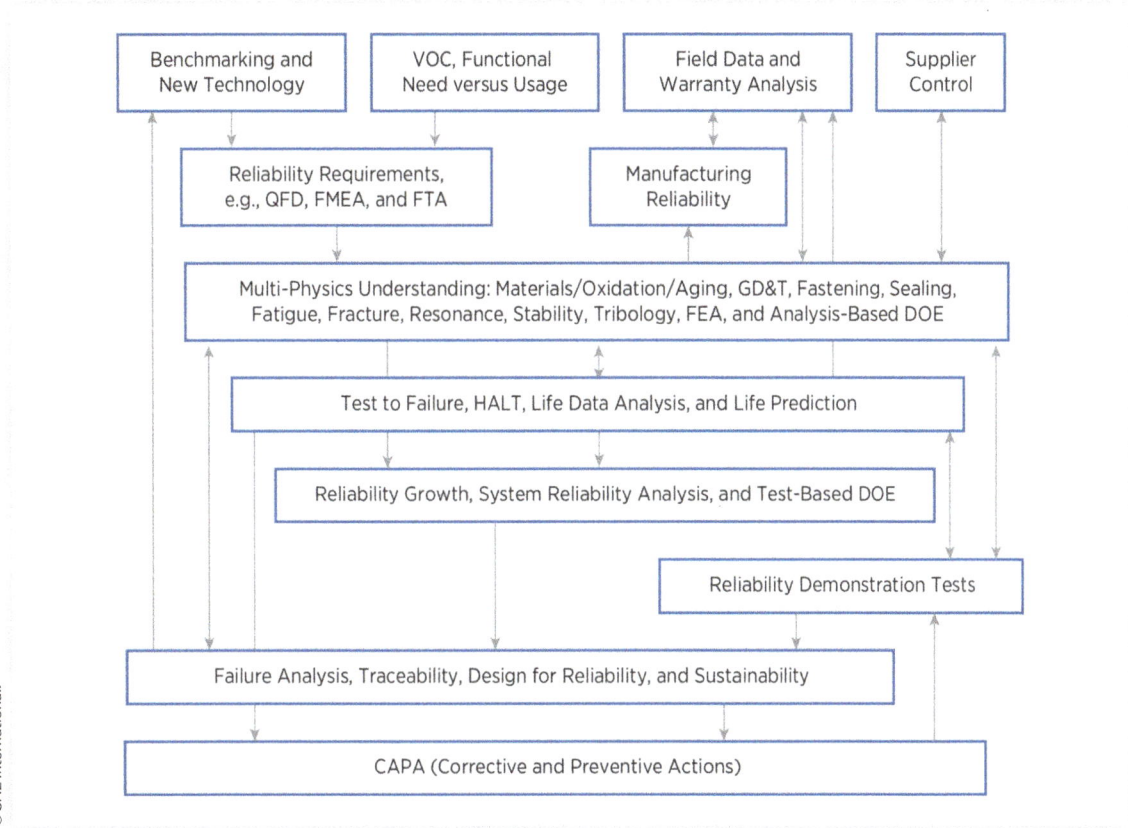

1.1.3. Tools for Reliability Realization

As a product progresses as designed, an idea that unifies all functional requirements is the concept of failure. Inventions are successful only to the extent that their creators properly anticipate how a device can fail to perform as intended [Petroski 1996]. It is crucial to understand the relationship among process control, component reliability, and system safety upfront in the design process. Historically, a variety of tools are employed to accomplish reliability realization by overcoming failures. There is a clear distinction between the goals and tools employed to assure quality versus those employed to analyze and improve reliability.

There is a significant contrast between the design for six sigma in quality engineering [Brue and Launsby 2003] and DFR in reliability engineering [Sadegh et al. 2006]. DFR is to exploit mainly the following specific fields:

1. Physics of failure, e.g., iron in rust and pathological driving behaviors.
2. Failure analysis, e.g., life data analysis and statistical tolerance analysis.
3. Life-cycle tests and prediction, e.g., stepwise accelerated life test.
4. Manufacturing reliability, e.g., manufacturing techniques and processes.
5. Supply chain reliability, e.g., inventory control and supplier selection.

6. Operational reliability, e.g., productive maintenance and autonomous driving.

7. Marketing reliability, e.g., warranty analysis and efficacy of advertisement.

8. Customer service conditions, e.g., usage, temperature, moisture, and degradation.

Activities in these fields prompt product reliability growth. Complementary tools available for growing product reliability include the following analytical techniques:

1. Voice of the customer (VOC) [Griffin and Hauser 1993].

2. Flow-down analysis [Dick and Jones 2012].

3. Quality function deployment [Hauser and Clausing 1988, Morrell 1987].

4. Failure modes and effect analysis (FMEA) [AIAG 2011].

5. Design for assembly and manufacture (DFAM) [Boothroyd et al. 2010].

6. Product variability, process capability, and gauge reliability and reproducibility (R&R) [AIAG 2008].

7. Test and control plans [Grant and Leavenworth 2004].

8. Measurement system analysis (MSA) [AIAG 2010].

9. Inferential mechanisms including physical laws of failure [Chiang 2019].

As supported by these tools, DOE summarizes the overall product performance against potential failure modes and has the final say about product reliability growth in the real world [Fisher 1925]. Experiments can be based on physical tests [Thomke 2003] and/or virtual simulations [Chiang and Tang 1995, Santner et al. 2018].

1.1.4. Advanced Reliability Planning

In a production endeavor, product reliability dominantly influences long-term business success. A product that continues to function as intended throughout its life guarantees happy customers who provide repeat business as well as referrals. A design and manufacturing company is strongly encouraged to initiate DFR activities at the concept feasibility stage in light of concurrent engineering. The stringent necessity in meeting today's sophisticated customer demands makes DFR more significant and valuable than ever for the following reasons:

(a) Product Differentiation: As automotive technology reaches maturity on many levels and the automotive market gets saturated, there are fewer opportunities to set products apart from the competition through traditional metrics (e.g., price, styling, and performance).

(b) Reliability Assurance: Each person needs a reliable vehicle, autonomous or not, to go to the workplace. The limited powertrain lives for electrical vehicles, more pleasing interior systems, sophisticated electronic modules, new component and material technologies, and less expensive parts make ensuring the reliability goal increasingly difficult.

(c) Affordability: The major cost benefit of implementing an advanced reliability plan is to promote early identification of required design, engineering, and manufacturing changes and thus avoid expensive "late changes," as shown in Figure 1.1.2. It eliminates redesign waste. Each part/assembly must have credible sources that can deliver consistent reliability (quality over time) in the volume as deliverable. The "rule of thumb" states that it may cost C times more to correct a major defect at the next stage of product development, as described below:

Completion level	Due cost of a corrective change
The part itself	Y
At subsystem assembly	CY
At final assembly	C^2Y
At the dealer/distributor	C^3Y
At the customer	C^4Y

where C is a constant that is greater than 1. Given that C = 10, a change after the product is introduced into the market because of unreliability may cost as high as 1,000 times as much as a preventive change at the research and development (R&D) stage.

(d) Preserving Profits: DFR helps products get to the market fast, preventing erosion of sales and market share and securing the brand name.

An automotive product development process can be divided into eight stages as exhibited in Table 1.1.1. An advanced product reliability goal must be built in the concurrent engineering—concurrently developing products and related manufacturing and marketing processes.

FIGURE 1.1.2 Numbers of design and manufacturing changes with and without advanced planning.

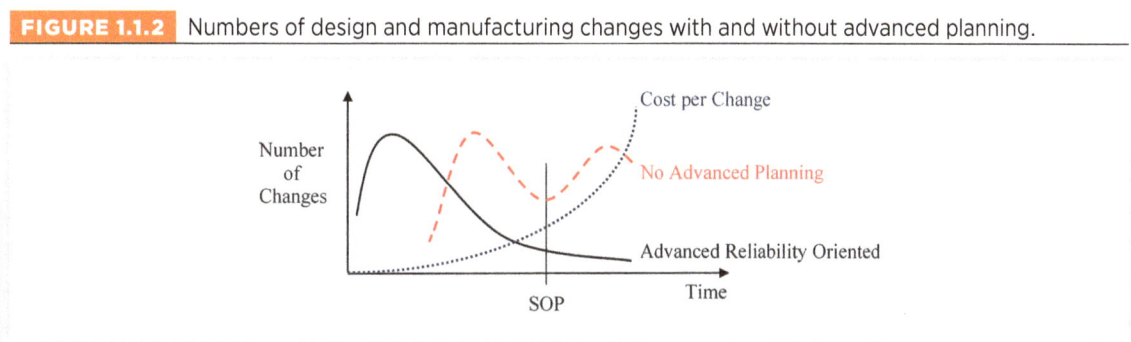

© SAE International.

TABLE 1.1.1 Development process for automotive products.

	Stage	Readiness review and approval
1	Market Research and Idea Generation	Idea Submission and Review
2	Conceptual Design and Feasibility	Project Charter
3	Conceptual Development and Projects	Business Plan
4	Product Design and Development	Design Readiness and Development Plan
5	Ramp-Up	Final Check and Manufacturing Readiness
6	Launch and Production Start-Up	Process Audit and Product Readiness
7	Production	Market Readiness
8	Marketing Strategy	Sales, Service, and Warranty

© SAE International.

Concurrent engineering hinges on contributions from multifunctional project teams in the product development process. For example, the "Design Readiness and Development Plan" review at Stage 4 (Table 1.1.1) includes the completion of the following activities:

(a) Functional performance

(b) Design for reliability

(c) Design for manufacture

(d) Design for sourcing

Work tasks deployed in these activities rely very much on DOE. The goal is to simultaneously optimize the product design against holistic reliability (Chapter 16, Volume IV), involving design factors, manufacturing variations, and latent variations in operation that may arise in the service environment. All these rely on the early flow-down analysis, including determining parameters that are statistically significant, the decomposition of system requirements of the product into requirements of subsystems and parts in an attempt to identify resources in need, and assigning resources to meet the need [Dick and Jones 2012].

1.1.5. Reliability Lauded in VOC

VOC is a decent in-depth process of capturing the customer's expectations, preferences, and even aversions [Griffin and Hauser 1993]. Activities related to VOC conducted at the start of any new product, process, or service design initiative provide a sound understanding of what the customer wants. A product leader is a good listener. For example, over a battery-powered electric vehicle, a consumer is concerned with purchase price (price differential relative to the gas-powered counterpart), limited range, speed of charge, battery life, recharging infrastructure, safety concern about the electric system, crashworthiness, vehicle size, limited top speed, and government incentive. A comparison between the VOC of Year 2015 potential buyers versus the VOC of Year 2020 potential buyers of electric vehicles is charted in Figure 1.1.3 via customer experience management analysis [Giffi et al. 2010]. It is quite clear that purchase price and reliability are the two leading factors among the following nine operating factors: price, reliability, quality, charging convenience, performance, utility, styling, resale/trade-in, and battery swapping. Therefore, the product development team shall ensure that the reliability issues are mitigated and properly resolved.

FIGURE 1.1.3 VOC of factors driving the adoption of electric vehicles.

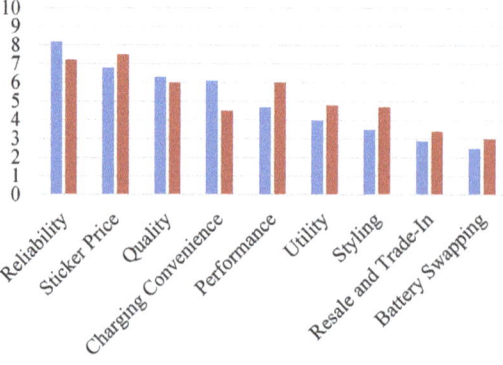

© SAE International.

1.1.6. **Product Reliability Growth**

Why do analysis and testing? The purpose is to examine a product to assure that it functions correctly and exhibits the properties and capabilities that meet its functional requirements. Areas of unreliability are exposed and corrective actions are made subsequently; thus, the product reliability will grow. A decrease in unreliability due to corrective actions is called the effectiveness factor (EF) which varies from one failure mode to another. As a measure of reliability growth, an EF of 80% means that a corrective action for a failure mode removes approximately 80% of the failure intensity, but 20% remains in the system.

Product reliability growth is the process of collecting, modeling, analyzing, and synthesizing data from reliability growth tests in the laboratory and analyses from the field. It aims at the improvement in product reliability over time owing to continuous improvements, originating from the design and manufacturing stages up to customers' feedback. The important factors for an effective reliability growth program are

(a) Develop a strategic management plan of potential failure modes.

(b) Study the effectiveness of corrective actions to be taken.

(c) Do testing and analysis that provide opportunities to identify problems.

(d) Check if the root causes of problems are verified.

(e) Implement corrections and check if functional requirements are validated.

(f) Listen to the customer's feedback (voice of the customer).

Testing and analysis to grow the product reliability are vital activities in the life cycle of a product. A latent failure mode is a type of failure that may not occur until the system has operated in the field for a certain period of time. For example, if the specifications for alignments, imbalances, electric motor circuit phase impedances, lubricant and coolant conditions and cleanliness, and vibrations and noises are not justified at the design stage, the electric vehicle sold will have a significant number of latent defects in the customer's hands. Because it is difficult to predict latent failures, product developers are urged to do the reliability growth management, starting from the R&D stage up to system maintenance and warranty services [Jin et al. 2010].

Nevertheless, it is now widely accepted that using designed experiments is the most effective way to grow product reliability. The situation leads to the question "What is an effective strategy for implementing this powerful tool?"

1.2. **Uncertainty of Measurement**

Measurement provides the fundamental basis for quantifying product performance, reliability, and sustainability. How to make sense of measured results in terms of reliability and their corresponding confidence level is essential to recognize the state of existence and quantify the potential influential factors [AIAG 2011]. Statistical methods applied to design for reliability and sustainability in measuring stochastic variables in the real world are far more complicated than the deterministic engineering principles. To see is to believe.

A statistic is anything that can be computed from the collected data. Each data constitutes one degree of freedom. For example, the mean of the measured data is a statistic with one degree of freedom (DOF). Assuming that the sample mean (also called average) is calculated from the N data, then there are still N − 1 DOF for other statistics. Statistical inferences drawn from the measured data derived from reliability demonstration tests and accelerated validation tests turn are viable for product development and research.

1.2.1. **Accuracy: Precision and Trueness**

Measurement is defined as the means of assigning a numerical value or an attribute to a characteristic of an object or event, which can be compared with other objects or events. The measurement of a property may be categorized by the following: characteristic (e.g., length), magnitude (e.g., 3.14156), unit (e.g., mm), and uncertainty (e.g., ±1.2). A measurement system is generally characterized by its precision and trueness, thus the resultant accuracy:

(a) Precision is the alias of dispersion, which indicates the "wide spread out" of the measured data, as shown in Figure 1.2.1. The wider the spread out the worse the precision.

(b) Trueness refers to the difference between the measured average and the reference mean, which is the deviation of the measured average relative to the theoretical value, as shown in Figure 1.2.1. Bias is calculated in reference to trueness as:

$$\text{Bias} = \frac{1}{N} \left(\sum_{n=1}^{N} t_n \right) - \text{Truth} \tag{1.2.1}$$

Accuracy is the combined effect of precision and trueness. The accuracy is high if its corresponding bias is zero and spread out is small.

FIGURE 1.2.1 Statistical meanings of accuracy, including spread out (precision) and bias (alias of trueness).

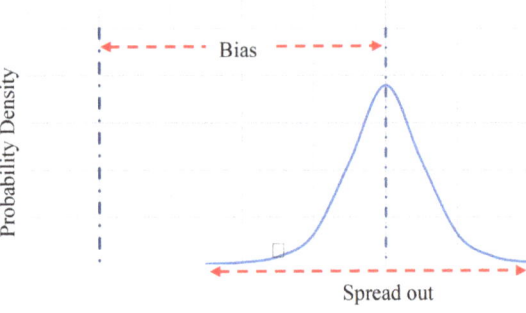

© SAE International.

The precision, including trueness and accuracy, of a measurement system is illustrated graphically in Figure 1.2.2. This is described in ISO 5725 which outlines the general principles to comply with regarding the accuracy (trueness and precision) of measuring equipment. It is concerned exclusively with measurement methods that work on a continuous scale and yield a single value as the test result. An accuracy of 100% means that the measured value is exactly the same as the theoretical value. To determine if it is adequate for the intended work tasks, a measurement system shall be assessed as follows:

(a) The measurement system must have discriminating readability and effective resolution over the range of measurement and can endure environmental factors.

(b) The measurement system sensitivity, i.e., the smallest input signal can result in a discernable output signal and must be guaranteed.

(c) The variability of the measurement system must be smaller than the production process variability with an acceptable statistical significance level (e.g., reliability = 90% and confidence = 90%).

(d) The variability of the measurement system must also be smaller than the specification limits with an acceptable statistical significance level.

If the statistical significance level is not available, one-tenth of the smaller of either the production process variability or the specification limits shall be the maximum allowable variability of the measurement system.

FIGURE 1.2.2   Precision and trueness of a measurement system.

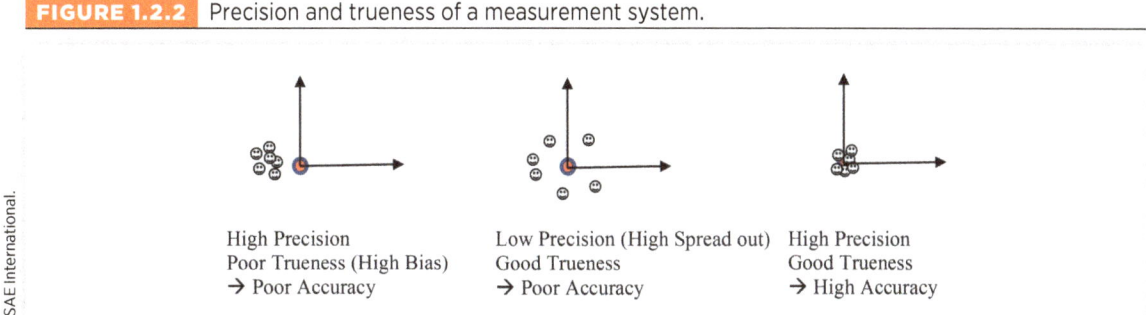

High Precision
Poor Trueness (High Bias)
→ Poor Accuracy

Low Precision (High Spread out)
Good Trueness
→ Poor Accuracy

High Precision
Good Trueness
→ High Accuracy

1.2.2. Essential Elements of Measurement

Measurement plays such a significant role in verifying the conceptual design, validating the product and process capabilities, and finalizing the product delivery. When a generalized measurement system is applied, the following essential elements, namely, SWIP^{3}E, constitute the major sources of variation in a measuring process:

- Standard: A standard shall poise itself on answering this question and provide an accepted basis for comparison in the physical sense and criteria for sampling acceptance including the statistical significance. Measurements that are traceable will agree closely to the same standard with a well-bounded sample variance. Traceability is defined by [ISO 2020] as "The property of a measurement or the value of a standard whereby it can be related to stated references, usually national or international standards, through an unbroken chain of comparisons all having stated uncertainties."

- Workpiece (Part or Assembly): Which workpiece is to be measured and what type of characteristics is to be measured? Both questions must be answered before doing any measurement. A workpiece characteristic, such as surface roughness, may be formed suffering from a dislocation of the mean value (biased average) and a wide dispersion of data (large sample standard deviation), an uneven distribution (sample skewness), and displaced peakedness (kurtosis).

- Instrument: The accuracy of measurement, also known as the theoretical value, is an objective real value that reflects the characteristics of something under certain conditions of time and space. With regard to the measurement facility, there is a certain deviation, more or less, between the measured result and the accurate value. This deviation in terms of precision and trueness is often the measurement error due to the applied equipment.

- Person: To err is human. Errors of measurement may accidentally stem from human mistakes. Getting bored with repetitive tasks, lack of training, and inattention to detail are also the root cause of measurement errors.

- Procedure: A procedure is developed to carry out the measurement based on a specific standard. It includes fixture arrangement, part orientation, part preparation, transducer location, automated versus manual, size, and expected range of conceived measurements. The measurement procedure has to be settled in advance of actually making any measurement, including the updated work instruction that is available at the machine operator's convenience.

- Policy: What purpose will the measured results be used for? When are the data needed? Who will do the tests and are they well-trained to work in the testing environment? How to present the data, especially to the customer? Company policy shall poise itself on answering these questions. Poor decision-making is also the root cause of measurement errors.

- Environment: Every measure is subject to environmental factors, including temperature, moisture, dirt, debris, ambient air movement, electromagnetic interference, air contamination, vibration, noise, and other laboratory conditions.

A measurement system is anticipated to stay in a statistically stable condition, i.e., the variation of the measurement system is merely subject to perceived common causes only. Common causes of defects should be eliminated to improve product reliability. Any special causes, which are not quantifiable or previously observed, shall be excluded, or corrected. The stability of a measurement system can be checked out using statistical process control (SPC) charts.

1.3. Unreliability

Product unreliability is the state of not meeting the intended objective or expectations of customers. It refers to the condition of being not reliable, untrustworthy, or undependable. Conventional wisdom has it that unreliability (i.e., failure) rather than reliability is dealt with in reliability engineering. Statistically, unreliability is a cumulative distribution function, and it can be expressed as:

$$U(t) \equiv 1 - R(t)$$

of which $R(t)$ is the reliability. Because the occurrence of a failure often appears stochastic, it is necessary to conduct designed experimental tests to determine if there is a statistically significant trend. A failure mechanism refers to the related physical or other process that leads to the failure. How to define a failure scientifically is intrinsic to the product of interest.

1.3.1. Physical and Functional Failures

The definition of failure is a complex matter, and it constitutes another measurement uncertainty in the real world. A failure mechanism may be addressed for one of the following observations:

(a) Physical Failure: The product has been physically damaged. For example, a part of the product breaks. Parts are made of materials that are expected to work under the thermomechanical state in their service environment [Chiang 2022]. Any fracture of a part is considered a failure by default. If a broken part has no influence on the system functionality as designed, it should be spared from the ordeal of appearing in the system.

(b) Functional Failure: The product fails to meet the functional requirement. Functional failure is the inability of a system to meet a specified performance standard or to function at the level of performance that has been specified as required. A failure mode is simply any event that causes a functional failure. Note that each product (e.g., equipment for measuring the temperature and relative humidity) often has multiple functions.

(c) Out of Specification: The product performance deviates from its normal state. When an operation deviates from its normal state, process operators should start to check instruments and analyze the process to find out the causes of the deviation. Thresholds are used to compare the sampled data of an attribute with predefined values. When a measured value gets lower (or higher) than the threshold value, it is considered a failure. This is typically applied in SPC.

One fundamental reason for the difficulties to deal with a potential failure mode of a large engineering product, such as an electric vehicle, is its inherent complexity. Complexity implies that different parts of the system are intercorrelated so that changes in one part may have effects on other parts. Complexity is generally a latent characteristic of large engineering products. More than one failure mechanism may be essentially the cause behind a failure mode.

1.3.2. Probability of Product Failure

The reliability of a manufactured product refers to the probability that it will perform satisfactorily for a specified period of time under stated use conditions. Product reliability predictions are often made possible in light of failure density, intensity, or rate, which is the probability expressed in failures per unit of time. There are four distinguishable types of failure rates defined in statistics:

1. Failure density f(t): It is defined as the probability per unit of time that the item experiences its first failure at time t, given that it starts to operate at time zero. This is the so-called probability density function (PDF) of failure. The integration or accumulation of f(t) over time, namely, F(t), is called the cumulative distribution function (CDF) of failure. F(t) is also called unreliability in contrast to reliability R(t). Note that R(t) + F(t) = 100%.

2. Hazard rate h(t): It is defined as the probability per unit time that the item experiences a failure at time t, given that it has survived up to time t (>0). In other words, h(t) = f(t)/R(t). Hazard rate h(t) is also called instantaneous failure rate—a theoretical measure of the risk of occurrence of a failure at a point in time.

3. Unconditional failure intensity ω(t): It is defined as the probability per unit time that the repairable item experiences a failure at time t (>0), given that it starts to operate at t = 0.

4. Conditional failure intensity λ(t): It is defined as the probability per unit time that the repairable item experiences a failure at time t (>0), but it has been repaired to be as good as new and is operating at time t. Conditional failure intensity λ(t) is also called the rate of occurrence of failures (ROCOF), which is normally expressed as failures per million hours (FPMH or 10^6 hr) or failures per billion hours (FPBH or 10^9 hr). For example, a component with a failure rate of 5 failures per million hours is expected to fail 5 FPMH.

It is important to distinguish between repairable and nonrepairable items when carrying out DFR activities.

Both instantaneous failure rate h(t) and conditional failure intensity λ(t) may be decreasing, increasing, or constant over the life of the product. Consequently, probabilistic engineering analysis is utilized to understand the uncertainty of the design, to identify high-risk areas for further improvement, to perform sensitivity analysis, and finally to conduct studies on reliability optimization.

1.3.3. Failure Measured by Time

The product development process takes place with a certain amount of uncertainty in a random environment. Failure metrics are set up to quantify the stochastic data required to effectively plan for and respond to system failures that are unavoidable. The top three widely applied failure metrics defined in the time domain are mean time to failure (MTTF), mean time between failures (MTBF), and mean time to repair (MTTR) and narrate as follows:

1. MTTF (Mean Time to Failure): It is a metric that measures the average amount of time a nonrepairable system operates before it fails. It is often used to measure the average life expectancy of an automotive component. A short MTTF means longer frequent downtime as the failing items need to be replaced.

2. MTBF (Mean Time between Failures): It is the expected time duration between two consecutive failures, i.e., the failure inter-arrival time, for a repairable system. The MTBF is a maintenance metric that reflects how long the facility (e.g., equipment) can operate without being disrupted. The ROCOF is often used to model the trend of failure, which can be constant, increasing, or decreasing, for the purpose of increasing the failure inter-arrival time.

3. MTTR (Mean Time to Repair): The expected time span ranging from a failure (or shutdown) to the repair (or maintenance completion) as defined for repairable systems. It refers to the amount of time required to repair a system and restore it to full functionality.

Because failures are important in managing downtime and its potential to harm people, it is essential to have a productive maintenance process in place to track all assets and equipment that are prone to failures.

1.4. Design for Reliability

Reliability is to deal with the fact on how a product performs the desired functional requirements over the expected lifetime of service, while durability is mostly concerned with how long a product can last despite the breakdowns it survives. DFR is therefore a procedure for assuring that a system performs the desired functional requirements over the expected lifetime under the customer's working conditions. It is ideally to be initiated at the conceptual design stage using virtual testing (e.g., finite element analysis [FEA]) before physical prototyping. DFR based on DOE is the most promising approach to grasp the ins and outs of cause and effect analysis and has been in practice in the automotive industry for years.

An effect of a failure is defined as the result of a failure mode on the function of the product or process as perceived by the customer. It should be described in terms of what the customer might see or experience should the identified failure mode occur. The application of DFR, including product reliability growth, can be divided into the following steps [Gardner 2006]:

1. Reference model and simulation: To understand the entire system.

2. Boundary diagram: To figure out subsystems and their interactions.

3. Parameter diagram (P-diagram): To disclose control and noise parameters.

4. Fishbone diagram: To characterize influential parameters.

5. Variation modes and effect analysis (VMEA): To prioritize parameters of interest in detail.

6. DOE: To quantify how sensitive the parameters are to the objective.

7. Optimization: To seek out the best solution.

Reliability is the probability that a product meets the intended functional requirements for a specified duration under expected operating conditions over time, while durability is defined as a product that has acceptable useful life without significant deterioration. A reliability goal based on survivability shall be set forth for each product at the early design stage. The reliability goal is statistically bound by a confidence level as the number of test specimens is always limited. For a component or system design, the norm in the automotive industry is 90% reliability with a 90% confidence level (R = 90% and C = 90%) over 10 years or 240,000 km, or so.

1.4.1. Reference Model and Simulation for System Identification

Reference models and simulation, including product sketches, three-dimensional (3D) solid models, differential equations, FEA, etc., will concretize the system visualization and fulfill the duty of product development. This is to abstract an applicable system, which can be a real product, prototype, 3D computer model, or drawing. For example, assume that there is a leak in the transmission gearbox. The investigation process may include the following intrusive and nonintrusive reference models and simulations: Visual observations; Drawing of leak paths; Bolt patterns; Bolt-fastening forces; Vibration monitoring and analysis; Tribology and lubrication; Thermal imaging and temperature measurement; Flow measurement of lubricant; Electrical testing and motor current analysis; Valve operation; Material creep assessment; Corrosion/oxidation monitoring and assessment; Structural rigidity; Process parameters; Abusive usage [Chiang 2019]. Evolution occurs when there is a lack of control over any of these factors that could negatively affect the mapping of system behaviors to functional requirements. Every sound system evolves from its failures. An idea that unifies all of engineering is the concept of failure. Inventions are successful only to the extent that inventors decently anticipate how a device can fail to perform as intended [Petroski 1996].

An understanding of processes evolvement is required before statistical tools such as DOE and SPC are applied. Selecting the appropriate experimental design is essential for effective experimentation. It is imperative to gain as much information as possible about the response-factor relationships while balancing against the cost of experimentation.

1.4.2. Boundary Diagram

A system is formed as a user-defined group of components, equipment, or facilities that support demanded operational requirements. These operational requirements will comply with mission criticality or with the environment, including product performance, reliability, health, safety, regulation, and other government- and business-defined requirements. Design implies the need for the ongoing activity of engineering control, including unaccounted variables emerging from the contradictions that are misidentified between system functionalities and product behaviors. It would be hard for an individual to know all the functions of a deployed complex system and thus required for making an evolutionary approach to design, but most systems can be divided into unique subsystems along user-defined boundaries. The boundaries are selected as a method of dividing a system into subsystems and components when its complexity makes analysis difficult. A boundary diagram is a graphical illustration of the system of interest in light of the relationships between its constituting subsystems, including assemblies, subassemblies, and components within the system, as well as the interfaces with the neighboring systems and environments.

An example of a boundary diagram is shown in Figure 1.4.1, which is a wireless power transmission (WPT) that has captured the interest of the automotive industry. It presents the interfaces of a charging system in the form of a block diagram. A WPT is a wireless charging system for electric vehicles using resonant coupled inductors. Interfaces between blocks are generally classified into the following three categories:

1. Functionality
2. Operation
3. Physical contacts (energy, material, signal, etc.)

FIGURE 1.4.1   Boundary diagram of a WPT [Cirimele et al. 2020].

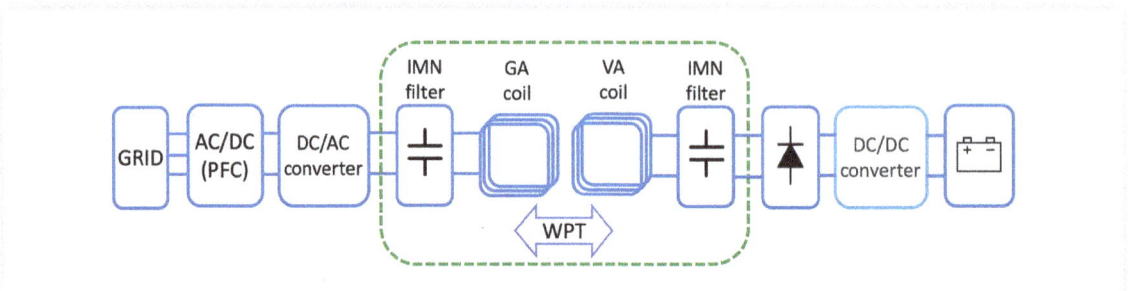

A system boundary or interface definition contains a description of the inputs and outputs of these items that may cross each boundary. A boundary diagram for the purpose of design may evolve into a reliability block diagram that models the essential functions of a complex system through the use of a series of "blocks" in which each block represents the work of a system component or subsystem. Reliability block diagrams as used in the automotive industry, as well as the defense industry, are frequently called P-diagrams (parameter diagrams), and the inputs and outputs are mostly parameterized.

Interfaces can be concretized further and components can be assigned with the help of an interface matrix. An interface matrix is employed to identify and clarify interfaces between subsystems and between the system environment, which are beneficial or detrimental to physical contacts, functional relationships, and operations. It also provides input-output connections and noise factors for deploying the follow-up P-diagram.

1.4.3. Parameter Diagram (P-Diagram)

A P-diagram takes the inputs from customer requirements and relates those input signals to the desired output functions of the design. It is a "thought map" relations diagram that can be utilized to identify the root causes of a complex problem. A thought map is a discussion aid that would yield a completed interrelationship diagram and serve as a useful memory jogger for the people that built the relationship map one arrow at a time. A P-diagram shall support the scope of the system defined in its related boundary diagram, and thus, it is essentially a schematic structured methodology.

Specifically, a P-diagram is a schematic diagram that encompasses signal factors, control factors, noise factors, and response variables that encompasses at least the following items [Juran and Gryna 1993]:

(a) Input Signals: Intended inputs, as a translation of purpose to functional requirements.

(b) Output Functions: Responses, functional requirements in terms of the customer's wants.

(c) Noise Factors: Unintended inputs (nuisances), e.g., service temperature and humidity.

(d) Control Factors: Measurables that are in control for effective transformation.

(e) Error States: Unintended outputs.

As a brainstorming tool, it supports downstream noise factor management strategies and verification methods, such as reliability demonstration matrix and design verification.

A typical P-diagram put in practice is a graphical tool employed to describe the operating environment in robustness-focused analysis, as demonstrated in Figure 1.4.2. The key lies in its simple, yet effective, belt-and-pulley design. Control factors are a list of factors that can be controlled such as materials, dimensions, and locations. Noise factors including both controllable and noncontrollable influences ought to be thoroughly examined [Ullman 1989].

FIGURE 1.4.2   Basic elements of a P-diagram.

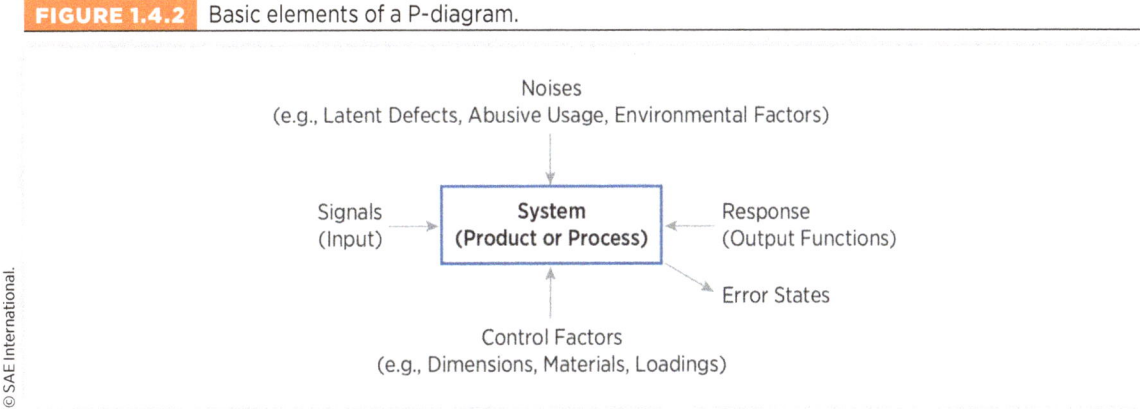

A P-diagram is also a brainstorming tool that supports downstream management strategies of noise factor and verification methods for product performance and reliability. Brainstorming is often applied for the identification of noise factors and quality history that may be used to supplement the identification of error states. Noise factors and error states are identified in terms of physics as negative interactions in the interface matrix, including:

(a) Demand-related noise, which is external to the design, such as piece-to-piece variation and changes over time.

(b) Capacity-related noises, which are internal to the design, such as customer usage, service environment, and system interactions.

It is shown that a continuous variable transmission (CVT) is more efficient than a simple gear train with a single reduction for battery electric vehicles (BEVs) [Ruan et al. 2015]. As exhibited in Figure 1.4.3, the P-diagram of a CVT that operates on an ingenious pulley system is shown for demonstration [Patel et al. 2005]. The sequential steps toward constructing a P-diagram are listed as follows [Ölme 2003]:

1. Determine the overall objective of the system as early as in the conceptual design phase or at the beginning of a problem-solving effort.

2. Identify the cross-functional team, including members who can provide expertise in the following areas:

   (a) Plan of the systemic approach.

   (b) Measurement and its know-how [Beckwith 1990].

   (c) How the system will be used.

   (d) Knowing sources of input variability [NIST/SEMATECH 2012].

   (e) Working knowledge of product performance and reliability, including testing methods.

   (f) Understanding of the design and manufacturing of the system to be optimized.

   (g) Statistical thinking [Ross 1987].

3. Review components and their functional requirements by getting all members of the cross-functional team familiar with the system and its related subsystems.

4. Clearly define boundaries of the system to easily tell the inputs from outputs.

5. Formulate each ideal function as characterizing the idealized relationship between an input signal and the functional response.

6. Verify control and noise factors [Taylor 1991].

   (a) Identifying parameters that affect the function of the system and classifying them into control factors and noise factors.

   (b) Parameterizing control factors as specified by the engineering team.

   (c) Parameterizing noise factors, which are related to design, manufacturing, usage, service, and environmental variabilities that the team does not want to change or cannot change.

FIGURE 1.4.3   P-diagram of a CVT transmission in search of fatigue life of shaft bearings.

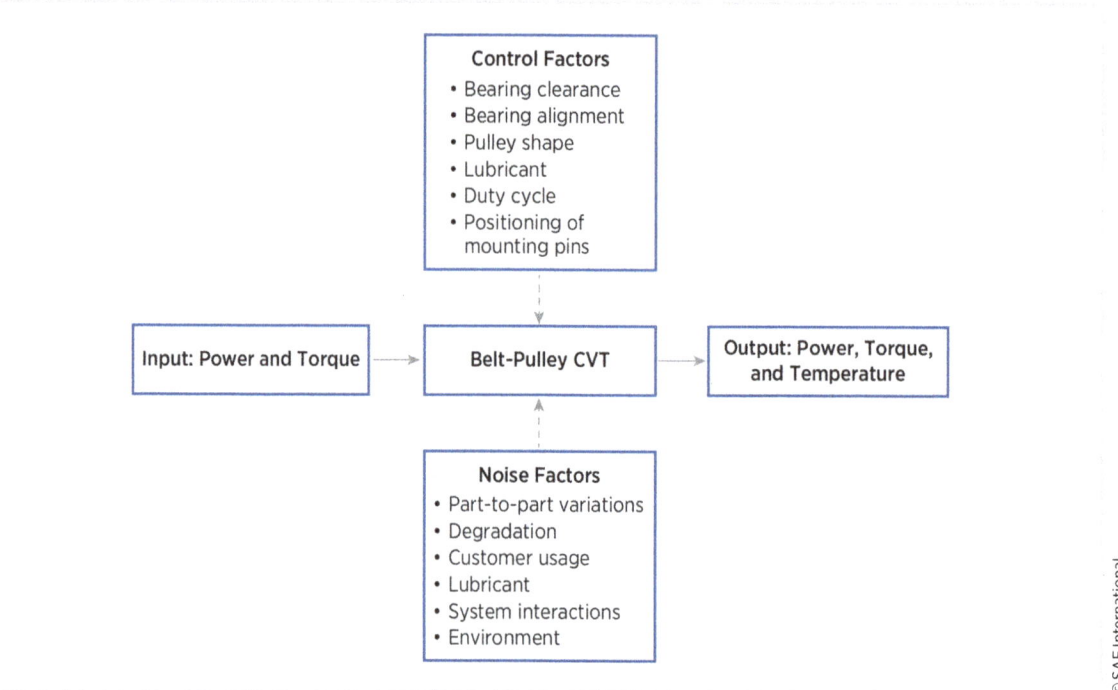

P-diagrams are commonly used for product design and quality defect prevention as guided by potential factors that cause an overall effect. A P-diagram is a visualization of the causes of a failure grouped into related categories. This is, in turn, used for problem analysis, root cause analysis, and quality improvement by identifying factors that have contributed to the problem. P-diagrams are usually constructed before conducting a DOE for a better understanding of potential influencing factors.

A P-diagram also allows the design engineer to aggregate component reliabilities to predict system reliability [National Research Council et al. 2015]. It can be used to optimize the allocation of reliability to individual components by considering the possible improvement of reliability and the associated costs subject to different design modifications. It is found to be profoundly useful when the product under study comes with many intersystem complexities and various operating conditions (e.g., electric vehicles), as the team members see the entire picture visually. Robustness is a concept that enters into statistics at different product development stages. At the design analysis stage, robustness refers to a technique that is not overly influenced by noises, i.e., achieving consistent performance by making the product/process insensitive to the influence of uncontrollable factors. Even if there is an outlier or bad data [Su and Tsai 2011], it is feasible to get the right answer regardless of who or what is involved in the process. Sometimes, it is also feasible to reduce the product deviation without actually removing the cause of variation.

1.4.4. Fishbone Diagram

A fishbone diagram, also called the Ishikawa diagram,[1] is one way to characterize parameters in the process of conducting a cause-and-effect analysis. Because it is often utilized to group the ideas (causes) after brainstorming and to classify them into their respective categories, it is also called a cause-and-effect diagram. The resultant effect forming the head of the fishbone and the potential causes forming the skeleton behind a fishbone diagram, which includes, but is not limited to, the following six M categories:

(a) Man (Personnel)

(b) Materials

(c) Methods

(d) Measurement

(e) Machines

(f) Mother Nature (Environment)

As an example, a fishbone diagram showing the specific cause-and-effect relationship between potential factors and the on-road energy consumption of BEVs is displayed in Figure 1.4.4.

FIGURE 1.4.4 Cause and effect analysis of on-road energy delivery of BEVs by fishbone diagram. Where Level 1 is Driver Assistance; Level 2 is Additional Assistance; Level 3 is Conditional Automation; Level 4 is High Automation; Level 5 is Full Automation.

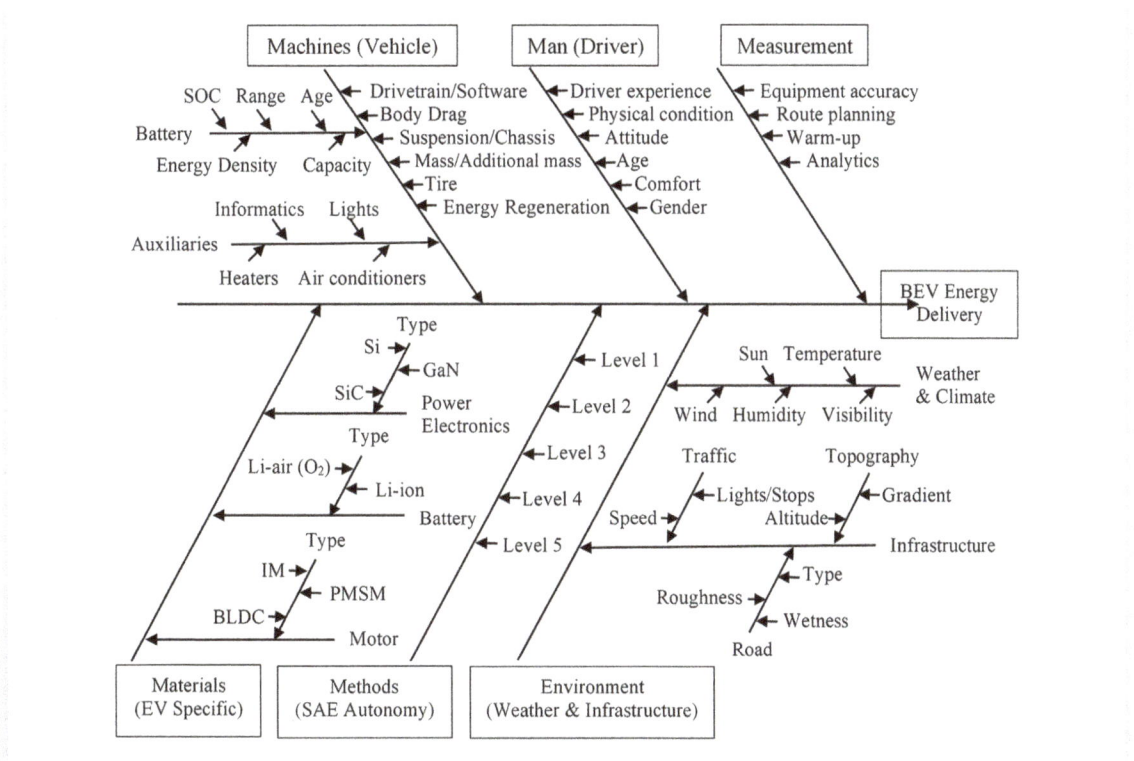

© SAE International.

---

[1] Ishikawa, K. (1968), The Fishbone Diagram Became Popular in the 1960s by Ishikawa as a Tool in the Quality Management Processes at Kawasaki Shipyards.

Brainstorming to generate ideas may help quickly organize them into affinity groups to stimulate more ideas for the exploration of potentially effective factors. It is straightforward and generally used for less complex linear problems. Planning for carrying out a DOE project requires structured brainstorming with different functional team members. The nature of discussions in the planning session is likely to vary from project to project, and it is best facilitated by someone who is well-versed in the technique and knows the problem.

1.4.5. Variation Modes and Effect Analysis

Variation mode and effect analysis is used to convert subjective knowledge about variations of potential influencing factors on an objective of interest, which can be a key product characteristic (KPC) or a certain failure mode into comparable values [Johansson et al. 2006]. It can be implemented through the following steps [Cronholm 2013]:

1. Breakdown of potential causes that affect the KPC: During problem-solving for a failure mode or performance enhancement process for a KPC, critical control items are first narrated by factors that affect product performance and/or reliability. A fishbone diagram can be helpful. Mathematically,

$$Y = f(X_1, X_2, ..., X_N) \tag{1.4.1}$$

where
  Y is the objective function, a KPC or failure mode
  $X_1, X_2,..., X_N$ are the N potential influencing factors

2. Sensitivity assessment: The sensitivity analysis can be done in the form of variances, formulated as:

$$\sigma_Y^2 = c_1 \, \sigma_1^2 + c_2 \, \sigma_2^2 + ... + c_N \, \sigma_N^2 + \text{Covariances} \tag{1.4.2}$$

where
  σ_Y^2 is the variance of the objective function
  $\sigma_1^2, \sigma_2^2, ...,$ and σ_N^2 are variances of the N potential influencing factors
  $c_1, c_2, ...,$ and c_N are coefficients

3. Variation size assessment: With all the coefficients in the above equation figured out by experience, brainstorming, and even an educated guess, it is feasible to decide which factors need to be put in control for detailed DOE.

VMEA is a technique conducted to prioritize parameters of interest in detail. VMEA is especially effective for brainstorming for a certain failure mode of a product. The following operational questions are proposed to find out the causes and effects [Equation (1.4.1)] of an event/failure of interest:

(a)   Subsystem or items under investigation

(b)   Modes of operation

(c)   Event that triggers the failure

(d)   Failure conditions

(e) Event that triggers the potential failure

(f) Potential failure

(g) Possible effect of the failure

(h) Measures taken to contain or prevent occurrences

(i) Classification of the severity of the failure

Potential failures due to design and manufacturing variations will be recorded in the DFMEA (Design Failure Modes and Effect Analysis) and PFMEA (Process Failure Modes and Effect Analysis), respectively. The final goal is to grow the product reliability as wrongdoings in the malfunction of a product are corrected.

1.5. Classification of DOE

DOE is a disciplined collective thinking of statistical methods and laws of physics with the goal of achieving the desired reliability at a comfortable confidence level. All the preceding steps presented in the last section (Section 1.4), i.e., reference model and simulation, boundary diagram, P-diagram, cause and effect (fishbone) diagram, and VMEA, are practically prepared for implementing DOE. DOE has been successfully applied worldwide in most business sectors, including research, design, manufacturing, supplier management, operation, service, and marketing [Weisman 2003].

1.5.1. Basic Types of DOE

Engineers perform experiments and analyze data as an integral part of their jobs. Whether or not engineers have learned statistics, they will use statistics. Reliability growth testing allows manufacturers to identify and eliminate critical failure modes throughout all stages of product development. Such practices become increasingly difficult to implement in today's fast-paced and highly-decentralized business practice, especially in meeting a wide range of fluctuating working conditions demanded by consumers. Firms start to bundle product performance with services by offering life-cycle reliability commitment under performance-based contracting on DOE. Basic DOE can be divided into five categories according to the applied statistical inference as follows:

1. DOE-t (Chapters 2 and 3): Traditional linear DOE, of which each factor has two levels and inferences are drawn from t-distribution, is usually used as a screening design.

2. DOE-F (Chapters 4 and 5): Traditional nonlinear DOE, of which at least one factor has more than two levels and inferences are drawn from F-distribution, is widely used as a response surface method (RSM).

3. DOE-LN (Chapter 8, Volume II): Logarithmic transformations of data for lognormal-distributed samples with skewed distributions and/or widely spread measures, such as fatigue life for some electronic components.

4. DOE-W and DOE-E (Chapter 9, Volume II): DOE based on Weibull distribution and exponential distribution, respectively. They are also in need of skewed distributions and/or widely spread measures.

5. Optimal Designs (Chapter 6): An incomplete DOE, of which the number of treatments is less than those required for meeting a traditional balanced orthogonal design (i.e., DOE-t and DOE-F designs), requires some optimization strategy to make it function like an affective DOE [Das and Lin 2011].

6. Mixture Design (Chapter 6): Mixture designs are applied when factors are mutually dependent. For example, it is desired to optimize the tensile strength of a fibrous composite, and the independent factors of interest are fibers, plastics, and additives. These factors are proportions of different components of a blend, and proportions of these factors must sum up to 100% in a mixture experiment. This complicated the experimental design and requires some optimization strategy to work it out. It can be further complicated by the mixture of different statistical distribution functions of failure modes and/or mechanisms, such as system reliability by mixed Weibull statistics (Chapter 16, Volume IV).

The statistical inference for DOE-t is based on the student's t-distribution while the statistical inference for DOE-F is based on the F-distribution [Box et al. 2005]. Reliability design of experiments (RDOE) is applied to reliability-related problems such as product life prediction/extension and product reliability growth. According to the inherent statistical distribution of the failure mode of interest, DOE-LN, DOE-W, and DOE-E are the top three frequently applied methodologies in the RDOE, for which the popular statistical inference of choice is the likelihood ratio. When the design gets involved with constrained mixture experiments that cannot be accomplished using balanced orthogonal designs or there are too many treatments to complete in time or beyond human effort, one may resort to optimal designs that have incomplete design matrices resulting from certain optimality algorithms.

1.5.2. Acceptance Sampling and Test Plans

Acceptance sampling, using statistical inferences based on reliability theories to determine whether to accept or reject a production lot of material, has been a common quality control technique applied in the automotive industry [Ross 1987]. A test plan unveils the strategy that will be utilized to verify and validate whether a product or process meets the functional requirements of the product. Acceptance sampling and test plans are extended to SPC that would screen out the potential latent defects in production, playing a role in product reliability growth. These techniques are applied across all product design and manufacturing stages, as gearing up for product reliability growth (Chapter 10, Volume II). Product life prediction based on RDOE will be legitimate and realistic when built on proper inferential mechanisms based on natural laws of conversion (Chapter 11, Volume II).

1.5.3. Statistical Tolerance Design

A dimension may vary from part to part because of the manufacturing procedure, working environment, material usage, tool degradation, and other stochastic factors. Each part dimension with the given tolerance range has its own individual statistic function inherently. The assembly of parts constitutes another statistical game by nature in the context of DOE (Chapter 7, Volume II).

In mechanical engineering, a product design is composed of multiple features, of which each comes with tolerance values that control the variable aspects of those features. Statistical tolerance analyses, statistical syntheses (i.e., synthesized allocations of tolerances), and deployment of algorithms for design for manufacturing and assembly (DFMA) are viable to product reliability. Statistical tolerance design, as a combination of statistical tolerance analysis and statistical tolerance synthesis, is focused on developing exact methods to reduce manufacturing costs and improve product reliability. In fact, it is a special case study on tolerancing dimensions and dimensional control based on DOE.

Statistical tolerance analysis is to take the variation of a set of inputs to calculate the expected variation of an output of interest [Bender 1968]. Two indices toward statistical tolerancing of mechanical parts are applied in parallel. They are process capability and distribution function zone, and both are disclosed and utilized when conducting a statistical tolerance analysis.

If the process capability is not met or cost advantage is not justified, statistical tolerance synthesis can be called on to duty in its manner to allocate dimensional tolerances that would meet the final output requirements in terms of both process capability and financial affairs.

1.5.4. Manufacturing Reliability

Relentless in the pursuit of manufacturing reliability using classical approaches, such as MTBF and MTTR, the technical capability may not yield the actual improvement in performance. Reliability metrics, such as MTBF and MTTR, tend to address lost time in production. Effective problem identification and root cause analysis are a necessity as the DOE can be deployed to discover the inherent reliability, which is recognized as the level of reliability embedded in the system as manufactured [Hayes and Wheelwright 1979]. Manufacturing reliability growth through DOE is an enormous deciding factor. One day of downtime is one day of lost production, unless a luxurious float is provided in the production line.

Case studies for DOE to grow manufacturing reliability are demonstrated using automotive applications, such as 3D printing, ball grid array packaging for electronic components, sealing by adhesive for both electric and electronic parts, effectiveness of gearbox sealing by bolts, durability of bolted joints, spot welding, material grinding, plastic molding, and coating. Process reliability involved in various mechanical-assembly processes is addressed in Chapter 12, Volume III. Manufacturing reliability may be evaluated by Weibull statistics in terms of production volume. Robust fabrication processes in tune with proper DOE exhibit an increasing demand for growing manufacturing reliability of mechanical components is the central theme of Chapter 13, Volume III. Manufacturing reliability of electronic components for electric vehicles via DOE techniques, including electronics packaging procedures and physics of failure, is addressed in Chapter 14, Volume III. 3D printing, as an additive manufacturing technology, enables a manufacturer to produce geometrically complex parts including shapes and textures, allowing the structural fabrication to use less material than subtractive manufacturing methods. How to build 3D-printed parts is addressed Chapter 15, Volume III.

1.5.5. Supply Chain Reliability

One of the automotive industry's most widely used international standards for quality management is ISO/TS 16949, which evolves from a series of publications of a new global industry standard by the International Automotive Task Force (IATF). Although "quality" has been rephrased again and again in ISO's and IATF's publications, most initiatives are consistently focused on product reliability, i.e., how to sustain product robustness over time via product development and process management involved with suppliers [Croom et al. 2000]. The supplier chain reliability is held in high regard and viable as addressed in Chapter 17, Volume IV.

1.5.6. Operational Reliability

System engineers in the area of operational reliability are at the core of creating system integration technology that will push the boundaries of market needs. As part of the product development team, a reliability engineer in the field will work on some unique and complex operating contexts to integrate product performance, including effectiveness, safety, efficiency, maintainability, testability, and ergonomics, into the overall configuration of a product and the related problem-solving skills. They do it faster than ever by drawing multiple information sources in collaboration with teams of diversified backgrounds to synthesize solutions that meet or exceed the customer's expectations. DOE for operational reliability plays such a viable role, as introduced in Chapter 18, Volume IV.

1.5.7. Marketing Reliability

Actions to enhance marketing reliability are anticipated before a product is launched. Product development tools such as VOC (Voice of the Customer), VEVA (Value Engineering and Value Analysis), DFMA, DVP&R (Design Verification Plan and Report), VSM (Value Stream Mapping), and CQI (Continuous Quality Improvement) utilize rational logic and functional analysis to identify relationships that increase product value. Benchmarking market value by data envelopment analysis is a tool for differentiating the VOC. DOE provides powerful statistical techniques aimed at quantitatively attributing the customer's decision-making to the product value offered using these product development tools. Nevertheless, no product is perfect. Hopefully, a reasonably accurate predictive method of expected product warranty returns based on either deterministic principles or stochastic simulations can be sought with short-term data in the break-in period. Continuous quality improvement can be successfully made in both break-in and useful life periods. Updated corrective and preventive actions in an on-demand customer's environment may be resolved in light of the 8D (8 Discipline) as bookkeeping tactics. It is customary for a manufacturer to adopt a FRACAS (Failure Reporting, Analysis, and Corrective Action System Software), or similar software, as a live document to keep track of the product reliability.

1.5.8. Holistic Approach

When a reliability study is put in gear toward various factors involved in a multifunctional organization that is involved with the design, manufacturing, quality control, supplier assessment, and market service, a holistic DOE can be taken to rationalize individual main effects from interactive complications among different functional departments, as delineated in Chapter 16, Volume IV. This often resorts to system reliability with mixed statistical distributions. A minimum aberration fractional factorial split-plot design is beneficial as it may increase the overall resolution with the smallest number of treatments (runs).

1.6. Software for DOE

DOE templates based on Excel can be set up easily to perform calculations and draw charts of main and interactive effects for most traditional experimental designs, e.g., 2^{K-P} and 3^{K-P} designs. Through Excel calculations and charts, an experimenter is able to understand the whole picture of the DOE analysis and synthesis. A significant number of practical problems encountered in the automotive industry can be solved quickly using Excel without commercial codes.

Nevertheless, commercial software packages for DOE do provide some proprietary features for advanced studies. As described by individual providers in their respective commercial advertisements, key points of some commercial software packages focused on DOE are addressed in alphabetical order as follows:

1. ***Design-Expert*** is a registered trademark of Stat-Ease, Inc., Minneapolis, MN, US.
   We can screen for vital factors and components, characterize interactions, and, ultimately, achieve optimal process settings and product recipes. Set flags and explore contours on interactive two-dimensional (2D) graphs and then visualize the response surface from all angles with rotatable 3D plots. Finally, maximize the desirability for all your responses simultaneously and overlay them to see the "sweet spot" meeting all specifications.

2. ***JMP*** is a subsidiary of SAS Institute.
   JMP software combines interactive visualization with powerful statistics. Example problems for DOE with JMP can be found in [Proust 2010].

3. ***Minitab*** is a registered trademark of Minitab, Inc., State College, PA, US.

   It is a powerful statistical software everyone can use. Predict, visualize, analyze, and harness the power of your data to solve your toughest business challenges from anywhere on the cloud. Example problems for DOE with Minitab can be found in [Mathews 2004].

4. ***MODDE*** is a registered trademark of Sartorius Data Analytics, Malmö, Sweden.

   Its built-in guidance and quality measures ensure users make the best experimental choices, so users get the most relevant and effective outcomes. MODDE is designed to help experimentalists get DOE right from the start. With an efficient DOE approach to problem-solving, users can significantly reduce experimental costs, de-risk projects and increase success rates, make the most of valuable samples/raw materials/human resources, accelerate progress and time to market while keeping within budget, achieve quality goals, and satisfy quality by design requirements.

5. ***ReliaSoft*** is a registered trademark of ReliaSoft, part of HBM Prenscia, part of Hottinger Bruel & Kjaer (HBK), Virum, Denmark.

   For durability and reliability (nCode and ReliaSoft), empowering data-driven confidence through accurate analysis and simulation enables customers to achieve success through failure prediction. In the field of product reliability growth, nCode is for the physics of failure and ReliaSoft for statistical analysis. Users will use ReliaSoft Weibull++ software and other ReliaSoft products with hands-on practice and through case studies to learn experiment design concepts and analysis methods.

6. ***SAS*** is a registered trademark of SAS Institute, Inc. Cary, NC, US.

   The SAS trademark is used in the following business: Computer consultation; Computer programming for others; Computer systems analysis; Computer services, namely, collection, analysis, management, and presentation of data and information for others; Providing temporary use of online non-downloadable software for data collection, analysis, and reporting; Maintenance of computer software; Computer graphics design; Designing and hosting the Websites of others on a computer server for an internal or global computer network. Example problems for DOE with SAS can be found in [Onyiah 2008] and [Atkinson et al. 2007].

7. ***SPSS*** is a registered trademark of IBM SPSS Statistics.

   Uncover data insights that can help solve business and research problems. It offers a user-friendly interface and a robust set of features that lets your organization quickly extract actionable insights from your data. Advanced statistical procedures help ensure high accuracy and quality decision-making. All facets of the analytics life cycle are included, from data preparation and management to analysis and reporting.

8. ***Statgraphics*** is a registered trademark of MathSoft, Inc., Cambridge, MA, US.

   It gives users the statistical tools to pursue excellence, gain understanding, and accomplish important business goals. A new link to Python functionality, a modernized graphical user interface with a convenient feature-locating ribbon bar, a procedure dashboard, and big data capability are included. Objectives: advancement of systems performance for quality, increased productivity, development of best practices, and optimization of policies and procedures, driving cost-saving efficiencies and controls.

9. ***S-Plus*** is a registered trademark of MathSoft, Inc., Cambridge, MA, US.

   It is an interactive programming environment for data analysis and graphics. It is based on the AT&T Bell Laboratories S program but has substantially greater statistical and graphic capabilities than the original S, particularly in the areas of time series analysis and regression methods.

10. ***Statistica (TIBCO)*** is now a registered trademark of TIBCO Software Inc., Palo Alto, CA, US. Statistica is a suite of analytics software products and solutions originally developed by StatSoft (Europe) GmbH. It was acquired by Dell in March 2014.

    TIBCO Statistica DOE lends experimental methods that can be used in agriculture, food and beverage, chemical, health sciences, manufacturing, marketing, power, and many other industries to optimize processes. Statistica is a data analysis and visualization program. Its data analysis capabilities cover thousands of algorithms, functions, tests, and methods.

In light of the first six chapters of the book, one would be able to comfortably apply the powerful DOE techniques to solving real-world problems using these statistical software packages, Excel, or computer programming (e.g., C/C++).

References

AIAG, *Advanced Product Quality Planning and Control Plan*, 2nd ed. (Southfield, MI: Automotive Industry Action Group, 2008).

AIAG (2010), *Measurement Systems Analysis*, 4th Edition, Automotive Industry Action Group, Southfield, MI, 240 pages; ISBN: 9781605342115.

AIAG, *Potential Failure Mode and Effects Analysis*, 4th ed. (Southfield, MI: Automotive Industry Action Group, 2011).

Atkinson, A.C., Donev, A.N., and Tobias, R.D., *Optimum Experimental Designs, with SAS* (Oxford, UK: Oxford University Press, 2007), ISBN:978-0-19-929660-6.

Beckwith, T.G., *Mechanical Measurements* (Reading, MA: Addison-Wesley Publishing Company, 1990).

Bender, A. Jr., "Statistical Tolerancing as it Relates to Quality Control and the Designer," SAE Technical Paper 680490 (1968), doi:https://doi.org/10.4271/680490.

Boothroyd, G., Dewhurst, P., and Knight, W., *Product Design for Manufacture and Assembly*, 3rd ed. (Boca Raton, FL: CRC Press, 2010).

Box, G.E.P., Hunter, J.S., and Hunter, W.G., *Statistics for Experimenters: Design, Innovation, and Discovery*, 2nd ed. (New York: John Wiley & Sons, 2005).

Brue, G. and Launsby, R.G., *Design for Six Sigma* (New York: McGraw-Hill, 2003).

Chiang, Y.J., *Mechanics and Design for Product Life Prediction* (Chongqing, China: Chongqing University Press, 2019), ISBN:978-7-5689-1917-6.

Chiang, Y.J., *Automotive Engineering Materials—Thermomechanical Properties* (Chongqing, China: Chongqing University Press, 2022), ISBN:978-7-5689-3293-6.

Chiang, Y.J. and Tang, C., "Accuracy Assessment to Applying 20-Node Solid Elements to Pressurized Composite Shells," *Finite Elements in Analysis and Design* 20 (1995): 219-231.

Cirimele, V. et al. (2016), "Uncertainty Quantification for SAE J2954 Compliant Static Wireless Charge Components," *IEEE Access*, September 2020, https://www.researchgate.net/publication/ 344340835.

Cronholm, K., "Design of Experiment Based on VMEA (Variation Mode and Effect Analysis)," *Procedia Engineering* 66 (2013): 369-382.

Croom, S., Romano, P., and Giannakis, M., "Supply Chain Management: An Analytical Framework for Critical Literature Review," *European Journal of Purchasing and Supply Management* 6 (2000): 67-83.

Das, R.N. and Lin, D.K.J., "On D-Optimal Robust Designs for Lifetime Improvement Experiments," *Journal of Statistical Planning and Inference* 141 (2011): 3753-3759.

Dick, J. and Jones, B. (2012), "On the Complexity of Requirements Flow-Down," Integrate Systems Engineering, Ltd., UK.

Fisher, R.A., "Theory of Statistical Estimation," *Mathematical Proceedings of the Cambridge Philosophical Society* 22 (1925): 700-725.

Gardner, R.H. (2006), "Tools for Design and Analysis of Experiments", *Statistical Considerations*, pp. 133-170.

Giffi, C. et al. (2010), "Gaining Traction: A Customer View of Electric Vehicle Mass Adoption in the U.S. Automotive Market," Deloitte Consulting LLP, Retrieved May 5, 2017 from www.cgiffi@deloitte.com.

Grant, E.L. and Leavenworth, R.S. (2004), *Statistical Quality Control*, Indian Edition, McGraw-Hill, New York; ISBN 13: 9780070435551.

Griffin, A. and Hauser, J.R., "The Voice of the Customer," *Marketing Science* 12, no. 3 (1993): 1-27.

Hauser, J.R. and Clausing, D. (1988), "The House of Quality," *Harvard Business Review*, May–June 1988, pp. 63-73.

Hayes, R. and Wheelwright, S.C. (1979), "Link Manufacturing Process and Product Life Cycles," *Harvard Business Review*, January–February 1979, pp. 133-140.

ISO (2020), "Accuracy (Trueness and Precision) of Measurement Methods and Results," ISO 5725-4:2020, ISO (the International Organization for Standardization is a Worldwide Federation of National Standards Bodies).

Jin, T. et al., "Reliability Growth Modeling for In-Service Electronic Systems Considering Latent Failure Modes," *Microelectronics Reliability* 50, no. 3 (2010): 324-331.

Johansson, P., Chakhunashvili, A., Barone, S., and Bergman, B., "Variation Mode and Effect Analysis: A Practical Tool for Quality Improvement," *Quality and Reliability Engineering International* 22 (2006): 865-876.

Juran, J.M. and Gryna, F.M., *Quality Planning and Analysis from Product Development through Use*, 3rd ed. (New York: McGraw-Hill, 1993), 256.

Mathews, P.G., *Design of Experiments with MINITAB: Homework Problems* (Milwaukee, WI: ASQ Quality Press, 2004).

Morrell, E.M., "Quality Function Deployment," SAE Technical Paper 870272 (1987), doi:https://doi.org/10.4271/870272.

National Research Council et al. (2015), "System Design for Reliability," in *Reliability Growth: Enhancing Defense System Reliability*, National Academies Press, Washington, DC, 266 pages; ISBN: 978-0309314749.

NIST/SEMATECH (2012), *e-Handbook of Statistical Methods*, April 2012, http://www.itl.nist.gov/div898/handbook.

Ölme, A. (2003), "How Can Design for Six Sigma Increase Product Reliability—SKF Experiences," in Creveling, C.M., Slutsky, J., and Antis, D. (eds), *Design for Six Sigma*, Prentice Hall PTR, Upper Saddle River, NJ.

Onyiah, L.C., *Design and Analysis of Experiments: Classical and Regression Approaches with SAS* (Boca Raton, FL: Chapman and Hall/CRC, 2008).

Patel, D., Ely, J., and Overson, M., "CVT Drive Research Study," SAE Technical Paper 2005-01-1459 (2005), doi:https://doi.org/10.4271/2005-01-1459.

Petroski, H., *Invention by Design How Engineers Get from Thought to Thing* (Cambridge, MA: Harvard University Press, 1996).

Proust, M., *Design of Experiments Guide* (Cary, NC: JMP, A Business Unit of SAS, 2010).

Ross, S.M., *Introduction to Probability and Statistics for Engineers and Scientists* (New York: John Wiley & Sons, 1987).

Ruan, J. et al. (2015), "Comparing of Single Reduction and CVT Based Transmissions on Battery Electric Vehicle," in *The 14th IFToMM World Congress*, Taipei, Taiwan, October 25–30, 2015.

Sadegh, P., Thompson, A., Luo, X., Park, Y. et al., "A Methodology for Predicting Service Life and Design of Reliability Experiments," *IEEE Transactions on Reliability* 55 (2006): 75-85.

Safie, F.M. (2010), "Understanding the Elements of Operational Reliability a Key for Achieving High Reliability," in *Trilateral Safety and Mission Assurance Conference (TRISMAC)*, Cleveland, OH, October 26–28, 2010.

Santner, T.J., Williamns, B.J., and Notz, W.I., *The Design and Analysis of Computer Experiments*, 2nd ed. (New York: Springer, 2018), ISBN:978-1493988457.

Su, X. and Tsai, C., *Outlier Detections* (New York: John Wiley & Sons, 2011).

Taguchi, G., *System of Experimental Design* (New York: UNIPUB, Kraus International Publications, 1987).

Taylor, W., *Optimization and Variation Reduction in Quality* (New York: McGraw Hill, 1991), ISBN:0-07-063255-3.

Thomke, S.H., *Experimentation Matters: Unlocking the Potential of New Technologies for Innovation* (Boston, MA: Harvard Business School Press, 2003).

Ullman, N.R., "The Analysis of Means (ANOM) for Signal and Noise," *Journal of Quality Technology* 21 (1989): 111-127.

Weisman, D.A., "Experimental Design with Applications in Management, Engineering and the Sciences," *Technometrics* 45, no. 1 (2003): 105.

Problems

P1.1: If hired to operate an automotive tier-one supplier of high-voltage batteries for electric vehicles, how would you deploy the reliability functionalities throughout the organization?

P1.2: Please prepare a fishbone diagram to elaborate on potential factors that would lead to a loosened bolted-joint.

2

Full Factorial Design 2^K

Design of experiments is employed to relate potential root causes to a functional requirement or failure mode of a complex system via variational analysis. The idea is to simultaneously vary all the factors using a statistical design, in which experiments are run for all combinations of levels for all the factors, instead of one factor at a time. Sustainable reliability based on DOE can be designed into a product or process by making it robust against noises during the three design stages: system design, parameter design, and tolerance design. Explicit interactive effects can be explored in addition to individual main effects in a higher-ordered nonlinear model. Consequently, quadratic or higher-ordered polynomials fit of the applied data encompassing response surface methods (RSM) can be optimized for the system robustness. DOE is so far the most promising method for problem-solving.

2.1. Why DOE

The DOE principles that were developed for agricultural experiments by Sir Ronald Fisher in England in the 1920s have been successfully adapted to automotive applications in the United States (US) since the 1970s. The idea is to vary all the factors at once using a statistical design, in which experiments are run for all combinations of levels for all the factors. One-Factor-At-a-Time (OFAT) is the approach to problem-solving using one factor at a time, basically as all the variables are constrained except one. There are two major advantages of conducting DOE over OFAT:

(a) Complete search of solution in the multidomain

(b) Tremendous reduction in the number of experimental tests

For example, DOE lends experimenters to conduct valid experiments in the presence of many naturally fluctuating conditions such as dimensional variation of parts/systems, road roughness, temperature variation, and other harsh environmental conditions. Such experiments provide the quantified relationship between performance and controllable factors that will enable engineers to identify the causes of product and process variations and reduce such variations by controlling key contributing parameters, thereby improving product reliability and quality.

Engineering design is a cognitive process to develop products and provide technical services based on knowledge and visualization. The fundamentals of knowledge are learned according to the taxonomy of the realization process of products and services. One way to characterize the development of a product and/or its manufacturing process is to do the factorization of the engineering design [Czitrom 1999]. Quality control is then conducted to improve the performance of the product or manufacturing process according to the identified factors. When quality control is done on the spot for immediate attention or closed-loop correction, it is called online quality control. On the other hand, it is called offline quality control. DOE is the major course in offline quality control.

2.1.1. Complete Search for Solution in the Multidomain

The viable reason why DOE should be applied instead of the OFAT method is to make the information of factorial effects available in the multidomain, in which the optimal solution can be searched for in all senses. This is to be explained using the follow-up example.

Let us assume that decision-making for the thickness of a specific pressure vessel is a function of two parameters: (1) strain (stress) and (2) cost of material. When it is thin, strain-induced failure is a problem. Too thick, the cost of material could be a major concern for the competitive market. An intuitive OFAT method has been in general practice for solving such a problem. It means one and only one of the available factors is varied at a time with the remaining factors held constant. In the example of a pressure vessel, the cost of material is fixed and stress is studied to find an allowed suitable stress level. This stress level is assumed the best choice, and then the stress is fixed at that level. The cost of material is then varied until a level of cost of material is found presumably to be the best choice. Finally, the choice of these two parameters is assumed to be the optimal design condition. However,

$$\text{Thickness} = f\left(\text{strain, cost of material}\right)$$

Note that the outcome (thickness) in terms of the two independent variables is a surface, which cannot be represented by only two curves searched on a surface for the optimal design solution as illustrated in Figure 2.1.1.

Instead, a systematic approach, based on DOE, to optimizing the thickness of the pressure vessel can be sought. DOE means an arrangement of a series of tests in which changes are made on purpose to the input variables for a product and/or process so that the reasons for changes in the output response can be identified and optimized. The main advantages of DOE include CQDR, as:

(a) C (Cost): Reducing cost

(b) Q (Quality): Improving product quality/process yield

(c) D (Delivery): Shortening development time

(d) R (Reliability): Reducing variability/long life span

A DOE tool proposes a test plan that covers the complete discretized parameters space of a certain problem. Experimental design involves (a) designing experiments for (b) testing hypotheses through controlling experiment factors so (c) to predict or establish a result (laboratory, theoretical, or numerical) (d) corresponding to the objective function based on (e) dependent and independent variables. Elements (a) to (e) comprise the key five actions that have to be taken into action. It pinpoints the sensitivity of each factor that has a potential influence on the outcome, and thus, engineers can have a design with high yield prior to going into production.

FIGURE 2.1.1   Search on a response surface.

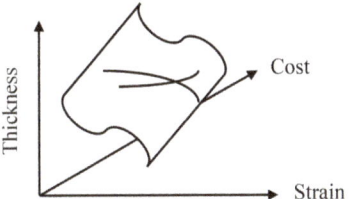

2.1.2. Tremendous Reduction in the Number of Experimental Tests

For simplicity, consider only the two-level factorial design in which each factor takes exactly two levels. The following provides a direct comparison between the DOE and OFAT method:

Factors	2-level factorial design	OFAT
2	4 runs to detect 3 effects	6 runs to detect 2 effects
3	8 runs to detect 7 effects	16 runs to detect 3 effects
5	16 runs to detect 15 effects; 32 runs to detect 31 effects	96 runs to detect 5 effects
7	64 runs to detect 63 effects; 128 runs to detect 127 effects	512 runs to detect 7 effects

If the OFAT method is applied, decision-making information on the factorial effects becomes available only after the entire experiment is completed. It takes too many tests as shown above. Nevertheless, DOE can be divided into fractions (blocked) and performed sequentially so that data from each fraction (block) of runs can be analyzed individually right after obtaining them.

2.2. What Is DOE

As argued above, DOE allows for multiple input factors to be manipulated to determine their effect on a desired output (response). By manipulating multiple inputs at the same time, DOE can identify important interactions between factors that may be missed when experimenting with one factor at a time. It can also resolve the confounding relationships between variables. All possible combinations can be investigated (full factorial) or only a portion of the possible combinations (fractional factorial).

(a) Full Factorial Design: 2^K (2 levels per factor) and 3^K (3 levels per factor)

(b) Fractional Factorial Design: 2^{K-P} (2 levels per factor) and 3^{K-P} (3 levels per factor)

Some special fractional factorial designs are mixed combinations of 2^{K-P} and 3^{K-P}. When 2^{K-P} and 3^{K-P} do not work for the problem, special algorithms for DOE such as DOE-LN, DOE-W, and DOE-E, and D-optimal design can be applied.

2.2.1. Application for DOE

The application of DOE has gained acceptance in the US automotive industry as an essential tool for improving the quality and reliability of product development, process design, operations, and services. This recognition is partially attributed to Dr. Genichi Taguchi, who promoted the use of DOE as a robust design tool such that the automotive parts/systems are relatively insensitive to environmental fluctuations. It is also due to user-friendly software packages (e.g., BBN, CADE, Design-Ease, JMP, Minitab, Quanterion Automated Reliability Toolkit, ReliaSoft, SAS, SPSS, Statistica, and Statgraphics) and the availability of training in the US automotive industry.

2.2.2. Experimental Design Process

The concept of product design or process validation using DOE shows how simple two-level factorial experimental designs can rapidly increase the engineer's knowledge about the behavior of the product or process being studied. An experimental design process is illustrated in the flowchart shown in Figure 2.2.1. It starts with defining the objective function that must be achievable and ends with carrying out implementable actions.

1. Define the problem or issue.
2. Determine the objective with goals.
3. Identify and select factors.
4. Set up DOE.
5. Conduct tests (physical or virtual with test validation).
6. Analyze data using a t-test (2^{K-P}), F-test (3^{K-P}), LR (likelihood ratio), etc.
7. Formulate the metamodel.
8. Do diagnostic checking.
9. Do RSM (DOE-F), DOE-LN, DOE-W, etc. as screening design if 2^{K-P} are not adequate.
10. Do RSM (DOE-F), DOE-LN, DOE-W, etc. based only on significant factors that are identified from the screening design.
11. Formulate the predictive equation and do optimization.
12. Take corrective and preventive actions.

A metamodel or surrogate model refers to an explicit presentation of the information that is presumably necessary to represent a system or process, while a predictive equation (predictive model) must be effective and adequate to represent a system or process. Designed experiments are science and thus whatever is done by an experimenter must be able to be duplicated by other experimenters, whether it supports or refutes the statistical hypothesis.

FIGURE 2.2.1 Experimental design process.

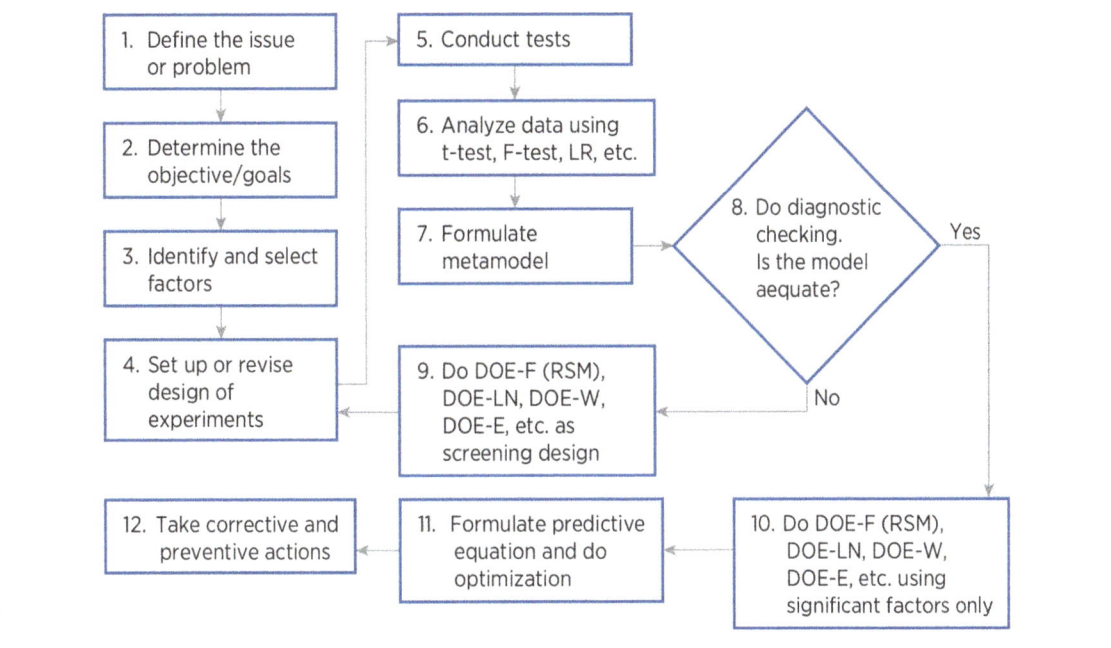

2.2.3. History of DOE

Various cause-and-effect doctrines have been narrated by school educators and religious people for more than two thousand years. Nevertheless, the relationship between causes and their effects on intended results had never been quantified scientifically until the DOE was unraveled in the early twentieth century. The important contributions to the theory and practice of statistical inference in both the 20th and 21st centuries have been those in DOE. The development timeline of DOE can be divided into the following four stages:

1. Theorization Stage (1918–1940s): DOE is first applied as an experimentation method for agriculture by R. A. Fisher—increasing the crop yield to feed the people after World War I and II. It had a tremendous influence on agricultural science as the DOE methodology explored by him and his co-workers unveiled the power of analysis of variance (ANOVA) [Fisher 1925]. As a man sows, so he shall reap.

2. Methodology Stage (1950s–1970s): More theoretical R&D by George Box (Fisher's son-in-law), Wilson, and other researchers [Cochran and Cox 1957] led to wide applications of the RSM in chemical and process engineering. The RSM was then soundly refined and generalized as an easy-to-use optimization tool in the areas of design and manufacturing.

3. Robust Engineering Stage (1970s–1990s): As the CQI for more reliable products was initiated in many manufacturing industries, product robustness became a key management goal, especially in the automotive industry as expected by Taguchi for the reduction in loss function. Reasonable robustness strategies in the product development stage have been implemented utilizing DOE to meet the VOC. Virtual tests also came into play in this period.

4. Economic Competitiveness Stage (2000s–): When economic competitiveness drives all business sectors in the 21st century, DOE has become a reliability tool widely accepted by most functional departments in more and more organizations, including manufacturing and service in addition to

the original functional needs in the department of product research and development. DOE has also been refined for meeting various product performance needs over time, and it has become an essential tool for developing product reliability growth throughout the entire organization in the 21st century—starting with VOC, design, R&D, engineering, manufacturing, operation, marketing, service, and organizational behavior and management. Specifically, the reliability design of experiments (RDOE) has been exploited for an objective function that is characterized by widely spread data, such as product creep-fatigue life cycles.

2.3. Objective Function and Goals

There will be one objective function, denoted by Y or y, which may consist of one goal only or a combination of several distinct goals. The appropriate measure has to be determined for the response (output) of the objective function. A variable measure is preferable, while an attribute measure with a "pass/fail" examination should be avoided. Furthermore, it must be ensured that the measurement system is stable and repeatable in pursuit of each objective function. For an attribute measure, the objective function as a response may be "graded" on a "variable" scale, e.g., ranging from 1 to 100, arbitrarily but still reasonably.

2.3.1. Quantifiable and Achievable

Each experiment aims at predicting the outcome (output) by varying the input variables, namely, independent factors. Each outcome must be quantifiable and achievable so that all possible combinations of the given factors investigated (full factorial) relating to the output can be mathematically expressed as γ:

$$Y = \gamma_0 + \sum_{i=1}^{K} \gamma_i x_i + \sum_{i=1}^{K} \sum_{j=1}^{K} \gamma_{ij} x_i x_j + \sum_{i=1}^{K} \sum_{j=1}^{K} \sum_{m=1}^{K} \gamma_{ijm} x_i x_j x_m + \dots + \varepsilon \qquad (2.3.1)$$

where

   Y is the dependent variable
   x_i is the independent variables
   γ_0 is a constant, the intercept
   γ_i is the coefficient for linear effects
   γ_{ij} is the coefficient for quadratic effects
   γ_{ijm} is the coefficient for three-factor interactive effects
   K is the total number of variables
   i, j, and m are the indexes, where $1 \le i \le K$, $1 \le j \le K$, and $1 \le m \le K$
   ε is the residual

The main credit in experimental design consists in the establishment of a valid, reliable, and replicable relationship between the independent outcome and input variables. Assume that factor one (x_1) has been found hypothetically to have a significant influence on outcome Y according to the test result intuitively; nevertheless, another variable (namely, x_2) is truly the one that leads to Y, while x_1 is not the true cause at all. Then x_2 is called a spurious variable. When the second variable (x_2) is involved and has not been controlled for, the relation is said to be a zero-order relationship. There are two possibilities for generating a zero-order relationship:

1. Intervening variable: Variable x_2, which lies between the suspected causing factor x_1 and outcome y, is the true causing factor. Then x_2 is an intervening variable.

2. Anteceding variable: Variable x_2, which comes into effect prior to the suspected causing factor (x_1) that consequently influences the test result, is the true causing factor. Then x_3 is an anteceding variable.

It is necessary for the experimenter to eliminate the effects of spurious, either intervening or anteceding, variables, otherwise to put them under control. Randomization, replication, and blocking are three techniques to reduce these undesired effects [Kensler et al. 2015].

2.3.2. RSM: Regression Using DOE

The RSM is often used to refine models after influential factors using screening factorial designs have been found out, especially when it is suspicious that curvatures exhibit in the response surface. Regression analysis is a form of inferential statistics. One example of response surface is the tensile strength of a die-cast aluminum part expressed as a function of die preheating pressure and squeeze pressure as shown in Figure 2.3.1.

FIGURE 2.3.1   Response surface of depth of wear versus load and frequency.

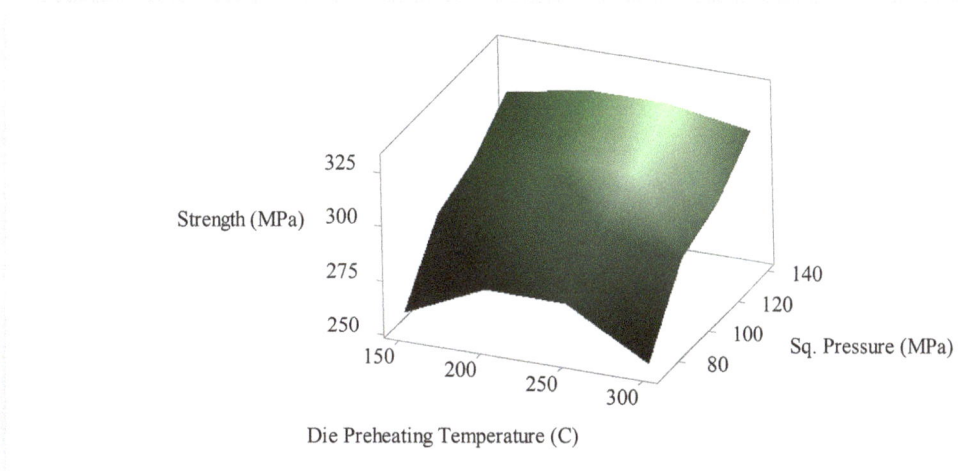

Since the experimental design at each stage may depend on the result of the previous one, a sequential series of such experimental designs is recommended to assure the robustness of the design. This is a general practice when using the RSM. A typical series of experiments consist of the following:

(a) Screening design, usually based on linear fractional factorial designs, is intended to identify potentially significant factors. In many applications, the number of factors that potentially affect the objective function is too great to study all factors in detail. The usual goal of a screening design is to identify the most important factors that affect the outcome.

(b) A more sophisticated quadratic (or even full factorial) RSM is intended to fully characterize the effects of important variables. This is to do optimization experiments to seek out the relationships among the most important factors and the objective function.

(c) A follow-up with confirmation runs is intended to verify the results. This is to confirm settings that improve precision, linearity, and bias.

When looking for Y_p as an approximation to outcome Y in Equation (2.3.1) with no residuals, such as:

$$Y_p = \gamma_0 + \sum_{i=1}^{K} \gamma_i x_i + \sum_{i=1}^{K} \sum_{j=1}^{K} \gamma_{ij} x_i x_j + \sum_{i=1}^{K} \sum_{j=1}^{K} \sum_{m=1}^{K} \gamma_{ijm} x_i x_j x_m \qquad (2.3.2)$$

one looks forward to following the search of the sequential series of experiments. Linear response surfaces for screening design with follow-up quadratic response surfaces for optimization are the commonly used strategies in practical applications.

Design for Linear Response Surface: Designed experiments should be executed iteratively so that information learned in one experiment can be applied to the next. Rather than running a very large experiment with many factors and using up most resources, one may start with a smaller experiment and then build on the results. A good guideline is not to invest more than one-quarter of your budget in the screening design. The equation for rendering a linear response surface can be obtained after truncating the higher-order terms in Equation (2.3.1) as:

$$Y = \gamma_0 + \sum_{i=1}^{K} \gamma_i x_i + \varepsilon \qquad (2.3.3)$$

The main effects are explored using the screening design. In general, if the main effect of a variable is not significant, its interactions with other factors are weak.

Design for Quadratic Response Surface: In a second-order quadratic model, typically used in response surface DOE with suspected curvatures, the three-way interactions and higher-order terms are omitted. The second-order equation can be then written as:

$$Y = \gamma_0 + \sum_{i=1}^{K} \gamma_i x_i + \sum_{i=1}^{K} \sum_{j=1}^{K} \gamma_{ij} x_i x_j + \varepsilon \qquad (2.3.4)$$

Sometimes, one begins with a smaller design to check whether square terms or interactions are present before you add more runs to estimate those terms. Foldover designs and composite designs (CD) with axial runs are two strategies in sequential experimentation. Optimization can then be carried out based on such as a robust regression given above. When additional runs are conducted at various center points, it is called central composite design (CCD) (Chapter 5). RSMs have been widely used in combination with finite element methods (FEM) and fatigue theories, replacing complex sensitivity analyses, in the field of structural design optimization and optimal product life prediction.

The reduction of the number of coefficients from Equations (2.3.1) to (2.3.4) for quadratic response design, and further to Equation (2.3.3) for linear RSM, is listed in Table 2.3.1. A metamodel is hereupon defined as an approximation model of the response such as Equation (2.3.3), and it may be described in terms of regression models or other techniques from a mathematical point of view. Based on the verification and validation using some simulation samples, a matching metamodel can be constructed and can be used later to predict response values at new factor settings. The construction of a metamodel is called fitting or learning in the realism of computer simulations.

CHAPTER 2 Full Factorial Design 2^K 35

TABLE 2.3.1 Reduction of unknown coefficients from the full model to linear and quadratic response designs.

No. of factors	Full model	Linear terms only	Up to quadratic terms	
2	$2^2 = 4$	3: $\gamma_0, \gamma_1, \gamma_2$	4:	$\gamma_0, \gamma_1, \gamma_2, \gamma_{12}$
3	$2^3 = 8$	4: $\gamma_0, \gamma_1, \gamma_2, \gamma_3$	7:	$\gamma_0, \gamma_1, \gamma_2, \gamma_3, \gamma_{12}, \gamma_{13}, \gamma_{23}$
4	$2^4 = 16$	5: $\gamma_0, \gamma_1, \gamma_2, \gamma_3, \gamma_4$	11:	$\gamma_0, \gamma_1, \gamma_2, \gamma_3, \gamma_4,$ $\gamma_{12}, \gamma_{13}, \gamma_{14}, \gamma_{23}, \gamma_{24}, \gamma_{34}$
5	$2^5 = 32$	6: $\gamma_0, \gamma_1, \gamma_2, \gamma_3, \gamma_4, \gamma_5$	16:	$\gamma_0, \gamma_1, \gamma_2, \gamma_3, \gamma_4, \gamma_5,$ $\gamma_{12}, \gamma_{13}, \gamma_{14}, \gamma_{15},$ $\gamma_{23}, \gamma_{24}, \gamma_{25},$ $\gamma_{34}, \gamma_{35},$ γ_{45}
6	$2^6 = 64$	7: $\gamma_0, \gamma_1, \gamma_2, \gamma_3, \gamma_4, \gamma_5, \gamma_6$	22:	$\gamma_0, \gamma_1, \gamma_2, \gamma_3, \gamma_4, \gamma_5, \gamma_6,$ $\gamma_{12}, \gamma_{13}, \gamma_{14}, \gamma_{15}, \gamma_{16},$ $\gamma_{23}, \gamma_{24}, \gamma_{25}, \gamma_{26},$ $\gamma_{34}, \gamma_{35}, \gamma_{36},$ $\gamma_{45}, \gamma_{46},$ γ_{56}
7	$2^7 = 128$	7: $\gamma_0, \gamma_1, \gamma_2, \gamma_3, \gamma_4, \gamma_5, \gamma_6, \gamma_7$	29:	$\gamma_0, \gamma_1, \gamma_2, \gamma_3, \gamma_4, \gamma_5, \gamma_6, \gamma_7,$ $\gamma_{12}, \gamma_{13}, \gamma_{14}, \gamma_{15}, \gamma_{16}, \gamma_{17},$ $\gamma_{23}, \gamma_{24}, \gamma_{25}, \gamma_{26}, \gamma_{27},$ $\gamma_{34}, \gamma_{35}, \gamma_{36}, \gamma_{37},$ $\gamma_{45}, \gamma_{46}, \gamma_{47},$ $\gamma_{56}, \gamma_{57},$ γ_{67}
8	$2^8 = 256$	7: $\gamma_0, \gamma_1, \gamma_2, \gamma_3, \gamma_4, \gamma_5, \gamma_6, \gamma_7, \gamma_8$	37:	$\gamma_0, \gamma_1, \gamma_2, \gamma_3, \gamma_4, \gamma_5, \gamma_6, \gamma_7, \gamma_8,$ $\gamma_{12}, \gamma_{13}, \gamma_{14}, \gamma_{15}, \gamma_{16}, \gamma_{17}, \gamma_{18},$ $\gamma_{23}, \gamma_{24}, \gamma_{25}, \gamma_{26}, \gamma_{27}, \gamma_{28},$ $\gamma_{34}, \gamma_{35}, \gamma_{36}, \gamma_{37}, \gamma_{38},$ $\gamma_{45}, \gamma_{46}, \gamma_{47}, \gamma_{48},$ $\gamma_{56}, \gamma_{57}, \gamma_{58},$ $\gamma_{67}, \gamma_{68},$ γ_{78}

© SAE International.

2.4. Factors

An experimenter not only talks about the response to the experiment but also thinks about relating factors to the response. A factor is just any categorical independent variable. Any two factors are expected to be independent of each other. Orthogonality, which means that each of the variables is independent, must be guaranteed in the selection of factors.

A treatment is defined as a particular combination of levels of the factors in an experimental test. A treatment constitutes a set of run conditions made of all design parameters at the selected levels, respectively. In brief, the term "run" is often used instead of "treatment." Note that "order" means the order to carry out all the treatments, experimentally or analytically, and thus an order number differs from its corresponding treatment number.

2.4.1. Types of Factors

Two types of factors may be involved in an experiment: design factors and nuisance factors. Design factors are those selected for study while nuisance factors do not present interest in the designated experiment. Typical nuisance factors are environmental temperature and humidity. Nuisance factors may be controllable, i.e., one may control their variations, or uncontrollable, i.e., parameters cannot be controlled during product use. Nuisance factors can be classified and dealt with as follows:

Nuisance characteristics	Example	How to deal with
Unknown and uncontrollable	Experimenter bias	Randomization
Known and uncontrollable, but measurable	Weight, volume, etc.	Experience counts
Known and controllable	Time, gender, etc.	Blocking

2.4.2. Effects of Factors and Their Interactions

A quantitative understanding of the factors that influence problem resolution, linearity, precision, and accuracy is integral to applying DOE. Problem-solving consists of using generic or ad hoc methods in an orderly manner to find solutions to problems. Some of the problem-solving techniques deployed rely on how to identify the significant factors and how to relate the significant factors to the response, such as closed-form solutions, DOE, D-optimal method, and neural networking. DOE is the first priority after the closed-form solutions, if the experiments can be designed and fully developed; if not, the D-optimal method may be applied; and if not, the methodology of (artificial) neural networks will be the final and last choice. Assume that

$$Y = f(A, B, C) \tag{2.4.1}$$

then the infinitesimal change of dependent variable y in response to the simultaneous infinitesimal changes of all the independent variables is

$$dY = (\partial f / \partial A)\, dA + (\partial f / \partial B)\, dB + (\partial f / \partial C)\, dC \tag{2.4.2}$$

When the variation of each variable is within a piecewise linear range, the above equation can be obtained as:

$$\Delta Y = (\partial f / \partial A)\, \Delta A + (\partial f / \partial B)\, \Delta B + (\partial f / \partial C)\, \Delta C \tag{2.4.3}$$

or

$$\Delta Y = (\partial f / \partial A)\,(A^+ - A^-) + (\partial f / \partial B)\,(B^+ - B^-) + (\partial f / \partial C)\,(C^+ - C^-) \tag{2.4.4}$$

where

A^+, B^+, and C^+ are high levels of variables A, B, and C, respectively
A^-, B^-, and C^- are low levels of variables A, B, and C, respectively
ΔA, ΔB, and ΔC are contrasts between high and low levels of variables A, B, and C, respectively

If the contribution of any factor (e.g., factor A) is not significant, substituting any value (ranging from A^- to A^+) into Equation (2.4.1) will yield the same Y value, of course with an insignificant random error. This constitutes the null hypothesis for the DOE. Nevertheless, the significance level has to be identified and proven statistically. The effect of a factor is conventionally described as the change in the response when the factor of concern goes from a low level to a high level.

(a) A positive effect means that going from a low level to a high level of a factor increases the response.

(b) A negative effect means that going from a low level to a high level of a factor decreases the response.

Conventionally, if the goal is to maximize a response, all factors with a positive effect would be set to operate at their high levels and all factors with a negative effect would be set to operate at their low levels. If the goal is to minimize the response, all factors with a positive effect would be set at low levels and all with a negative effect would be set at high levels.

2.4.3. Selections—Doable and Controllable

Level is defined as the value assigned to, and thus assumed by, a factor in an experiment. Factorial designs let an experimenter or analyst take a comprehensive approach to study all potential input variables. Removing a factor from the experiment slashes the chance of determining its importance to zero. For each factor (parameter), the experimenter is encouraged to determine the extrema, but realistic, high and low levels as wished to be investigated. Extreme levels of each factor selected should be realistic, not absurd. It is not encouraged to have extreme levels that go beyond what is now in use as it may lead to an impractical solution.

2.4.4. Dimensionless Format

A design matrix is created for the factors under investigation in the DOE. The design matrix will show all possible combinations of high and low levels for each input factor. These high and low levels can be generically coded as −1 and +1, e.g., (−1, 1) for a two-level design and (−1, 0, 1) for a three-level design. If A is a variable, its dimensionless variable a, which varies between −1 and 1, can be represented as:

$$-1 \leq a = \frac{A - A_{average}}{A_{range/2}} \leq 1 \tag{2.4.5}$$

of which A is the variable of interest. The average and range can be calculated respectively as:

$$A_{average} = \tfrac{1}{2}(A_+ + A_-) \tag{2.4.6}$$

and

$$A_{range/2} = \tfrac{1}{2}(A_+ - A_-) \tag{2.4.7}$$

where A_+ and A_- are variable A evaluated at high and low levels, respectively. Select the widest applicable level range $(A_+ - A_-)$ if practically feasible.

2.4.5. Design Contrasts and Effects

Contrasts can be represented by vectors and sets of orthogonal contrasts that are uncorrelated and independently distributed if the data are normal distributed statistically. Because of this independence, each orthogonal treatment provides different information to the others, as shown in Figure 2.4.1 for a DOE with three factors—A, B, and C. Orthogonality assures that the forms of comparison (contrasts) can be legitimately and efficiently carried out. Assume that a, b, and c are the three dimensionless coded variables (varying between −1 and 1) for factors A, B, and C, respectively. Design contrasts of the individual main effects contributed by individual factors can be described as follows:

1. Factor A: All the nodes with a = 1 versus all the nodes with a = −1
2. Factor B: All the nodes with b = 1 versus all the nodes with b = −1
3. Factor C: All the nodes with c = 1 versus all the nodes with c = −1

Assume that Y = f(A, B, C) for the study case with the design matrix 3^3 as shown in Figure 2.4.2. Then, one can search for the solution in the entire 3D "solid" space, and the corresponding design contrasts of individual main effects are mathematically described as:

$$\text{Contrast of A} = [Y(1, -1, -1) + Y(1, -1, 1) + Y(1, 1, -1) + Y(1, 1, 1)] / 4 -$$
$$[Y(-1, -1, -1) + Y(-1, -1,1) + Y(-1, 1, -1) + Y(-1, 1, 1)] / 4 \tag{2.4.8}$$

$$\text{Contrast of B} = [Y(-1, 1, -1) + Y(-1, 1, 1) + Y(1, 1, 1) + Y(1, 1, -1)] / 4 -$$
$$[Y(-1, -1, -1) + Y(-1, -1, 1) + Y(1, -1, 1) + Y(1, -1, -1)] / 4 \tag{2.4.9}$$

and

$$\text{Contrast of C} = [Y(-1, -1, 1) + Y(1, -1, 1) + Y(1, 1, 1) + Y(-1, 1, 1)] / 4 -$$
$$[Y(-1, -1, -1) + Y(1, -1, -1) + Y(1, 1, -1) + Y(-1, 1, -1)] / 4 \tag{2.4.10}$$

FIGURE 2.4.1 Design contrasts as illustrated using the 2^3 design.

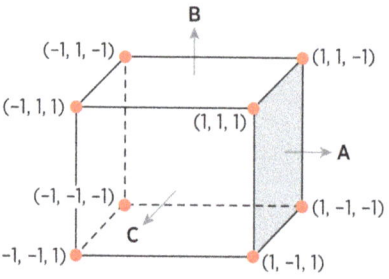

© SAE International.

When the design contrast of factor A is calculated, the variation subjected to factor B is neutralized simultaneously (i.e., −1 + −1 + 1 + 1 = 0 for a = 1, while −1 + −1 + 1 + 1 = 0 for a = −1) so is the variation subjected to factor C neutralized. Design contrasts of factors B and C are also calculated by that token. Thus, the main effects of factors A, B, and C are defined as follows:

$$\text{Main effect of factor A} = \tfrac{1}{2} \text{ Contrast of factor A} \tag{2.4.11}$$

$$\text{Main effect of factor B} = \tfrac{1}{2} \text{ Contrast of factor B} \qquad\qquad (2.4.12)$$

and

$$\text{Main effect of factor C} = \tfrac{1}{2} \text{ Contrast of factor C} \qquad\qquad (2.4.13)$$

2.4.6. Statistical Significance

Each DOE should include a clear statement proposing the analysis-related statistical distribution and DOFs to be undertaken before a "significant" statistical inference can be drawn. It typically involves the manipulation of the process of statistical analysis and the DOFs until they return a figure below a certain threshold level of statistical significance, e.g., $\alpha = P(t_v) \leq 5\%$, as shown in Figure 2.4.2, of which $P(t_v)$ is the so-called probability value (p-value). One must be cautious of adopting the p-value since dubious conclusions often result from exercising the experimenter's own confirmation bias. The p-value indicates the degree to which the data conform to the trending predicted by the test hypothesis including all other assumptions used in the test (the underlying statistical model). For example, p-value = 1% means that the data are not close to what the statistical model (including the test hypothesis) predicted; while p-value = 30% means that the data are much closer to the model prediction, allowing for chance variation.

FIGURE 2.4.2 P-value locating unlikely observations of a statistical distribution.

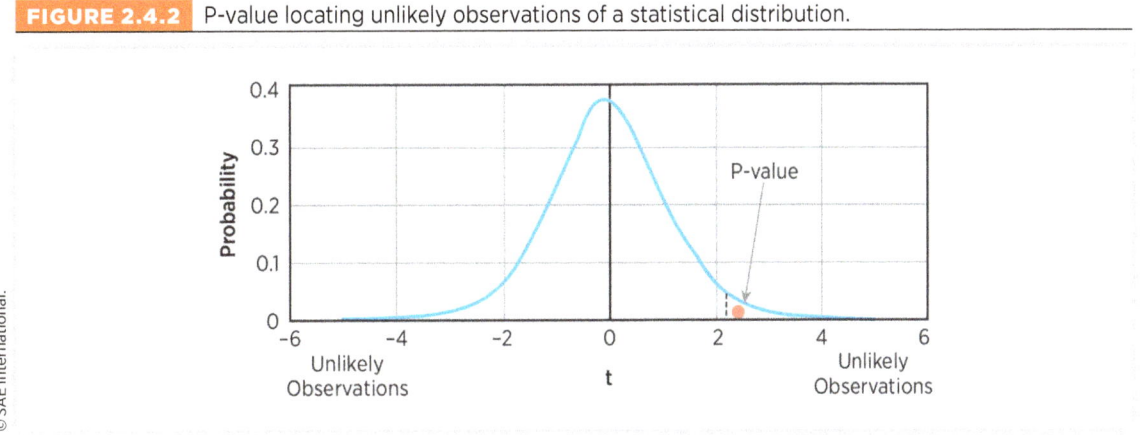

The conclusion of a statistical inference is a statistical proposition. Both t-test and F-test are widely applied statistics, based on which statistical inferences are drawn from DOE. The t-test with null hypothesis is the most convenient statistic for a design matrix with factors of two levels, while the F-test is generally aimed at a design involving three levels or higher. Note that F-test is based on the assumption that the test data are of the population that has a normal distribution.

In general, no statistical significance can be drawn because of the following mistakes [Anderson and Whitcomb 2015]: studying the wrong problem, measuring the wrong response, not having adequate precision, studying the wrong factors, and having too many runs outside the region of operability.

2.5. Statistical Inference by Student's t-Distribution

Student's t-distribution, also simply the t-distribution, is a member of a family of continuous probability distributions that appears when estimating the mean of a normally distributed population in situations such as conducting DOE, of which the sample size is limited and the population standard deviation is unknown. The probability density function (PDF) of a t-distribution is symmetric and bell shaped, but it has elongated tails, meaning that it tends to produce extreme values that fall far from its mean in contrast to the normal distribution function. Because of the complexity of numerical calculation, the probability of cumulative distribution function (CDF) of the student's t-distribution is listed in Table 2.5.1 for reference.

TABLE 2.5.1 Probability points of accumulative t (student's t) distribution (probability = $1 - \alpha$; one-sided).

DOF	75%	80%	85%	90%	95%	97.5%	99%	99.5%	99.9%	99.95%
1	1.000	1.376	1.963	3.078	6.314	12.710	31.820	63.660	318.300	636.600
2	0.816	1.080	1.386	1.886	2.920	4.303	6.965	9.925	22.330	31.600
3	0.765	0.978	1.250	1.638	2.353	3.182	4.541	5.841	10.210	12.920
4	0.741	0.941	1.190	1.533	2.132	2.776	3.747	4.604	7.173	8.610
5	0.727	0.920	1.156	1.476	2.015	2.571	3.365	4.032	5.893	6.869
6	0.718	0.906	1.134	1.440	1.943	2.447	3.143	3.707	5.208	5.959
7	0.711	0.896	1.119	1.415	1.895	2.365	2.998	3.499	4.785	5.408
8	0.706	0.889	1.108	1.397	1.860	2.306	2.896	3.355	4.501	5.041
9	0.703	0.883	1.100	1.383	1.833	2.262	2.821	3.250	4.297	4.781
10	0.700	0.879	1.093	1.372	1.812	2.228	2.764	3.169	4.144	4.587
11	0.697	0.876	1.088	1.363	1.796	2.201	2.718	3.106	4.025	4.437
12	0.695	0.873	1.083	1.356	1.782	2.179	2.681	3.055	3.930	4.318
13	0.694	0.870	1.079	1.350	1.771	2.160	2.650	3.012	3.852	4.221
14	0.692	0.868	1.076	1.345	1.761	2.145	2.624	2.977	3.787	4.140
15	0.691	0.866	1.074	1.341	1.753	2.131	2.602	2.947	3.733	4.073
16	0.690	0.865	1.071	1.337	1.746	2.120	2.583	2.921	3.686	4.015
17	0.689	0.863	1.069	1.333	1.740	2.110	2.567	2.898	3.646	3.965
18	0.688	0.862	1.067	1.330	1.734	2.101	2.552	2.878	3.610	3.922
19	0.688	0.861	1.066	1.328	1.729	2.093	2.539	2.861	3.579	3.883
20	0.687	0.860	1.064	1.325	1.725	2.086	2.528	2.845	3.552	3.850
21	0.686	0.859	1.063	1.323	1.721	2.080	2.518	2.831	3.527	3.819
22	0.686	0.858	1.061	1.321	1.717	2.074	2.508	2.819	3.505	3.792
24	0.685	0.857	1.059	1.318	1.711	2.064	2.492	2.797	3.467	3.745
26	0.684	0.856	1.058	1.315	1.706	2.056	2.479	2.779	3.435	3.707
28	0.683	0.855	1.056	1.313	1.701	2.048	2.467	2.763	3.408	3.674
30	0.683	0.854	1.055	1.310	1.697	2.042	2.457	2.750	3.385	3.646
40	0.681	0.851	1.050	1.303	1.684	2.021	2.423	2.704	3.307	3.551
50	0.679	0.849	1.047	1.299	1.676	2.009	2.403	2.678	3.261	3.496
60	0.679	0.848	1.045	1.296	1.671	2.000	2.390	2.660	3.232	3.460
120	0.677	0.845	1.041	1.289	1.658	1.980	2.358	2.617	3.160	3.373
∞	0.674	0.842	1.036	1.282	1.645	1.960	2.326	2.576	3.090	3.291

2.5.1. t-Distribution

The student t-test is one of the simplest statistical tests to perform and provides insight into the nature of statistical comparisons. The significance level of an effect or interaction can be obtained by checking the t-distribution, of which the PDF is [Ross 1987]:

$$f(t) = \frac{\Gamma[\frac{1}{2}(v+1)]}{(v\,\pi)^{1/2}\,\Gamma[\frac{1}{2}\,v]}\,(t^2/v+1)^{-(v+1)/2} \qquad (2.5.1)$$

where

t is the student's t variable and $-\infty < t < \infty$
v is the DOFs, $v \geq 1$ or the t-distribution is not defined
Γ is the Gamma function

When $v = \infty$, PDF f(t) in the t-distribution becomes a normal distribution function. Its curves resemble the PDF of normal distribution very much, as shown in Figure 2.5.1.

FIGURE 2.5.1   PDFs of t-distribution with different v DOFs.

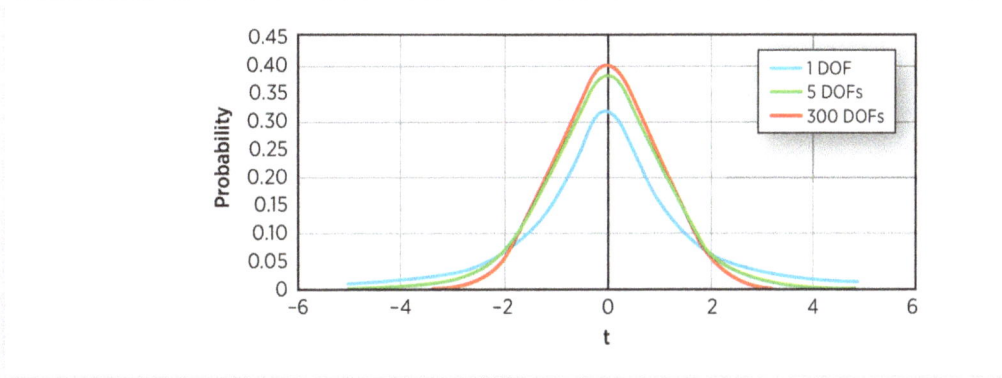

The numerical value for a student's t-statistics is approximated by the difference between the calculated effect and mean value as measured by the sample standard deviation, i.e.,

$$t = (\text{Calculated effect} - \text{Mean value})/\text{Sample standard deviation}$$

Or say that

$$t_v = (\gamma_{\text{effect}} - \gamma_{\text{null}})/S \qquad (2.5.2)$$

where

γ_{effect} is the effect, main or interactive
γ_{null} is the mean value, i.e., the target or expected value, which is "zero" by null hypothesis
S is the sample standard deviation

When conducting a DOE, one may assume that $\gamma_{\text{null}} = 0$ because it is the null hypothesis in nature—no effect until it is proven to be effective.

2.5.2. Null Hypothesis

In statistical inference on observational data, the null hypothesis refers to a general statement or default position that there is no difference between two measured phenomena—before and after the treatment. During the experiment, the experimenter is trying to either reject or fail to reject this hypothesis. Rejecting (or disproving) the null hypothesis means to conclude that there are grounds for believing that there is a difference between two phenomena, e.g., the input variables potentially having a measurable effect on the response variable in the DOE.

The t-test (in its simplest form) is used to compare an observed or experimentally determined mean value from some perfectly known expected value (typically the null hypothesis of no treatment effect). Therefore, the null hypothesis, denoted by H_o, is a statement of "no effect" or "no difference." On the other hand, the difference between the effect and the target is significant if $P(t_v) = \alpha$ is small (e.g., $\alpha \leq 10\%$), as long as the difference goes to an extremity. It tells that the influence of the factorial effect on the response is valid. Thus, in inferential statistics, the null hypothesis and alternative hypothesis are respectively:

- H_o: There is no difference between the effect and target if $p(t_v)$ is large, e.g., $p(t_v) > 10\%$

and

- H_1: There is a difference between the effect and target if $p(t_v)$ is small, e.g., $p(t_v) < 10\%$

Note that $p(t_v)$, namely, p-value, is the probability for the difference between the effect and its related target as denominated by the sample standard deviation, as shown in Figure 1.2.1. It is also called asymptotic significance for t-distribution [Stockburger 2007]. In statistical hypothesis testing, a p-value is the probability for a given statistical model that the statistical inference would be the same as or of greater magnitude than the actual observed result when the null hypothesis is true.

In case the null hypothesis is rejected, an alternative hypothesis is accepted in its place. A p-value (i.e., α) can be used to locate the unlikely observations on one side of a statistical distribution for some cases or to locate the unlikely observations on both sides for other cases. When α is small (usually $\alpha \leq 10\%$ or 5%), one is confident that observed differences are unlikely subjected to chance alone and that repeating the experiment will give similar results. It is analogous to the legal practice in the US in which a defendant is presumed to be innocent (null hypothesis) until proven guilty beyond reasonable doubt (alternative hypothesis). The reasonable doubt is drawn from a statistical inference by the significant level (p-value).

2.5.3. Determine the Experimental Error

Reliability can be affected by the validity of how the experiments are conducted. One may ask whether it tests what is meant to test. This will be answered by observing the sample random errors, also called experimental errors, i.e., residual ε given in Equation (2.3.1). The experimental error will be used for checking the statistical significance of each main effect or interaction on the response. There are several ways to determine the experimental error. Among the options are

1. Taking multiple independent measurements under the same treatment
2. Pooling the "insignificant" effects
3. Making sense of the potential engineering error
4. Using dummy factors
5. Selecting the center point of the experimental levels and make several experimental runs at that point
6. Replicating the entire experimental matrix

2.5.4. Experimental Error with Replications by Sample Variance Sample Standard Deviation

Given Q sample variances resulting from Q treatments in a DOE, the unbiased estimate of overall sample standard deviation (i.e., random error) can be obtained by combining individual sample variances S_q^2 associated with treatment q ($1 \leq q \leq Q$) and their associated DOFs ν_q ($1 \leq q \leq Q$) as:

$$S = \left(\frac{\nu_1 S_1^2 + \nu_2 S_2^2 + \dots + \nu_q S_q^2 + \dots + \nu_Q S_Q^2}{\nu_1 + \nu_2 + \dots + \nu_q + \dots + \nu_Q}\right)^{1/2} \tag{2.5.3}$$

where
S is the sample standard deviation of the combined data of all treatments
S_q is the sample standard deviation of the qth treatment in DOE; q = 1, 2, ..., Q

Note that ν_q is the DOF of data within treatment q and the denominator of the above equation is the total DOFs (ν) that:

$$\nu = \nu_1 + \nu_2 + \dots + \nu_Q \tag{2.5.4}$$

How is each S_q^2 calculated? Consider Y_{qn}, where subscript q is the treatment and subscript n is the number of replications within the qth treatment, representing a single piece of data obtained from physical tests. Note that a treatment is a run condition with all the involved factors being set up according to the DOE. Then, each individual sample standard deviation, i.e., S_q, is:

$$S_q^2 = \left[\sum_{n=1}^{N_q} (y_{qn} - \bar{y}_q)^2\right] / (N_q - 1) \tag{2.5.5}$$

and

$$\nu_q = N_q - 1 \tag{2.5.6}$$

where
S_q^2 is the sample variance of treatment q
y_{qn} is an individual data at replication n under treatment q
$\bar{y}_q$ is the average of replicated experimental results, abiding within treatment q
N_q is the number of replications under treatment q
n is the index of replication
ν_q is the DOF corresponding to treatment q

The number of replications for individual treatment may be different. Each piece of data possesses one DOF inherently. The sample variance is calculated after taking the average, which is a statistic that accounts for 1 DOF. Therefore, there are $N_q - 1$ DOFs associated with the qth sample variance, given that there are N_q replications under the qth treatment.

2.5.5. Experimental Error with No Replications

The algorithm for taking multiple independent measurements under the same treatment (i.e., $N_q \geq 2$) is described by Equation (2.5.3). When there is only one data for each treatment ($N_q = 1$), there will be no replicated error. The alternate estimate of the overall sample standard deviation can be obtained by pooling the insignificant main and interactive effects, designated as E_m^2 ($1 \leq m \leq M$), where m is the number of insignificant main and interactive effects. By the same token as Equation (2.5.4), the sample standard deviation (i.e., experimental random error) of all the data can be calculated as:

$$S = \left(\frac{E_1^2 + E_2^2 + \ldots + E_M^2}{M} \right)^{1/2} \tag{2.5.7}$$

where

$E_1, E_2, \ldots, E_M$ is the insignificant main and interactive effects
M is the total number of insignificant effects

Of these six approaches (Section 2.5.3), pooling the insignificant main and interactive effects is not always favored, but it can be effectively applied to "measuring" the data that are obtained using numerical methods (e.g., FEM) with no random-error message in the nature of things.

2.5.6. Engineering Error

Another approach to the experimental error is to make engineering sense of the situation. In engineering practice, a 2% error is usually considered tolerable, and thus an effect (main or interactive) that is less than 2% of the grand average (i.e., the constant of the predictive equation) is excluded from the metamodel for formulating a reasonable predictive equation. In fact, the LR, as proposed in Chapters 7, 8, and 9 (Volume II), has a similar concept of operations.

2.5.7. Test for Outliers

During a DOE in the product development process, another question people ask frequently "Are the extreme observations outliers? Are they due to anomalous causes?" These questions can be translated into statements in the statistical language [Stockburger 2007] as:

$$H_o \text{ (null hypothesis):} \qquad \text{Extremity} = \mu \tag{2.5.8}$$

or

$$H_1 \text{ (alternative hypothesis):} \qquad \text{Extremity} \neq \mu \tag{2.5.9}$$

of which "Extremity" and μ are the extremity of interest and performance mean, respectively. When an extremity is an observation distant from other observations, it is called an outlier. An extremity can be too small or too large.

The modified Thompson tau test is a sustainable practice based on the student's t-distribution, of which the rejection zone is defined by the following equation:

$$\text{Rejection Zone} = \frac{(N - 1) \, t_{\alpha/2}}{[N \, (N - 2 + t_{\alpha/2}^2)]^{1/2}} \tag{2.5.10}$$

where

N is the sample size

$t_{\alpha/2}$ is the critical value in the t-distribution with the associated DOFs.

Whether a specific variable, say variable y, is an extreme can be checked using the following equation:

$$\text{If } |(y - \bar{y}) / S_y| > \text{Rejection Zone} \tag{2.5.11}$$

x is an outlier; otherwise, x is not. Substituting Equation (2.5.10) into Equation (2.5.11) leads to:

$$t_{\alpha/2}^2 < \frac{N(N-2)\,|(y-\bar{y})/S_y|^2}{(N-1)^2 - |(y-\bar{y})/S_y|^2} \tag{2.5.12}$$

This test can be used to evaluate multiple possible outliers. One may start with the farthest outlier; an outlier can be discarded, or even noticeably included, in further analysis [Adikaram et al. 2014].

EXAMPLE 2.5.1

The following data were obtained from an environmental durability test: $\bar{y} = 70{,}950$ cycles (sample average), $S_y = 31{,}030$ cycles (sample standard deviation), and Q = 7 (sample size). Given that $y_3 = 10{,}151$ cycles, is the third data an outlier?

Solution:

Given $|(y_3 - \bar{y}) / S_y| = |(10151 - 70950) / 31030| = 1.96$

then $t_{\alpha/2}^2 < Q(Q-2)\,|(y_3 - \bar{y})/S_y|^2 / [(Q-1)^2 - |(y_3 - \bar{y})/S_y|^2]$

$= 7(7-2)(1.96)^2 / [(7-1)^2 - (1.96)^2] = 4.178$

i.e., $t_{\alpha/2} < 2.044$.

$P(t6, \alpha/2 < 2.044) \approx 5\%$. In other words, $\frac{1}{2}\alpha \approx 5\%$. Thus, the statistical confidence for the first test of the third data to be an outlier is 95%.

2.5.8. Process Bias

Bias is defined as the difference between the true value and the observed average of the measurements, as described by Equation (1.2.1), which presents a relocation of the mean. Accuracy means that the process is not biased, while reliability means that the process has low variability. Inaccuracy results from systematic error, while unreliability is subject to random errors. The misunderstanding between the reliability and accuracy of experimental tests can be clarified as follows:

	Reliable	Unreliable
Accurate	Answers are consistently correct with little variation between them.	Answers are consistently correct, but they vary from one to another.
Inaccurate	Answers are consistently incorrect with little variation between them.	Answers are consistently incorrect, and they vary from one to another.

The accuracy of an experiment is how close the measured result is to the correct accepted value. Bias may be eliminated or reduced by calibration of standards and/or instruments. Errors that contribute to bias can be present even when all equipment and standards are properly calibrated and under control in the beginning. Temperature, vibration, and moisture probably have a high potential for introducing bias into measurements.

2.5.9. Process Stability

Process stability is defined as the total variation (drift) in the measurement obtained with a measurement system on the same part characteristic over an extended period. It is the change in bias over time. The stability of a generalized measurement system can be tested like a production process using statistical process control charts.

2.6. Full Factorial Design with Two Levels

The study hereon is on how to provide a wise decision support when facing different types of DOE for the right product reliability growth. Let us start with the 2^K design, which is defined as the $2 \times 2 \times \cdots \times 2$ design with K factors. Each factor level of every factor appears the same number of times in such a balanced design. Orthogonality is guaranteed for any pair of factors, each possible level combination appears the same number of times in the design.

2.6.1. 2^3 Factorial Design with No Replication

Here is a simple and practical example that walks engineers through the basic ideas behind DOE. Three different test procedures for assessing nail quality are documented in ASTM F680, Methods of Testing Nails: (1) impact bend angle test, (2) flexural yield strength test, and (3) hardness test. Extensive tests to differentiate these three test procedures by three distinct physical parameters (factors), i.e., carbon content, nominal shank diameter, and hardening, were done by [White et al. 1990], as given in Table 2.6.1. Subject to the inhomogeneity of the constitutive materials of a nail and the stochastic nature of the manufacturing process, the influence of these three physical properties on any of the three test procedures cannot be formulated deterministically based on physical laws in the mechanics of materials.

Given the nonlinear variation of the flexural yield strength versus carbon content [White et al. 1990], the analysis is divided into two blocks, of which each is made of a 2^3 design. Relevant factors and their corresponding design levels are listed in Table 2.6.1. The example data of impact bend angle in the low carbon region will be used for illustrating the analysis procedure, while leaving the analysis in the high carbon region as homework practice for students. The statistical model of a 2^3 factorial design can be represented by:

$$Y = \text{mean} + \gamma_a\,a + \gamma_b\,b + \gamma_c\,c + \gamma_{ab}\,a\,b + \gamma_{ac}\,a\,c + \gamma_{bc}\,b\,c + \gamma_{abc}\,a\,b\,c \qquad (2.6.1)$$

where

Y is the response, e.g., impact bend angle, flexural yield strength, or Vickers hardness mean is the sample mean, i.e., average

a is the dimensionless variable for A, a = −1 for SAE 1010 steel and a = 1 for SAE 1022 steel

b is the dimensionless variable for B, b = −1 for B = 2.67 mm and b = 1 for B = 3.43 mm

c is the dimensionless variable for C, c = −1 for no hardening and c = 1 for "hardened"

γ_a, γ_b, and γ_c are individual main effects of factors A, B, and C, respectively, γ

γ_{ab}, γ_{ac}, and γ_{bc} are two-factor interactions, i.e., interactive effects of AB, AC, and BC, respectively

γ_{abc} is three-factor interactions, i.e., interactive effects of ABC

TABLE 2.6.1 Design factors for impact bend angle of nails in the low carbon region [Chiang and Chu 1993].

Factor	Level (−)	Level (+)	Variate (−1, 1)
A: Carbon content (SAE #)	SAE 1010	SAE 1022	a = A
B: Shank diameter (mm)	2.67	3.43	b = (B − 3.05)/0.38
C: Hardening	No	Yes	c = C

Used with permission of ASTM International, from "Quality Assessment Testsfor Nails," Chu, MH; Chiang, YJ; Petersen, DR, 21, 1993; permission conveyed through Copyright Clearance Center, Inc.

Test data resulting from the impact bend angle based on the full factorial design for three factors are listed in Table 2.6.2. As usual, the first column is reserved for the treatment number, i.e., the run number for simplicity. Elements of the test matrix are assigned according to the next three columns, as exhibited in Table 2.6.2 for carrying out the physical tests. Note that there are only three independent factors, i.e., A, B, and C. Here are the algorithms for assigning low- and high-leveled working conditions to each variable in a column-wise order:

(a) "−1" and "+1" are assigned alternatively to factor A (Column 2).

(b) "−1 −1" and "+1 +1" are assigned alternatively to factor B (Column 3).

(c) "−1 −1 −1 −1" and "+1 +1 +1 +1" are assigned alternatively to factor C (Column 4).

(d) AB, AC, BC, and ABC are to be calculated according to the multiplication rule.

For example, if A = +1 and B = −1, then AB = (+1) (−1) = −1 within the same treatment.

TABLE 2.6.2 The 2^3 factorial design for impact bend angle of nails in the low carbon region [Chiang and Chu 1993].

Run	Order	a	b	c	ab	ac	bc	abc	Y (°)	Y$_p$# (°)
1	3	−1	−1	−1	1	1	1	−1	66	62.6
2	6	1	−1	−1	−1	−1	1	1	52	50.9
3	1	−1	1	−1	−1	1	−1	1	26	27.6
4	7	1	1	−1	1	−1	−1	−1	28	27.4
5	4	−1	−1	1	1	−1	−1	1	70	73.4
6	8	1	−1	1	−1	1	−1	−1	39	40.1
7	2	−1	1	1	−1	−1	1	−1	40	38.4
8	5	1	1	1	1	1	1	1	16	16.6
Contrast		−16.75	−29.25	−1.75	5.75	−10.75	2.75	−2.25		
Effect		−8.375	−14.625	−0.875*	2.875	−5.375	1.375*	−1.125*	42.125	(Average)
Error		1.143	1.143	1.143	1.143	1.143	1.143	1.143		
t-ratio		7.325	12.791	0.765	2.514	4.701	1.203	0.984		
t$_{3,1-\alpha}$ = t$_{3,90\%}$		1.638	1.638	1.638	1.638	1.638	1.638	1.638		
Significant		Yes	Yes	No	Yes	Yes	No	No		

Notes:

Run (or treatment): Index for numbering the eight test conditions

Order: The order for carrying out the eight experimental tests

A, B, C: Main effects of factors A, B, and C, respectively

AB: Interaction between factors A and B

AC: Interaction between factors A and C

BC: Interaction between factors B and C

ABC: Interaction among factors A, B, and C

t-ratio: t-ratio = effect/error

t$_{3,1-\alpha}$: From t-distribution (Table 2.5.1) at α = 10%, when the t-ratio is smaller than the preset criterion t$_{3, 1-10\%}$ = t$_{3, 90\%}$, the effect is not significant

*: Small effects pooled together to estimate the sample standard error using Equation (2.5.7)

#: Predicted value Y$_p$ based on the criterion that α = 10%

Response variable Y is used to denote the observations obtained from experiments, such as physical tests or finite element analysis (FEA), and its corresponding variable Y_p stands for the predictive value for variable Y. An experimental design matrix can be arranged in two different formats, without or with interactions:

(a) Q×K (number of rows × number of columns) matrix, where index Q is the number of treatments (run conditions) and index K the number of factors.

(b) Occasionally, interactive terms are appended to the main factors, and the number of columns is increased accordingly, as demonstrated in Table 2.6.2 for the design matrix 2^3.

After the design matrix is defined, one can conduct the experimental tests randomly. It means that the order for carrying out tests differs from the order of run conditions (treatments).

Calculation of the design contrast for each factor and interaction is to be illustrated here. Let us consider factor A (2nd column of Table 2.6.2) first. As shown in Figure 2.4.1, all the corner nodes with positive A are on the right of the A-axis, while all the corner nodes with negative A are on the left of the A-axis. The design contrast between the nodes with positive A and those with negative A is then calculated as:

$$A_{contrast} = \tfrac{1}{4} [(1 \times 52) + (1 \times 28) + (1 \times 39) + (1 \times 16)]$$

$$+ \tfrac{1}{4} [(-1 \times 66) + (-1 \times 26) + (-1 \times 70) + (-1 \times 40)]$$

The above equation can be rearranged to:

$$A_{contrast} = \tfrac{1}{4} [(-1 \times 66) + (1 \times 52) + (-1 \times 26) + (1 \times 28)$$

$$+ (-1 \times 70) + (1 \times 39) + (-1 \times 40) + (1 \times 16)]$$

$$= -16.75$$

In other words, the design contrast attributed to factor A is equal to one-quarter ($\tfrac{1}{4}$) of the sum of multiplying "column A" by "column Y" one by one. Note that $\tfrac{1}{4}$ is applied since there are four sets of contrast data in this study. Note that contrast is the relative change between −1 and +1, while the effect is the relative change from 0 (the origin) to +1 (or −1). Thus, the main effect of factor A, i.e., η_A denoted in Equation (2.6.1), is half of the contrast as:

$$\gamma_A = \tfrac{1}{2} A_{contrast} = -8.375$$

By the same token, the other main and interactive effects can be calculated, as given in Table 2.6.2. The mean is to be estimated using the grand average of all the test data as:

$$\text{Grand Average} = (66 + 52 + 26 + 28 + 70 + 39 + 40 + 16) / 8 = 42.125$$

In summary, the estimated main effects, multifactor interactions, and the grand average can be listed as follows:

- Main effect: $\gamma_A = -8.375$, $\gamma_B = -14.625$, $\gamma_C = -0.875$
- Two-factor interaction: $\gamma_{AB} = 2.875$, $\gamma_{AC} = -5.375$, $\gamma_{BC} = 1.375$
- Three-factor interaction: $\gamma_{ABC} = -1.125$
- Average: $\gamma_o = 42.125$

The next concern will be which main effect or interaction listed above is statistically significant. The individual sample variance corresponding to each run listed in Equation (2.5.3) is not available because there is one and only one data per treatment (run). However, in light of the calculated results for main effects and interactions, one can tell some data are significantly larger than others by one order. The only method for calculating the overall sample variance is to pool the following three "insignificant-looking" effects together:

$$S_C = -0.875, S_{BC} = 1.375, S_{ABC} = -1.125$$

Given that M = 3, Equation (2.5.7) reduces to:

$$S = \left(\frac{S_1^2 + S_2^2 + S_3^2}{M}\right)^{1/2} = \left[\frac{(-0.875)^2 + (1.375)^2 + (-1.125)^2}{3}\right]^{1/2} = 1.143$$

of which the total DOFs ν is the total number of data pooled, i.e., M = 1 + 1 + 1 = 3. The t-ratio for each run is obtained using Equation (2.5.2) by setting $\gamma_{null} = 0$,

$$t_\nu = t_3 = \frac{\gamma_{effect} - \gamma_{null}}{S} = \eta_{effect} / S$$

FIGURE 2.6.1   Pareto plot of main effects and interactions on the impact bend angle.

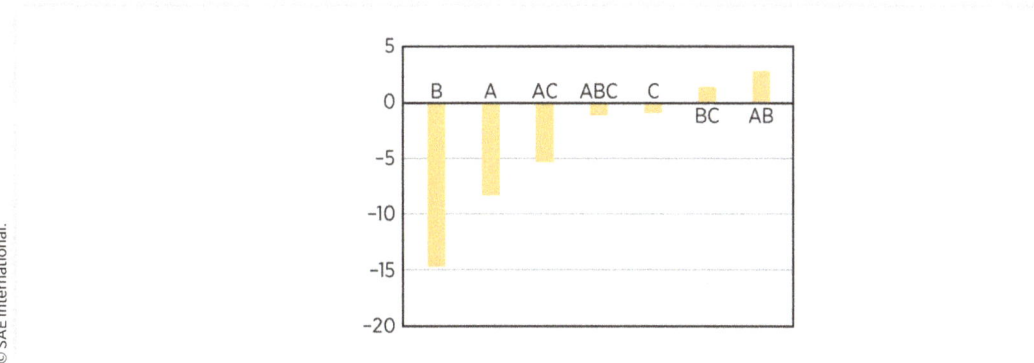

When the effects are arranged in the order of relative size as shown in Figure 2.6.1, it is called the Pareto plot. Based on the parting line of significance at α = 10% that $t_{3,1-\alpha} = t_{3,90\%} = 1.638$ (from Table 2.5.1), one has the following regression model for this case study on the impact bend angle:

$$Y_p = 42.125 - 8.375\, a - 14.625\, b + 2.875\, a\, b - 5.375\, a\, c \qquad (2.6.2)$$

or

$$Y_p = 42.125 - 8.375\, A - 14.625\, (B - 3.05) / 0.38 + 2.875\, A\, [(B - 3.05) / 0.38] - 5.375\, A\, C$$

$$= 159.51 - 31.45\, A - 38.487B + 7.5658\, A\, B - 5.375\, A\, C \qquad (2.6.3)$$

since b = [B - (3.43 + 2.67) / 2] / [(3.42 - 2.67) / 2] = (B - 3.05) / 0.38 as given in Table 2.6.1.

In light of the above equation that one has a confidence interval (CI) of 90% ($\alpha = 10\%$), it can be concluded that both carbon content (factor A) and nominal shank diameter (factor B) have strong individual main effects on reducing the impact bend angle: the more the better, but their interaction (AB) has a minor drawback. Although hardening has no effect, its interaction with the carbon content (factor A) is effective in the reduction of bend angle—tending to help SAE 1022 steel reduce the impact bend angle. Note that SAE 1022 steel has higher carbon content than SAE 1010 steel. In other words, the influence of hardening increases with increasing carbon content.

The predicted values (Y_p) based on Equation (2.6.3) for all eight treatments (runs) are plotted against the test values (Y) in Figure 2.6.2. The strong correlation, R = 99.4%, shows that the DOE modeling is valid, and thus, conclusions drawn from Equation (2.6.3) will be used as the quality assessment guideline for the impact bend angle test of nails within the operating ranges of the relevant parameters given in Table 2.6.1.

FIGURE 2.6.2   Correlation between the test data (Y) versus predicted values (Y_p).

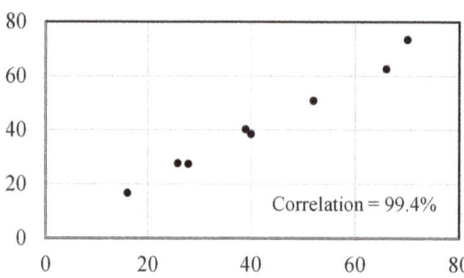

Correlation = 99.4%

© SAE International.

2.6.2. 2^K Factorial Design

One can readily generalize the design matrix 2^3 to a two-level full factorial having K factors, 2^K. The generation of the full factorial design matrix can be obtained thru the following steps:

1. The first column containing variable A starts with −1 and then +1, alternating in sign for all 2^K runs.

2. The second column containing variable B starts with −1 repeated twice, then two repeats of +1s, alternating with every two repeats in sign for all 2^K runs.

3. The third column containing variable C starts with −1 repeated four times, then four repeats of +1s, alternating with every four repeats in sign for all 2^K runs.

4. And so on.

In general, the ith column containing variable x_i (A for x_1, B for x_2, C for x_3, etc.) starts with $2i − 1$ repeats of −1 followed by $2i − 1$ repeats of +1. For example, the design matrix of 2^3 is given in Table 2.6.1. Note that the run order listed in the table is not necessarily the order to carry out the experiments. The order to carry out the experiments has to be decided randomly to eliminate the unnecessary "order effect."

It would be better to conduct experimental tests after the product or process under investigation is stabilized and statistically in control. Randomization, replication, and blocking are three techniques that allow for resolving ambiguities that could be encountered when conducting designed experiments.

2.6.3. Replication

Replication refers to repeating a treatment under the same run conditions but with different subjects and/or experimenters, while repetition involves repeated measures such as measuring the sample multiple times. Measurements are usually subject to variation and uncertainty. It offers a more precise estimate of the treatment effects and obtains an estimate of the error effect. In general, measurements are repeated and experiments are replicated to help identify the sources of variation and quantify uncertainty. The number of replications (sample size) is the number of experimental tests that receive each treatment (run condition). Replication increases the DOFs in statistics and reduces the uncertainty in the outcome. The experimenter will have more confidence in the result. In other words, replication lets the experimenter have a better estimate of the true effects of treatments that further strengthen the reliability and validity of the experiment. The more replications the better, although six replications are generally accepted as a practice norm.

The estimate of effects in a DOE with replications is herein illustrated using an example of the travel time of a car model (300 grams) based on the design matrix 2^3. Three inherent operating parameters (A, weight; B, tire traction; and C, road roughness) and their design levels are listed in Table 2.6.3. The objective is to find out the influences of these three factors on the travel time (Y) after an initial push (i.e., winding torque) is applied. The experimental design is herein to be analyzed using DOE-t with two replications for each treatment (Table 2.6.4).

TABLE 2.6.3 Operating factors and their design levels.

Factor	Level (−)	Level (+)	Coded var. (−1, 1)
A: Vehicle mass	Car (300 g)	Car (300 g) + phone (200 g)	a = A
B: Tire traction	High	Low	b = B
C: Road roughness	Coarse (cement)	Smooth (hardwood)	c = C

© SAE International.

TABLE 2.6.4 The 2^3 factorial design for evaluating the race time of the model car.

Run	A	B	C	AB	AC	BC	ABC	Y_1	Y_2	Y	Y_p	ε
1	−1	−1	−1	1	1	1	−1	6.6	6.5	6.55	6.506	−0.044
2	1	−1	−1	−1	−1	−1	1	7.0	7.0	7.0	7.006	0.006
3	−1	1	−1	−1	1	1	1	5.4	5.6	5.5	5.531	0.031
4	1	1	−1	1	−1	−1	−1	6.4	6.4	6.4	6.406	0.006
5	−1	−1	1	1	−1	−1	1	6.0	6.1	6.05	6.056	0.006
6	1	−1	1	−1	1	1	−1	6.8	7.0	6.9	6.931	0.031
7	−1	1	1	−1	−1	−1	−1	5.4	5.5	5.45	5.456	0.006
8	1	1	1	1	1	1	1	6.0	6.0	6.0	5.956	−0.044
Contrast	0.688	−0.788	−0.263	0.037	0.013	0.037	−0.188	—	—	—	—	
Effect	0.344	−0.394	−0.132	0.0185	0.0065	0.0185	−0.094	—	—	6.231	(Ave).	
Error	0.071	0.071	0.071	0.071	0.071	0.071	0.071					
t-ratio	4.86	5.57	1.856	0.265	0.088	0.265	1.326					
$t_{8,1-90\%}$	1.397	1.397	1.397	1.397	1.397	1.397	1.397					
Significant	Yes	Yes	Yes	No	No	No	Minor					

© SAE International.

Notes:
Y: Average of two experimental test results (Y_1 and Y_2) of each treatment
Y_p: Predicted value of Y, based on the criterion that α = 10% (assumedly)

There are two replicated tests for each run condition (treatment). Thus, based on Equation (2.5.3), individual sample variances corresponding to treatments herein can be used for computing the sample variance for each effect as:

$$S = \left(\frac{v_1 S_1^2 + v_2 S_2^2 + v_3 S_3^2 + v_4 S_4^2 + v_5 S_5^2 + v_6 S_6^2 + v_7 S_7^2 + v_8 S_8^2}{v_1 + v_2 + v_3 + v_4 + v_5 + v_6 + v_7 + v_8}\right)^{1/2}$$

$$= \left(\frac{1 \times 0.005 + 1 \times 0 + 1 \times 0.005 + 1 \times 0.02 + 1 \times 0 + 1 \times 0.005 + 1 \times 0.005 + 1 \times 0}{1 + 1 + 1 + 1 + 1 + 1 + 1 + 1}\right)^{1/2}$$

$$= 0.0707$$

where, for example, $S_1^2 = [(6.6 - 6.55)^2 + (6.5 - 6.55)^2]/(2 - 1) = 0.005$; DOF $v_1 = 2 - 1$ for Treatment 1 has two individually and independently replicated data available. Similarly, the sample variances and DOFs for the other seven treatments can be obtained.

The relative contribution from all the main and interactive effects is depicted in Figure 2.6.3. Based on the parting line of significance at $\alpha = 10\%$ such that $t_{8,1-\alpha} = t_{8,90\%} = 1.397$ (from Table 2.6.1), one has the following regression model for this case study on the processing time:

$$Y_p = 6.2315 + 0.344\,a - 394\,b - 0.131\,c - 0.094\,a\,b\,c \tag{2.6.4}$$

Note that the contribution from the three-factor interaction is minor when measured by $t_{8,90\%} = 1.397$, but its engineering ratio to the mean is 1.51% (0.094/6.2315) which can be regarded as a minor effect. The predicted values (Y_p) based on the above equation for all the eight treatments are plotted against the test values (Y) in Figure 2.6.4(a), with a 90% confidence ($\alpha = 10\%$). For this case study, tire traction has the highest impact on travel time and then followed by road roughness and car mass.

FIGURE 2.6.3   Frequency plot and Pareto plot of effects.

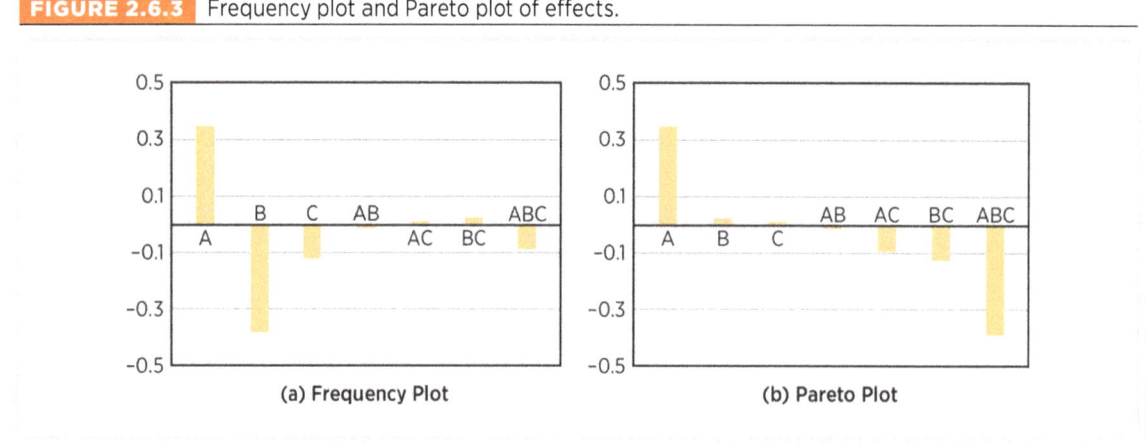

(a) Frequency Plot   (b) Pareto Plot

© SAE International.

The eight residual values can then be calculated using the following equation for the eight treatments:

$$\varepsilon = Y_p - Y \tag{2.6.5}$$

Since all three operating parameters are attributes, Equation (2.6.4) can be rewritten as:

$$Y_p = 6.2315 + 0.344\ A - 394\ B - 0.131\ C - 0.094\ A\ B\ C \qquad (2.6.6)$$

As exhibited in Figure 2.6.4, the correlation plot, i.e., defined as the test data versus the predicted values, with a correlation R = 99.88% and the residual plot, i.e., defined as the residuals versus the predicted values, shaped like a meatball with a correlation R = 0% suggest that the predictive equation given above is both effective and adequate.

FIGURE 2.6.4   Model checking with (a) correlation plot and (b) residual plot.

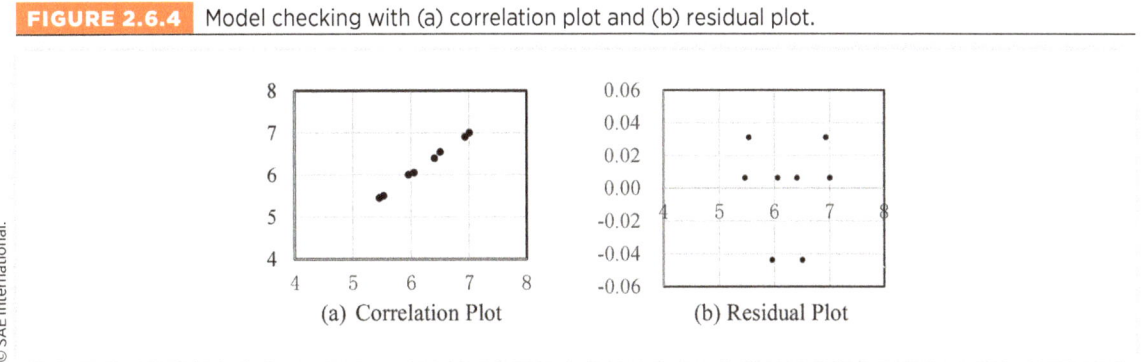

© SAE International.

2.6.4. Randomization

Randomization means the random allocation of factors and the order of experiments. The order of running experiments shall be randomized to avoid influence by uncontrolled variables [Ganju and Lucas 1999]. Randomly assigning test variables and sequences in a deliberate test process such as DOE is a critical step that helps eliminate potential biases from the conclusions. Random assignment of treatments distinguishes a rigorous, "true" experiment from an adequate, but less-than-rigorous, "quasi-experiment." For example, the "Order" numbers provided in Table 2.6.2 is the order to conduct the experimental tests, and these may be obtained from random numbers generated by computer simulations to randomize the potential "latent" influence of the "Run" numbers on the test results. One may explicitly randomize using computer software, coins, dice, cards, etc. which later ends with the proper use of the probability theory which gives a solid foundation for statistical inference.

Three statistically related factors can impair the successful matching of the independent variable means of treatments assembled using randomization. These are the sample size, variance of the data, and number of treatments (groups) being populated. Well-matched treatments yield a measurement much closer to the nominal measurement of pure randomization, such as average or mode. When the treatment (group) size is large relative to the variability, randomization robustly generates treatments that are well matched with any statistics. However, when a treatment (group) size is small, the expected discrepancy in any covariate under randomization can be surprisingly large, hindering statistical inference [Bertsimas et al. 2015].

2.6.5. Blocking

Blocking is a technique, which refers to arranging similar factors in a set of treatments. If all tests are treated as a big group, the within-group variability may be very large. After dividing the experimental conditions into several "blocks," the analyst narrows down the error variance in each block, and thus, the within-group variability is smaller. For example, in an experiment, an analyst collected the data in two days, which may have different working conditions for conducting the tests. An experimenter worries that this might produce uncontrollable noise and thus included the date of run as a block factor in the design [Montgomery 2019]. Factors are preferred to independent variables whose values are under control.

Blocking, also called stratification, is the arrangement of experimental units into groups (blocks/lots) consisting of units that are similar to one another. Blocking screens out noise caused by known sources of variation, such as raw material batch, shift changes, or machine differences. For example, suppose that some experimental measurements are made in the morning and some in the afternoon. If a difference between morning and afternoon measurements is anticipated, the analyst has to ensure that within each period there are equal numbers of subjects in each treatment group and to take account of the difference between morning and afternoon measurements in the analysis. In other words, stratification itself becomes a "block factor." Thus, a reduction is known, but irrelevant sources of variation between units allow greater precision in the estimation of the source of variation under study. By dividing experimental runs into homogeneous blocks and then arithmetically removing the difference, the sensitivity of the DOE is enhanced. Therefore, blocking improves the precision with which comparisons among the factors of interest are made.

By using block factors, one can avoid biases that might occur because of differences between the subjectively allocated treatments, and as a way of estimating the noise in the experiment. In summary, blocking is an algorithm used for containing nuisance factors that may contribute to undesirable variation and make it difficult to assess the systematic effects of a treatment. Taking the block effect into consideration, one can rewrite Equation (2.3.4) as [Myers et al. 2010]:

$$Y = \gamma_0 + \sum_{i=1}^{K} \gamma_i x_i + \sum_{i=1}^{K} \sum_{j=1}^{K} \gamma_{ij} x_i x_j + \sum_{m=0}^{M} \gamma_m b_m + \ldots + \varepsilon \qquad (2.6.7)$$

where
   γ_m is the block effect
   b_m is the block factor
   m is the index for block, $1 \leq m \leq M$

The metamodel given above assumes that the block factors do not interact with any treatment, as well as are not correlated with any other variable factor. Block factor b_m given in the above equation is only an indicator of its existence, referring to $m \neq 0$ (i.e., blocks identified). Nevertheless, if one can neither fix nor stratify a variable—e.g., age would be set as a block factor for finding out the influences of sugar, milk (cream), and coffee on the coffee taste, but it is hard to stratify it—then the experimenter can just randomize it instead. If the effects of a block factor must be separately extracted, then one has to choose a full factorial design on the block factor. The effect of a block factor is illustrated using an example (Section 2.9.1).

2.7. Diagnostic Checking

For further use of the metamodel to relate the response to the design factors and optimize the system, proper error analyses such as effectiveness, adequacy, and validation must be performed. The error term includes the fitting residuals and the effects of the nuisance factors. The fitting residuals are caused by the terms that are not included in the metamodel. The effects of the nuisance factors are noncontrollable or unknown, and therefore, they are not accommodated in the metamodel. Once the estimated metamodel is obtained, an experimenter can deploy the following tools to do the diagnostic checking to see if the effectiveness, adequacy, and validation of the fitted metamodel are met:

1. Correlation between the test data and predicted values.
2. Correlation between the residuals and predicted values.
3. Error percentage using an extra test point at the center (a = 0, b = 0, c = 0, …) compared with the corresponding predicted value from the predictive equation.
4. Uniform probability plot of residuals.
5. Normal probability plot of effects, if the objective function is normally distributed in statistics.
6. Lack-of-fit plot (Chapter 4).

To err is human. Human errors that can occur during experiments include failure to follow directions, mishaps in measuring, contamination of materials, miscalculations of data, and underestimating the impact of environmental factors. To further use the metamodel so to relate the response to the design factors and optimize the system, proper error analyses such as effectiveness, adequacy, and validation must be performed. Simultaneously, to facilitate efficient estimates of effects, the following four desirable characteristics of an experimental design are required: orthogonality, rotatability, constant variance, and minimum bias.

2.7.1. Checking Effectiveness

The effectiveness of a DOE model can be checked using the plot of test results versus the predicted values and the correlation between them on a linear regression—a straight line with a 45° slope if the vertical and horizontal axes are commonly scaled. This is called a prediction plot. When both data sets are commonly scaled, the shape of the correlation plot will also tell the correlation as follows:

(a) Straight line: Good correlation
(b) Ball-shaped figure: No correlation (when commonly scaled)

If the DOE model is effective, the statistical correlation between the predicted values and their corresponding test data of all treatments must be greater than some value, e.g., R ≥ 90% as a general practice.

2.7.2. Checking Adequacy

The adequacy of a DOE metamodel can be checked using the plot of residuals versus the predicted values (ε–Y_p plot) and the correlation between them on a linear regression—a horizontal line. This is called a residual plot. If the metamodel is correct, the correlation between the predicted values and the corresponding residuals must be very small. A plot of residuals versus predicted values should look like a spotted soccer ball. The correlation can be calculated explicitly for both. A violation of the constant variance assumption can be detected with a residual plot by noting that the variation in the vertical direction seems to differ significantly at different points along the horizontal axis. Residual plots are widely used for checking model adequacy in linear regression analyses and maximum likelihood methods. They are used to diagnose whether a model or a distribution can fit the data well. Residual plots can also help experimenters find outliers in the data set. By examining the pattern of residual plots, one can identify whether there are additional variables that should be included in the regression model.

Another way to check the model adequacy is the location-spread plot ($|\varepsilon|^{1/2}$–Y_p plot). Similar to the residual plot, the location-spread plot also indicates a potential violation of the constant variance assumption if the vertical locations of the $|\varepsilon|^{1/2}$ data seem to differ significantly at different points along the horizontal axis.

2.7.3. **Model Validation**

In practice, a final run has to be conducted so to confirm the effectiveness of modeling using DOE. Jumping into a paradox should be avoided. Generally speaking, model validation can be justified via the following two approaches:

1. The "central point" of all the selected factors, i.e., (0, 0, 0, …, 0), as each factor is parameterized between –1 and 1. The central (nominal) point will be replicated to ensure that the random error (noise) of the measurement is small enough.

2. The current design configuration is a good choice to be employed for model validation since its market information is generally available, including product validation data and the customer's feedback.

Once a metamodel is validated, it becomes a predictive equation, which can be then used to predict new responses that are not included in the run conditions (treatments) conducted to formulate the predictive equation. A predictive equation is also called a predictor. How to optimize the system response based on the predictor will be the follow-up work after a DOE is validated.

2.8. **Statistical Inference by Normal Probability Plots**

Measurement accuracy is statistically characterized by the population mean and its related standard deviation. Let f(x) and F(x) be the PDF and CDF of the continuous random variable x, respectively. The statistics of variable x can then be characterized by the following:

1. Mean: A measure of central tendency (expected value), as $\mu \equiv \int x\, f(x)\, dx$.
2. Median: Variable $x_{50\%}$ at which $f(x_{50\%}) = 50\%$.
3. Mode: Variable x_{mode} at which $f(x_{mode})$ is the maximum among all f(x) values.
4. Variance: A measure of dispersion (also called spread) of f(x).

Note that a statistical function must be independent and identically distributed. Each set of data must be checked if it fits a certain type of statistical distribution.

2.8.1. **Normal Distribution**

Errors of measurement are observed to be able to follow a symmetric bell-shaped diagram as shown in Figure 2.8.1 when the sample size is huge. It is conventionally called normal distribution or Gaussian distribution after Mr. Gauss (1809), of which the continuous random variable x ranges from $-\infty$ to ∞. The PDF of a normal distribution with a mean μ_x and variance σ_x^2 is respectively described by the following equation [Ross 1987]:

$$f(x) = \frac{1}{(2\pi)^{1/2}\, \sigma_x} \exp[-\tfrac{1}{2}(\frac{x - \mu_x}{\sigma_x})^2] \tag{2.8.1}$$

where
  x is a variable
  μ_x is the mean, which is the location parameter
  σ_x is the standard deviation, which is the scale parameter

FIGURE 2.8.1 PDF of a normal distribution.

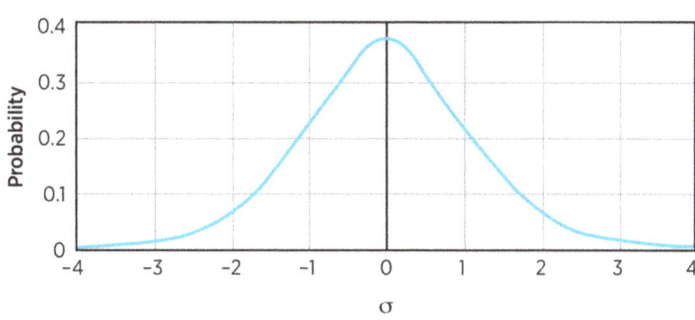

The PDF of the corresponding standard normal distribution with a mean of 0 and variance of 1, namely, $N(0, 1)$, is then

$$\phi(z) = \Phi\left(\frac{x - \mu_x}{\sigma_x}\right) = \frac{1}{(2\pi)^{1/2}} \exp(-\tfrac{1}{2} z^2) \qquad (2.8.2a)$$

and

$$z \equiv \frac{x - \mu_x}{\sigma_x} \qquad (2.8.2b)$$

where
    z is the standardized variable
    ϕ is the statistical PDF

The cumulative normal distribution function is the integration of the normal density function given above from the negative infinitive to the value of interest, i.e.,

$$F(x) = \int_{-\infty}^{x} \frac{1}{\sigma_t (2\pi)^{1/2}} \exp\left[-\tfrac{1}{2}\left(\frac{t - \mu_t}{\sigma_t}\right)^2\right] dt \qquad (2.8.3)$$

The notation for random variable x being normally distributed with mean μ_x and variance σ_x^2 is denoted as $x \sim N(\mu_x, \sigma_x^2)$, while the standard normal distribution function is $\Phi \sim N(0, 1)$. Random variable x may range from $-\infty$ to ∞, so does the mean. The variance is always positive, i.e., $\sigma_x^2 > 0$, where σ_x is named the standard deviation. The probability at $1\sigma_x$, $2\sigma_x$, $3\sigma_x$, and beyond can be assessed by integrating the PDF (Figure 2.8.1) according to Equation (2.8.3). Note that the mean of a normal distribution function is also its median and mode because of its bell-shaped PDF, which is symmetric.

The sample mean and variance of discrete random variable x can be calculated from the obtained data as follows:

$$\mu_x = \frac{1}{N}\left[\sum_{n=1}^{N} x_n\right] \qquad (2.8.4)$$

and

$$S_x^2 = \frac{1}{N-1} \left[\sum_{n=1}^{N} (x_n - \mu_x)^2 \right] \tag{2.8.5}$$

They are called sample mean and sample variance. S_x is called the sample standard deviation, sample error, or error. If mean μ_x and variance S_x^2 are obtained from the entire population group, they can be called population mean and population variance, which do not exist virtually in real-world engineering applications.

2.8.2. Normal Probability Plots (Normal Quantile Plots)

A normal probability plot is a graphical presentation of the corresponding CDF for assessing whether or not a data set is approximately normal distributed. A normal PDF f(x) is plotted in Figure 2.8.1, and its corresponding CDF is:

$$F(z) = p \tag{2.8.6}$$

The associated normal probability plot of its distribution function is demonstrated in Figure 2.8.2, which is called the quantile-quantile plot (q-q plot) in statistics. The word quantile comes from the word quantity, and it refers to dividing a probability distribution into areas of equal probability. The term quantile function is a synonym for the inverse distribution function (or percent point function) that can be written as:

$$F^{-1}(p) = z \tag{2.8.7}$$

A q-q plot is a probability plot, which is a graphical method for comparing two probability distributions by plotting their quantiles against each other. It is a graphical tool to help statisticians assess if a set of data plausibly originate from a theoretical distribution function. For example, the probability distribution of a set of raw or reduced experimental data can be plotted against $N(0, 1)$, which is the set of data with the solid line (mean = 0, variance =1) given in Figure 2.8.2, to check if the data are truly normal distributed.

FIGURE 2.8.2   Cumulative distributions of different normal probability functions.

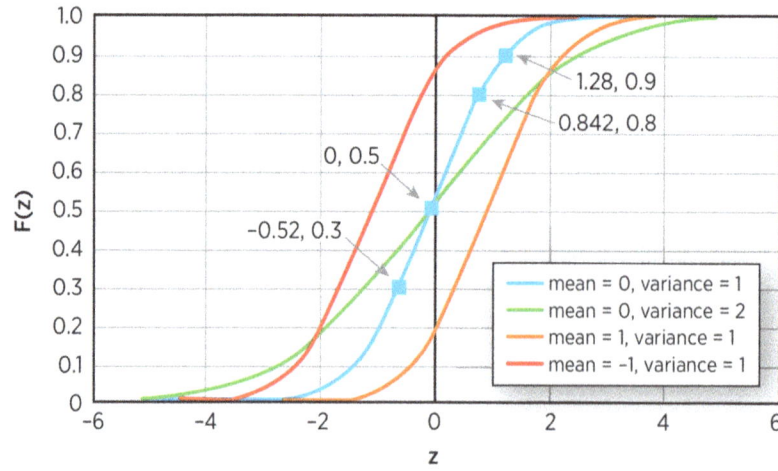

© SAE International.

The average of n samples taken from any distribution with finite mean and variance will have a normal distribution as long as there are a lot of data according to the central limit theorem. Statistically insignificant experimental data like random error will lie close to a straight line as demonstrated by $N(0, 1)$, while statistically significant experimental data will depart from the straight line and be displaced at both ends. This is not only true for a normal-distributed sample data set but also for sample data that are consistent with one of the following three distributions: uniform distribution, student's t-distribution, and exponential distribution. The data points plotted against the corresponding three CDFs also lie close to a straight line, as demonstrated using q-q plots in Figure 2.8.2. Quantile statistics provide information on whether the values of a set of data are clustered or do not follow a specific statistical pattern:

(a) The points shown in a q-q plot comprising factors with small and/or insignificant effects on the response will constitute (roughly) a straight line on the plot as denoted by the null hypothesis.

(b) The points for that with large and significant effects, main and interactive, will visually fall off of the straight line at both ends, as illustrated in Figure 2.8.2.

If all the main effects and interactions are not statistically significant, it will stay close to the normal distribution curve following the null hypothesis that H_o: $\gamma = 0$, of which γ can be one of the following: γ_0, γ_i, γ_{ij}, etc.

Q-q plots have been in use for graphically comparing two probability distributions by plotting their quantiles against each other. A normal probability plot, in the form of a q-q plot, is made accordingly to graphically verify its normality assumption for effects derived from a univariate population that are mutually independent and identically distributed [Bechhoffer and Dunnett 1988]. The use of a normal probability plot of residuals for the analysis of a factorial design with no replications is in need when it lacks a formally estimated sample standard deviation (experimental error).

2.8.3. Making Normal Probability Plots Using Excel

Excel provides several worksheet functions for working with normal distributions. Assume that there are n data available for making a normal probability plot. The normal probability plot is a chart of effects in ascending order versus z-values that resemble the CDF curve of $N(0, 1)$. One may take the following steps to create a normal probability plot:

1. Putting the n observations in an ascending order, namely, $\eta_1 \leq \cdots \leq \eta_i \leq \cdots \leq \eta_n$. These observations are herein the main and interactive effect resulting from the DOE design matrix, and subscript i is the respective ascending order number of each effect. These data are the outcomes in the CDF, and they are to be identified on the ordinate (vertical axis) as nonparametric quantile regression resulting from a cumulative normal distribution curve.

2. Assigning a rank-ordered quantile point Φ_i to each corresponding η_i, as presented in Table 2.8.1. For example, given that there are n effects and interactions to be identified and $n \leq 10$, then Φ_i can be expressed as [Minitab]:

$$\Phi_i = (i - 0.375) / (n + 0.25), \text{ for } i = 1, 2, 3, ..., n \quad (n \leq 10) \qquad (2.8.8)$$

of which Φ_i is the normal order statistic medians, ranging from 0 to 1. Note that Φ is the cumulative normal distribution function (Figure 2.8.2), and it is also called the standard normal quantile function. If $n > 10$, it is to be estimated using the following equations [Minitab]:

$$\Phi_i = (i - 0.5) / n, \text{ for } i = 1, 2, 3, ..., n \qquad (n > 10) \qquad (2.8.9)$$

3. Calculating $z_i = \Phi_i^{-1}$, i.e., inverting Φ_i. This inversion is implemented in Microsoft Excel as $z_i = \text{NORMSINV}(\Phi_i)$.

4. Making an η_i–z_i chart, which is then the normal probability plot.

The calculation of Φ_i has to get involved with sophisticated manipulations of the standard normal distribution function. Fortunately, one can obtain it utilizing a Microsoft Excel function, e.g., $z_i = \text{NORMSINV}(\Phi_i)$. Data for Φ_i and z_i, as listed in Table 2.8.1, are provided here for 7 effects (for 8 runs) and 15 effects (for 16 runs) for convenience. These are the two cases frequently used for typical applications of DOE in practice.

The normal probability plot made this way is a q-q plot, and it can be also used to determine the process capability.

TABLE 2.8.1 Data required for the normal probability plot for DOE with 8 runs (7 effects) and 16 runs (15 effects), as $z_i = \text{NORMSINV}(\Phi_i)$ implemented in Excel.

7 effects		15 effects		
$\Phi_1 = 0.0862$	$z_1 = -1.364$	$\Phi_1 = 0.0333$	$z_1 = -1.83$	
$\Phi_2 = 0.2241$	$z_2 = -0.758$	$\Phi_2 = 0.1000$	$z_2 = -1.28$	
$\Phi_3 = 0.3621$	$z_3 = -0.353$	$\Phi_3 = 0.1667$	$z_3 = -0.97$	
$\Phi_4 = 0.5000$	$z_4 = 0$	$\Phi_4 = 0.2333$	$z_4 = -0.73$	
$\Phi_5 = 0.6380$	$z_5 = 0.353$	$\Phi_5 = 0.3000$	$z_5 = -0.52$	(Illustrated in Figure 2.8.2)
$\Phi_6 = 0.7759$	$z_6 = 0.758$	$\Phi_6 = 0.3667$	$z_6 = -0.34$	
$\Phi_7 = 0.9138$	$z_7 = 1.364$	$\Phi_7 = 0.4333$	$z_7 = -0.17$	
		$\Phi_8 = 0.5000$	$z_8 = 0.00$	(Illustrated in Figure 2.8.2)
		$\Phi_9 = 0.5665$	$z_9 = 0.17$	
		$\Phi_{10} = 0.6333$	$z_{10} = 0.34$	
		$\Phi_{11} = 0.7000$	$z_{11} = 0.52$	
		$\Phi_{12} = 0.7667$	$z_{12} = 0.73$	
		$\Phi_{13} = 0.8333$	$z_{13} = 0.97$	
		$\Phi_{14} = 0.9000$	$z_{14} = 1.28$	(Illustrated in Figure 2.8.2)
		$\Phi_{15} = 0.9667$	$z_{15} = 1.83$	

It is assumed that all the effects should be normally distributed with a mean value around zero based on the given hypothesis that they are not significant. Thus, one can check the normal probability plot for significant effects, which depart from the straight line. When certain effects are significantly different from "zero departure," they should be considered significant. This is called the Lenth method [Lenth 1989]. The assumptions of normality and constant variance in a linear model, such as DOE-t and ANOVA regressions, are quite robust to show these departures [Shapiro and Francia 1972]. It means that even if the assumptions are not met perfectly, the resulting p-values will still be reasonable estimates.

A sample size of 15 or 7 is too small to truthfully reveal the normal pattern, and thus, a right-skewed distribution that deviates slightly from the linear straight line is unavoidable, as exhibited in Figure 2.8.3. Note that the following is not normally distributed:

(a) Censored or truncated data, including time-to-event variables

(b) Discrete counts

(c) Ordinal

(d) Product life (e.g., km or cycles)

A proper application of Φ_i for the corresponding distribution function has to be utilized for diagnostic checking of DOE models [Filliben 1975]. If a set of raw data fits a lognormal distribution function well, it may show a concaved curve in a normal probability plot (i.e., η_i–z_i plot). For such a case, a logarithmic transformation of the original raw data should be taken before doing a DOE analysis.

EXAMPLE 2.8.1

Construct a normal probability plot of the 15 effects (as listed below) resulting from a DOE design matrix with 16 runs to study the adherence of adhesive joints (Section 12.6, Volume III). Determine the effects that are statistically significant. Do the data follow an approximately normal distribution when the influential factors are excluded?

Solution:

Rearrange both the main and interactive effects in ascending order, listed as follows:

FIGURE 2.8.3 Normal probability plot.

Effect	η_i	Φ_i (percentile)	z_i
D	−73.1	0.0333	−1.83
AD	−23.8	0.1	−1.28
BD	−11.9	0.1667	−0.97
ABCD	−6.25	0.2333	−0.73
AB	−5	0.3	−0.52
AE	−3.13	0.3667	−0.34
C	4.38	0.4333	−0.17
AC	5.0	0.5	0
ACD	5.0	0.5667	0.17
CD	6.88	0.6333	0.34
BCD	10.6	0.7	0.52
ABC	11.3	0.7667	0.73
ABD	17.5	0.8333	0.97
A	51.3	0.9	1.28
B	61.8	0.9667	1.83

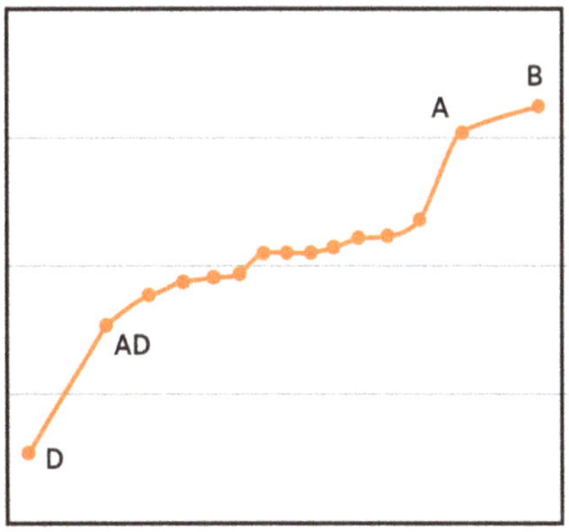

© SAE International.

Both Φ_i and z_i given above are obtained from Equation (2.8.9). The η_i–z_i plot is depicted in **Figure 2.8.3**, by which one can see that factors A, B, and D have statistically significant influences on the adherence of glue. Factor D has a negative effect on the left extremity of the figure, while factors B and A have positive effects on the right extremity. Since all data points, except A, B, and D, lie close to a straight line, the data do approximately follow a normal distribution.

2.8.4. Half-Normal Probability Plots

The half-normal distribution is a fold at the mean of an ordinary normal distribution with a mean of zero. In other words, a half-normal distribution is the distribution of the abs(z), which falls in the range (0, ∞), with z having a normal distribution with a mean of zero. A half-normal probability plot resembles its normal probability plot counterpart except that the absolute value of an effect (main or interactive) is the value plotted on the horizontal axis, while the shape or color of the data points indicates whether the original effect is positive or negative.

2.9. Full Factorial Design Examples

Though obviously already less risky than resolution III designs, resolution IV designs are not even as risky as they might appear. As a consequence, out of the many factors investigated, it is to be expected that only a few are statistically significant and the experimenter is fully informed of interactions with factors whose main effects are themselves significant. A full factorial design with four, five, or six factors takes advantage of being able to separate individual effects, including the main effects and multifactorial interactions. Every treatment in the design matrix accounts for measuring the response variable so that the analysis meets the need of making a large demand on effects with no confounding issues but meaningful statistical inferences. Nevertheless, one general understanding of DOE in practice is called the sparsity effect—usually, only a few of the many factors investigated will eventually turn out to be important as the "Pareto rule" reigns. A Pareto plot reveals the information for determining where to begin with the corrective actions.

2.9.1. 2^4 Factorial Design—Case Study on Tasting Coffee

A group of students are invited to a coffee party to find out a great recipe for making tasty coffee. The factors that would have an impact on the taste of the coffee have been identified and listed in Table 2.9.1. The experiments were conducted in the classroom and one student was invited to grade the coffee made on the spot. The grade of each treatment ranges from 0 to 100. The first three factors (i.e., sugar, milk, and coffee) are the key ingredients; factor D is the waiting time used as a block (block factor) while expecting the answer to the question "Does cold coffee taste the same as hot coffee?" Since one anticipates a difference in measurement between fresh coffee (2 minutes after coffee is made) and cold coffee (42 minutes after coffee is made), one has to ensure that there are equal numbers of subjects in each treatment group within each period. It means creating a block factor (waiting time after making) to take account of the difference between periods (2 minutes versus 42 minutes).

TABLE 2.9.1 Factors and their design levels identified for making tasty coffee.

Factor	Level (−)	Level (+)	Variate (−1, 1)
A (Sugar dose)	2 teaspoon	3 teaspoon	$a = (A − 2.5)/0.5$
B (Milk dose)	1 teaspoon	2 teaspoon	$b = (B − 1.5)/0.5$
C (Coffee dose)	2 teaspoon	3 teaspoon	$c = (C − 2.5)/0.5$
D (Waiting time after the coffee was made)	2 minutes	42 minutes	$d = (D − 22)/20$

© SAE International.

The 16 run treatments for full factorial design matrix 2^4 are listed in Table 2.9.2(a). The calculated main effects are also given in Table 2.9.2(a) while the multifactor interactions are in Table 2.9.2(b). Contributions of all the effects to the demanded objective function are plotted in Figure 2.9.1(a), which is often called a frequency plot. After being rearranged in the order of relative size as shown in Figure 2.9.1(b), it is called a Pareto plot.

TABLE 2.9.2 A 2^4 factorial design for making tasty coffee.

(a) Main effects							
Run	a	b	c	d	Y	Y_p	ε
1	−1	−1	−1	−1	70	71.875	1.875
2	1	−1	−1	−1	80	79.375	−0.625
3	−1	1	−1	−1	75	75.000	0.000
4	1	1	−1	−1	85	82.500	2.500
5	−1	−1	1	−1	75	72.500	−2.500
6	1	−1	1	−1	60	62.750	2.500
7	−1	1	1	−1	65	66.875	1.875
8	1	1	1	−1	75	74.375	−0.625
9	−1	−1	−1	1	70	67.500	−2.500
10	1	−1	−1	1	75	75.000	−0.000
11	−1	1	−1	1	70	70.625	0.625
12	1	1	−1	1	75	78.125	3.125
13	−1	−1	1	1	65	68.125	3.125
14	1	−1	1	1	60	58.125	−1.875
15	−1	1	1	1	65	62.500	−2.500
16	1	1	1	1	70	70.000	0.000
Contrast	3.125	3.125	−8.125	−4.375			
Effect	1.563	1.563	−4.063	−2.188		70.9375 (*Average*)	
Error	0.699	0.699	0.699	0.699			
t-ratio	2.236	2.236	5.814	3.130			
$t_{8,1-\alpha}$	1.397	1.397	1.397	1.397			
Significant?	Yes	Yes	Yes	Yes			

(b) Interactions											
Run	ab	ac	ad	bc	bd	cd	abc	abd	acd	bcd	abcd
1	1	1	1	1	1	1	−1	−1	−1	−1	1
2	−1	−1	−1	1	1	1	1	1	1	−1	−1
3	−1	1	1	−1	−1	1	1	1	−1	1	−1
4	1	−1	−1	−1	−1	1	−1	−1	1	1	1
5	1	−1	1	−1	1	−1	1	−1	1	1	−1
6	−1	1	−1	−1	1	−1	−1	1	−1	1	1
7	−1	−1	1	1	−1	−1	−1	1	1	−1	1
8	1	1	−1	1	−1	−1	1	−1	−1	−1	−1
9	1	1	−1	1	−1	−1	−1	1	1	1	−1
10	−1	−1	1	1	−1	−1	1	−1	−1	1	1
11	−1	1	−1	−1	1	−1	1	−1	1	−1	1
12	1	−1	1	−1	1	−1	−1	1	−1	−1	−1
13	1	−1	−1	−1	−1	1	1	1	−1	−1	1
14	−1	1	1	−1	−1	1	−1	−1	1	−1	−1
15	−1	−1	−1	1	1	1	−1	−1	−1	1	−1
16	1	1	1	1	1	1	1	1	1	1	1
Contrast	4.375	−4.375	−0.625	0.625	−0.625	0.625	4.375	−1.875	1.875	1.875	−1.875
Effect	2.188	−2.188	−0.313	0.313	−0.313	0.313	2.188	*−0.938*	*0.938*	*0.938*	*−0.938*
Error	0.699	0.699	0.699	0.699	0.699	0.699	0.699	0.699	0.699	0.699	0.699
t-ratio	3.130	3.130	0.447	0.447	0.447	0.447	3.130	1.342	1.342	1.342	1.342
$t_{8,1-\alpha}$	1.397	1.397	1.397	1.397	1.397	1.397	1.397	1.397	1.397	1.397	1.397
Significant?	Yes	Yes	Yes	No	No	No	Yes	No	No	No	No

Notes:
Y: Data obtained from physical tests
Y_p: Predicted value of Y, based on the criterion that α = 10%
ε: Residuals

FIGURE 2.9.1 Relative size of effects: (a) frequency plot and (b) Pareto plot.

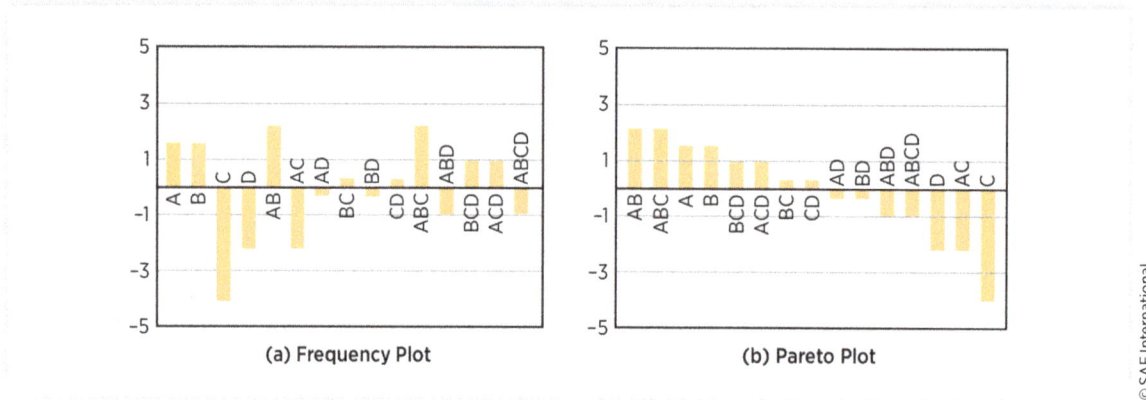

(a) Frequency Plot   (b) Pareto Plot

© SAE International.

Since there is no replication, the small effects are pooled together to estimate the sample random error according to Equation (2.5.3). It is shown that in addition to the four main effects, which have different levels of influence on the result, two two-factor interactions and one three-factor interaction are also statistically significant, using the statistical inference on the p-value that $\alpha = 10\%$, i.e., $t_{8,90\%}$ listed in Table 2.9.1. Thus, the predictive equation can be written as:

$$Y = 70.9375 + 1.5625\, a + 1.5625\, b - 4.0625\, c - 2.1875\, d$$

$$+ 2.1875\, a\, b - 2.1875\, a\, c + 2.1875\, a\, b\, c$$

(2.9.1)

FIGURE 2.9.2 Correlation between test data (Y) and predicted values (Y_p) and residual plot.

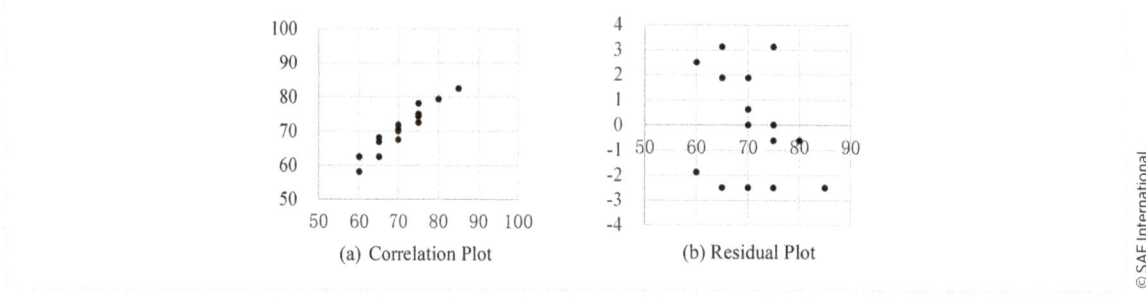

(a) Correlation Plot   (b) Residual Plot

© SAE International.

Model checking on the effectiveness and adequacy is demonstrated by the correlation plot in Figure 2.9.2(a) and the residual plot in Figure 2.9.2(b), respectively. The strong correlation (R = 95.5%) between the test data and the predicted values exhibits the effectiveness of the model. Simultaneously, the low correlation (R ≈ 0) between the residuals and the predicted values renders the adequacy of the model; as the residual plot looks so "random," it is shaped like a spotted soccer ball.

2.9.2. 2^5 Factorial Design—Case Study on Finite Element Accuracy

Finite element methods (FEM) have become a powerful tool for numerical analysis of composite structures. Applications range from stress-strain analysis, vibrations, heat transfer, mold flow, and electromagnetic field to their coupled effects because of the need to accommodate the following work tasks:

(a) Calculations of stress intensity factors for structural flaws or cracks.

(b) High interlaminar stress gradients invited by free-edge effects.

(c) Curved edges and surfaces of complex geometry.

(d) Less limitation on skew angles in meshing.

(e) Flexibility in handling incompressible materials.

(f) Implementing multi-integration schemes, including $2 \times 2 \times 2$, $3 \times 3 \times 3$, and 14-point algorithms [Cook et al. 2001].

Twenty-node isoparametric solid (brick) elements have become a popular general-purpose element to model composite structures.

It is well accepted by researchers to predict the behaviors of composite solids by 20-node isoparametric elements, but the application to composite shells is dubious. The lack of adequate understanding can lead to puzzling and misleading results. On the finite element accuracy, a factorial analysis is conducted to explore the impact of the following five factors:

(a) Aspect ratio: Ratio of length (average of four sides) to width (average of four sides) of an element.

(b) Radius-to-thickness ratio: Ratio of radius (on the centroidal surface, averaged) to thickness (averaged) of an element; its inverse can be used as a measure of curvature.

(c) Material anisotropy: Ratio of Young's modulus in the fiber direction to that in the transverse direction.

(d) Fiber angle: Fiber alignment measured from the longitudinal axis of the shell.

(e) Integration order: Reduced integration ($2 \times 2 \times 2$) and full integration ($3 \times 3 \times 3$) in Gauss quadrature formulae.

The case study is done on the application of 20-node isoparametric solid elements to a clamped composite cylindrical shell (baseline data: axial length of 508 mm, radius of 508 mm, and thickness of 25.4 mm) subjected to an internal pressure (p = 0.02758 MPa). The full factorial design 2^5, with no confounding patterns, is deployed and their corresponding design levels are listed in Table 2.9.3. The objective is to check the finite element accuracy as measured by the radial displacement at the central cross-section when the cylinder is subjected to factorial variations.

TABLE 2.9.3 Factors and levels used for the study on FEA on composite shells [Chiang and Tang 1995].

Factor	Level (−)	Level (+)
A: Aspect ratio	1.25	5.00
B: Radius-to-thickness ratio	20	200
C: Material anisotropy	Glass/epoxy*	Graphite/epoxy#
D: Fiber angle (degrees)	90°	0°
E: Integration order	$2 \times 2 \times 2$	$3 \times 3 \times 3$

Notes:
*Glass/epoxy: E_{11} = 51.7 GPa, E_{22} = E_{33} = 13.8 GPa
    G_{12} = G_{13} = G_{23} = 8.62 GPa; ν_{12} = ν_{13} = ν_{23} = 0.25
#Graphite/epoxy: E_{11} = 159.6 GPa, E_{22} = E_{33} = 10.4 GPa
    G_{12} = G_{13} = G_{23} = 6.42 GPa; ν_{12} = ν_{13} = ν_{23} = 0.25

The design matrix and the test results for the finite element error are given in Table 2.9.4. All the finite element calculations, including pre-processors and post-processors, were performed on an in-house finite element computer program developed by Chiang and Tang [1995], namely, CADOCS

(Computer-Aided Design Of Composite Structures). Solutions to obtain simultaneous equations were derived according to the column-wise skyline algorithm [Cook and Bretl 1979]. The validity of the finite element program has been confirmed repeatedly when it was originally developed for tire modeling and body panels by the first author of [Chiang and Tang 1995]. Note that the skew angle was kept zero throughout the experimental design. When the integration order of 3 × 3 × 3 is applied, the calculated radial displacement at the central cross-section under internal pressure p = 0.02758 MPa is 0.00928 mm, which correlates extremely well with the exact solution of 0.00932 mm, derived from the closed-form solution [Kraus 1967].

TABLE 2.9.4 Full factorial design 2^5 for finite element accuracy in modeling composites [Chiang and Tang 1995].

Run	a	b	c	d	e	Y (FEA error)
1	−1	−1	−1	−1	−1	−0.01631
2	1	−1	−1	−1	−1	0.08015
3	−1	1	−1	−1	−1	0.18485
4	1	1	−1	−1	−1	1.10771
5	−1	−1	1	−1	−1	−0.00256
6	1	−1	1	−1	−1	0.20353
7	−1	1	1	−1	−1	0.38727
8	1	1	1	−1	−1	0.96373
9	−1	−1	−1	1	−1	−0.01135
10	1	−1	−1	1	−1	−0.01956
11	−1	1	−1	1	−1	−0.00446
12	1	1	−1	1	−1	−0.00459
13	−1	−1	1	1	−1	−0.00446
14	1	−1	1	1	−1	−0.00459
15	−1	1	1	1	−1	0.17143
16	1	1	1	1	−1	0.79577
17	−1	−1	−1	−1	1	−0.01593
18	1	−1	−1	−1	1	−0.01715
19	−1	1	−1	−1	1	−0.00215
20	1	1	−1	−1	1	0.10799
21	−1	−1	1	−1	1	0.00953
22	1	−1	1	−1	1	−0.00412
23	−1	1	1	−1	1	0.00523
24	1	1	1	−1	1	0.11101
25	−1	−1	−1	1	1	−0.01140
26	1	−1	−1	1	1	−0.01135
27	−1	1	−1	1	1	−0.01190
28	1	1	−1	1	1	0.10676
29	−1	−1	1	1	1	−0.00464
30	1	−1	1	1	1	−0.00451
31	−1	1	1	1	1	−0.00178
32	1	1	1	1	1	0.10811

Calculations on statistical inferences are performed according to contrast coefficients using the DOE [Box et al. 2005]. Since there is no replication, the statistical sample standard error is formed by pooling the small four- and five-factorial interactions together. Significance levels for main effects and interactions are presented in Table 2.9.5. The final assessment equation for the finite element error based on the balanced DOE for this case study can be summarized as:

$$Y_p = 0.163 + 0.11\ a + 0.153\ b - 0.025\ d - 0.140\ e$$

$$+\ 0.093\ a\,b - 0.084\ a\,e - 0.123\ b\,e + 0.024\ de - 0.064\ a\,b\,c \tag{2.9.2}$$

This case study proves that 20-node isoparametric solid elements can be used to model thin-shelled solids, e.g., all tire components including liners. Although shell finite elements can be also applied, they are not able to predict the interlaminar normal stress. On average, the radial displacement at the central cross-section predicted by 20-node isoparametric solid elements is 16.3%. Some interesting findings are listed as follows [Chiang and Tang 1995]:

1. Increasing either the aspect ratio or the radius-to-thickness ratio tends to damage the finite element accuracy. When both are increased, their interaction does more damage to the finite element accuracy. However, the effectiveness of the three-factor interaction (aspect ratio × radius-to-thickness ratio × integration order) shows that the 3 × 3 × 3 integration will relieve some of the damage done by increasing both aspect ratio and radius-to-thickness ratio.

2. It seems that the anisotropy of material has little influence on the finite element accuracy, at least within the range of $E_{11}/E_{22} = 3.75$ and $E_{11}/E_{22} = 14.60$. This is contradictory to what was claimed by [Noor and Camin 1976] in which the OFAT method is applied. However, aligning fibers along the curved edges makes the finite element accuracy worse. This can be explained by the fact that successive coordinate transformations made from the material axes of the natural (body) coordinates, then to the global coordinate, have damaged the finite element accuracy over a subtended angle of a curve.

3. Integration order is also a significant main effect, as proven mathematically. The change of integration order from 2 × 2 × 2 to 3 × 3 × 3 very much improves the finite element accuracy. In light of two-factor interactions, as given in equation 2.9.2, the 3 × 3 × 3 integration order can recover some damage done by a high aspect ratio and high radius-to-thickness ratio.

TABLE 2.9.5 Significance levels for effects on FEA error in modeling composites [Chiang and Tang 1995].

Variable	Effect	Error*	t-ratio	Significance level
A	0.2212	0.01805	12.25	<2.5%
B	0.3056	0.01805	16.93	<2.5%
C	0.0149	0.01805	0.83	—
D	−0.0503	0.01805	2.79	<2.5%
E	−0.2808	0.01805	15.56	<2.5%
AB	0.1863	0.01805	10.32	<2.5%
AC	−0.0201	0.01805	1.11	—
AD	−0.0179	0.01805	0.99	—
AE	−0.1675	0.01805	9.28	<2.5%
BC	−0.0115	0.01805	0.64	—

(Continued)

TABLE 2.9.5 (Continued) Significance levels for effects on FEA error in modeling composites [Chiang and Tang 1995].

Variable	Effect	Error*	t-ratio	Significance level
BD	−0.0117	0.01805	0.65	—
BE	−0.2452	0.01805	13.58	<2.5%
CD	−0.0269	0.01805	1.49	—
CE	−0.0056	0.01805	0.31	—
DE	0.0472	0.01805	2.61	<2.5%
ABC	−0.0333	0.01805	1.84	—
ABD	0.0191	0.01805	1.06	—
ABE	−0.1289	0.01805	7.14	<2.5%
ACD	0.0003	0.01805	0.02	—
ACE	0.0169	0.01805	0.94	—
BCD	−0.0094	0.01805	0.52	—
BCE	0.0077	0.01805	0.43	—
BDE	0.0096	0.01805	0.53	—
CDE	0.0240	0.01805	1.33	—
ABCD*	—	—	—	—
ABCE*	—	—	—	—
ABDE*	—	—	—	—
ACDE*	—	—	—	—
BCDE*	—	—	—	—
ABCDE*	—	—	—	—

*Interactive effects pooled together to calculate the statistical error based on Equation (2.5.3).

2.9.3. 2^6 Factorial Design—Case Study on Simple-Stranded Wire Cable

The functional requirement of a wire cable is to connect parts together by transmitting forces from one part to another in a flexible way that distributes stresses uniformly over every cable cross-section, except at the cable end [Ayub et al. 1997]. Although analytic equations [Costello and Butson 1983] can be used to predict the stress concentration of simple-stranded wire cables, most of them are frictionless contact and valid only for cross-sections far away from cable ends where the boundary effect exists; the thin rod theory used to account for interwire friction may not be a good assumption [Raoof and Kraincanicm 1994]. Therefore, FEM and DOE are blended together to characterize the failure of 1 × 7 simple-stranded wire cables, which have been in use for automotive side-door latch systems, subjected to axial loading at both ends [Chiang 1996; SAE 1991].

The case study is done on the application of eight-node isoparametric solid elements to 1 × 7 wire cables, as depicted in Figure 2.9.3. In the virtual experiment based on FEA, one end of the cable strand is held fixed in all three directions, while the other end is subjected to a uniform displacement (working as the axial loading) which is either fixed or set free to rotate according to the variations in design factors. Torsional, bending, and axial deformations develop simultaneously when a stranded cable is axially loaded. FEM turns out to be an ideal tool for modeling the cable in the 3D space. The design levels for the six design parameters listed in Table 2.9.6 are feasible for various design combinations.

FIGURE 2.9.3   A 1 × 7 simple-stranded wire cable for automotive side-door latching systems [Chiang 1996].

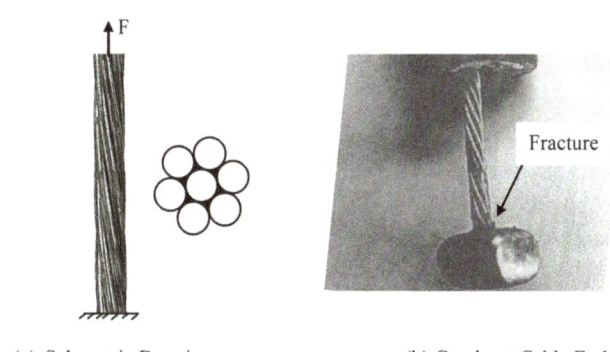

(a) Schmetaic Drawing.   (b) Cracks at Cable End.

Used with permission of Elsevier Science & Technology Journals, from Finite Elements in Analysis and Design: The International Journal of Applied Finite Elements and Computer Aided Engineering, Young J. Chiang, 1996; permission conveyed through Copyright Clearance Center, Inc.

TABLE 2.9.6   Factors and levels used for the study on characterizing wire cables [Chiang 1996].

Factor	Level (−)	Level (+)
(A) Radius of core wire (mm)	1.97	2.47
(B) Radius of each helical wire (mm)	1.865	1.365
(C) Helical angle, 90-H* (degree)	9.23	17.03
(D) FEA boundary condition at the loaded end	Fixed	Free to rotate
(E) Length of cable modeled (mm)	7.67	9.971
(F) Contact condition between strands	Sliding	Non-sliding

Used with permission of Elsevier Science & Technology Journals, from Finite Elements in Analysis and design: The International Journal of Applied Finite Elements and Computer Aided Engineering, Young J. Chiang, 1996; permission conveyed through Copyright Clearance Center, Inc.

The full factorial design 2^6, with no confounding patterns, is deployed and their corresponding design levels are listed in Table 2.9.7. Because of the high gradient of stress variation over each cross-section and the requirement of axial loading, eight-node isoparametric solid (brick) elements were employed to calculate all six stress (strain) components in the 3D space [Chiang 1996]. There are 292 elements and 579 nodes in each model for each of the 64 runs (treatments). The computation is carried out by the commercial finite element program ANSYS. The mechanical properties of all the constituent steel are assumed to be linear elastic with Young's modulus E = 197.9 GPa and Poisson's ratio ν = 0.28. A uniform axial displacement of 0.001 mm was applied at one end and the other end is fixed. The contact between the core wire and surrounding helical wires is assumed to be in two extreme situations: sliding and non-sliding (adhesive). All six helical wires are of the same material and size. A uniform finite element mesh pattern was applied to all 64 treatments as intended to eliminate the effect of mesh quality. The finite element accuracy has been verified and validated as shown in [Chiang 1996] and will not be described in detail here. Since the outer helical wires usually produce a higher stress level than the core wire, attention in terms of reliability is paid to the stress rise in these helical wires. The nominal axial stress and strain are respectively defined as:

$$\sigma_{nom} = F\,/\,Area$$

and

$$\varepsilon_{nom} = \delta\,/\,L$$

where

σ_{nom} (MPa) is the nominal stress

ε_{nom} is the nominal strain

F (N) is the reaction force on one end of the cable where a uniform displacement is applied

Area is the cross-sectional area

δ (mm) is the applied uniform displacement

L (mm) is the length

Three indices are employed to present the structural performance of the wire cable. They are stiffness, stress rise at the middle cross-section of the cable, and stress rise at the cable end. The stiffness is defined as:

$$\text{Stiffness} \equiv \sigma_{nom} / \varepsilon_{nom}$$

The stress rises at the fixed end and at the middle cross-section are respectively defined as follows:

$$K_e \equiv \sigma_{e,high} / \sigma_{nom} - 1 \qquad \text{(Stress rise at the fixed end)}$$

and

$$K_m \equiv \sigma_{m,high} / \sigma_{nom} - 1 \qquad \text{(Stress rise at the middle cross-section)}$$

where

K_e is the stress rise at the cable end

K_m is the stress rise at the middle cross-section

$\sigma_{e,high}$ (MPa) is the highest stress at the cable end

$\sigma_{m,high}$ (MPa) is the highest stress at the middle cross-section

TABLE 2.9.7 Full factorial design 2^6 for FEA of wire cables [Chiang 1996].

Run	A	B	C	D	E	F	K_e	K_m	Stiffness
1	−1	−1	−1	−1	−1	−1	36.33	18.26	194.8
2	1	−1	−1	−1	−1	−1	35.70	17.74	195.6
3	−1	1	−1	−1	−1	−1	33.86	11.82	193.7
4	1	1	−1	−1	−1	−1	32.84	17.24	195.1
5	−1	−1	1	−1	−1	−1	61.02	27.05	179.3
6	1	−1	1	−1	−1	−1	58.69	25.16	181.6
7	−1	1	1	−1	−1	−1	55.95	27.98	177.1
8	1	1	1	−1	−1	−1	52.41	25.53	180.9
9	−1	−1	−1	1	−1	−1	70.02	20.13	178.2
10	1	−1	−1	1	−1	−1	68.58	19.17	179.7
11	−1	1	−1	1	−1	−1	86.22	24.92	170.2
12	1	1	−1	1	−1	−1	82.63	22.53	173.6
13	−1	−1	1	1	−1	−1	121.70	44.78	146.5
14	1	−1	1	1	−1	−1	115.30	40.85	150.9
15	−1	1	1	1	−1	−1	141.60	59.87	132.5
16	1	1	1	1	−1	−1	126.20	50.71	140.9
17	−1	−1	−1	−1	1	−1	35.33	19.06	192.6
18	1	−1	−1	−1	1	−1	34.56	18.38	193.6
19	−1	1	−1	−1	1	−1	31.86	18.89	191.9

(*Continued*)

TABLE 2.9.7 (Continued) Full factorial design 2^6 for FEA of wire cables [Chiang 1996].

Run	A	B	C	D	E	F	K_e	K_m	Stiffness
20	1	1	−1	−1	1	−1	30.75	17.85	193.4
21	−1	−1	1	−1	1	−1	60.41	30.16	174.0
22	1	−1	1	−1	1	−1	57.51	28.15	176.9
23	−1	1	1	−1	1	−1	53.47	31.40	172.5
24	1	1	1	−1	1	−1	49.33	28.24	176.9
25	−1	−1	−1	1	1	−1	86.81	25.85	168.6
26	1	−1	−1	1	1	−1	84.15	24.23	171.1
27	−1	1	−1	1	1	−1	104.60	34.31	157.8
28	1	1	−1	1	1	−1	98.38	30.49	162.9
29	−1	−1	1	1	1	−1	154.30	65.62	127.2
30	1	−1	1	1	1	−1	141.60	58.79	133.3
31	−1	1	1	1	1	−1	167.70	81.99	117.3
32	1	1	1	1	1	−1	145.40	70.33	124.2
33	−1	−1	−1	−1	−1	1	35.16	27.31	198.2
34	1	−1	−1	−1	−1	1	34.22	25.25	199.7
35	−1	1	−1	−1	−1	1	32.92	23.73	196.9
36	1	1	−1	−1	−1	1	31.66	22.00	198.9
37	−1	−1	1	−1	−1	1	60.14	37.55	182.2
38	1	−1	1	−1	−1	1	57.43	34.33	185.2
39	−1	1	1	−1	−1	1	55.46	34.57	179.6
40	1	1	1	−1	−1	1	51.73	31.20	183.9
41	−1	−1	−1	1	−1	1	62.64	24.10	181.5
42	1	−1	−1	1	−1	1	60.09	21.56	183.6
43	−1	1	−1	1	−1	1	66.47	24.91	177.5
44	1	1	−1	1	−1	1	141.60	58.79	133.3
45	−1	−1	1	1	−1	1	111.20	49.14	157.9
46	1	−1	1	1	−1	1	103.90	42.33	157.9
47	−1	1	1	1	−1	1	113.60	53.96	146.2
48	1	1	1	1	−1	1	100.60	43.19	154.5
49	−1	−1	−1	−1	1	1	34.64	26.39	195.3
50	1	−1	−1	−1	1	1	33.57	24.51	196.9
51	−1	1	−1	−1	1	1	31.43	22.90	194.2
52	1	1	−1	−1	1	1	30.11	21.37	196.12
53	−1	−1	1	−1	1	1	60.30	38.74	175.8
54	1	−1	1	−1	1	1	57.08	35.66	179.2
55	−1	1	1	−1	1	1	53.58	35.89	173.8
56	1	1	1	−1	1	1	49.25	32.33	178.6
57	−1	−1	−1	1	1	1	86.81	25.85	168.6
58	1	−1	−1	1	1	1	84.15	24.23	171.1
59	−1	1	−1	1	1	1	116.30	55.44	145.7
60	1	1	−1	1	1	1	62.81	24.67	176.3
61	−1	−1	1	1	1	1	126.90	66.57	139.4
62	1	−1	1	1	1	1	116.30	55.44	145.7
63	−1	1	1	1	1	1	116.60	69.14	138.1
64	1	1	1	1	1	1	101.20	62.67	145.8

Stress rise at the cable end, which is of more concern on the unreliability of cables than the other two indices, will be examined here. The other two cases are to be left to the students as homework problems. Calculations of statistical inferences are performed according to design contrasts based on the t-distribution [Box et al. 2005, Montgomery 2019]. Since there is no replication, the statistical sample standard error is formed by pooling the small four- and five-factorial interactions together. See detailed information on significance levels of main and interactive effects on the stress rise in [Chiang 1996]. The final assessment equation for the predictive equation of stress rise at the cable end based on the balanced DOE for this case study can be summarized as:

$$K_{e,p} = 72.44 \quad - 2.67\,a + 18.12\,c + 27.79\,d + 3.13\,e - 6.16\,f$$

$$- 1.39\,a\,c - 1.57\,a\,d - 1.04\,b\,c + 2.47\,b\,d - 2.11\,b\,f$$

$$+ 6.91\,c\,d + 3.83\,d\,e - 5.81\,d\,f - 1.78\,e\,f$$

$$- 2.19\,b\,d\,f - 1.06\,c\,d\,f - 1.93\,d\,e\,f$$

It means that the average axial stress rise at the loaded end is 72.44% for all the 64 treatments, when measured by the nominal axial stress. All the five design factors have significant individual main effects on the axial stress rise. The boundary conditions at the loaded end (factor B) and the helical angle (factor H) of helical wires are by far the two leading influential factors. A free rotation allowed at the loaded end or a large helical angle will increase the axial stress rise—doing more damage. The adhesive friction between the core and surrounding helical wires tends to reduce the stress rise. The larger core wire seems to reduce the axial stress rise. The length of the strand is also an influential factor, albeit only to a small extent. Nine two-factor interactions and three three-factor interactions also have a significant influence on the axial stress rise at the loaded end, albeit to a minor extent.

References

Adikaram, K., Hussein, M., Effenberger, M., and Becker, T. (2014), "Outlier Detection Method in Linear Regression Based on Sum of Arithmetic Progression," *The Scientific World Journal* 2014, p. 821623.

Anderson, M. J. and Whitcomb, P. J. (2015), *DOE Simplified: Practical Tools for Effective Experimentation*, 3rd Edition, Productivity Press, London, 268 pages; ISBN: 9780429258022.

Ayub, M., Lee, D. J., Chiang, Y. J., and Barkczynski, D. (1997), "Robustness Study of Cables and Rods for Automotive Door Latch Systems," *Proceedings of Total Product Development Symposium*, Dearborn, MI, November 5–6, 1997.

Bechhoffer, R. E. and Dunnett, C. W. (1988), "Percentage Points of Multivariate Student-t Distributions," in *Selected Tables in Mathematical Studies*, Vol. 11, American Mathematical Society, Providence, RI.

Bertsimas, D. et al. (2015), "The Power of Optimization over Randomization in Designing Experiments Involving Small Samples," *Operations Research*, 63, pp. 868-876.

Box, G. E. P., Hunter, J. S., and Hunter, W. G. (2005), *Statistics for Experimenters: Design, Innovation, and Discovery*, 2nd Edition, John Wiley & Sons, New York.

Chiang, Y. J. (1996), "Characterizing Simple-Stranded Wire Cables under Axial Loading," *Finite Elements in Analysis and Design*, 24, pp. 49-66.

Chiang, Y. J. and Chu, M. H. (1993), "Quality Assessment Test for Nails," *Journal of Testing and Evaluation*, 21(1), pp. 51-56.

Chiang, Y. J. and Tang, C. (1995), "Accuracy Assessment to Applying 20-Node Solid Elements to Pressurized Composite Shells," *Finite Elements in Analysis and Design*, 20, pp. 219-231.

Cochran, W. G. and Cox, G. M. (1957), *Experimental Design*, John Wiley & Sons, New York.

Cook, R. D. and Bretl, J. L. (1979), "A New Eight Node Solid Element," *Int'l Journal of Numerical Methods in Engineering*, 14(4), pp. 593-615.

Cook, R. D., Malkus, D. S., Plesha, M. E., and Witt, R. J. (2001), *Concepts and Applications of Finite Element Analysis*, 4th Edition, John Wiley & Sons, New York.

Costello, G. A. and Butson, G. J. (1983), "Stresses in Multilayered Cables," *Journal of Energy Resource Technology*, 105, pp. 337-340.

Czitrom, V. (1999), "One Factor at a Time versus Designed Experiments," *The American Statistician*, 53(2), pp. 126-131.

Filliben, J. J. (1975), "The Probability Plot Correlation Coefficient Test for Normality," *Technometrics*, 17(1), pp. 111-117.

Fisher, R. A. (1925), "Theory of Statistical Estimation," *Mathematical Proceedings of the Cambridge Philosophical Society*, 22, pp. 700-725.

Ganju, J. and Lucas, J. M. (1999), "Detecting Randomization Restrictions Caused by Factors," *Journal of Statistical Planning and Inference*, 81, pp. 129-140.

Kensler, J. L. K., Freeman, L. J., and Vining, G. G (2015), "Analysis of Reliability Experiments with Random Blocks and Subsampling," *Journal of Quality Technology*, 47(3), 235-251.

Kraus, H. (1967), *Thin Elastic Shells*, Wiley, New York.

Lenth, R. V. (1989), "Quick and Easy Analysis of Un-Replicated Factorials," *Technometrics*, 31, pp. 469-473.

Montgomery, D. C. (2019), *Design and Analysis of Experiments*, 10th Edition, John Wiley & Sons, Hobaken, NJ, 688 pages; ISBN: 978-1-119-49244-3.

Myers, R. H. et al. (2010), *Generalized Linear Models with Applications in Engineering and Sciences*, 2nd Edition, John Wiley & Sons, Hoboken, NJ.

Noor, A. K. and Camin, R. A. (1976), "Symmetry Considerations for Anisotropic Shells," *Computer Methods in Applied Mechanics and Engineering*, 9(3), pp. 317-335.

Raoof, M. and Kraincanicm, I. (1994), "Critical Examination of Various Approaches Used for Analyzing Helical Cables," *Journal of Strain Analysis*, 29, pp. 43-55.

Ross, S. M. (1987), *Introduction to Probability and Statistics for Engineers and Scientists*, John Wiley & Sons, New York.

SAE (1991), "Passenger Car Side Door Latch Systems," SAE J839 JUN91, SAE Handbook, Warrendale, PA.

Shapiro, S. S. and Francia, R. S. (1972), "An Approximate Analysis of Variance Test for Normality," *Journal of American Statistical Association*, 67, p. 215.

Stockburger, D. W. (2007), "Hypothesis and Hypothesis Testing," in Salkind, N. J. (Ed.), *Encyclopedia of Measurement and Statistics*, SAGE Publications, Thousand Oaks, CA.

White, M., McLain, T., and Pldla, D. (1990), "Relationship between the Results of Nail Impact Bend Angle and Selected Nail Material Properties," *Journal of Testing and Evaluation*, 18(3), pp. 219-226.

Problems

P2.1: Three engineering factors and the related design levels for the Vickers hardness test of nails that have low carbon content are given as follows [Chiang and Chu 1993]:

Factor	Level (−)	Level (+)	Coded variable
A: Carbon content (SAE #)	SAE 1010	SAE 1022	a = A
B: Shank diameter (mm)	2.67	3.43	b = (B − 3.05)/0.38
C: Hardening	No	Yes	c = C

The full factorial design 2^3 has been conducted, and the physical test results (Y values) of the Vickers hardness test are listed as follows:

Run#	a	b	c	ab	ac	bc	abc	Y	Z
1	−1	−1	−1	1	1	1	−1	219	297
2	1	−1	−1	−1	−1	−1	1	297	306
3	−1	1	−1	−1	1	1	1	252	260
4	1	1	−1	1	−1	−1	−1	249	315
5	−1	−1	1	1	−1	−1	1	298	429
6	1	−1	1	−1	1	1	−1	429	536
7	−1	1	1	−1	−1	−1	−1	189	457
8	1	1	1	1	1	1	1	489	493

Please derive the predictive equation based on the DOE.

P2.2: Three engineering factors and the related design levels for Vickers hardness test of nails that have high carbon content are given as follows [Chiang and Chu 1993]:

Factor	Level (−)	Level (+)	Coded variable
A: Carbon content (SAE #)	SAE 1022	SAE 1040	$a = A$
B: Shank diameter (mm)	2.67	3.43	$b = (B − 3.05)/0.38$
C: Hardening	No	Yes	$c = C$

The 2^3 full factorial design has been conducted, and the physical test results (Z values) are listed in P2.1 (above). Please derive the predictive equation based on the DOE.

P2.3: The Freescale Cup is a college student competition aimed at building a wireless-controlled high-speed intelligent race car (model) that follows the prescribed trajectory. In running the experiment, the race car model is equipped with various sensors and driven by the signals wirelessly sent by a remote controller. To quantify the potential influences of four operating parameters on the completion time of a predetermined running course, a group of undergraduate students (School of Automotive Engineering, Chongqing University) sets up a full factorial design 2^4 that take the design levels of the four operating parameters of a toy model. Their design levels used for the study on a remote-controlled toy vehicle are listed as follows:

Operating factors	Level (−)	Level (+)
A. Steering sensitivity	Adjusted	Unadjusted
B. Acceleration/deceleration	Acceleration	Deceleration
C. Tire-track interface	Wet	Dry
D. Control frequency (max. Steering angle)	150 Hz	155 Hz

Sixteen test data (in seconds), corresponding to the standard treatment sequence for full design matrix 2^4 as shown in Table 2.9.2, while in the same run (treatment) order, are given as follows: (1) 32.123, (2) 32.242, (3) 28.841, (4) 28.894, (5) 31.026, (6) 31.328, (7) 28.304, (8) 28.487, (9) 31.711, (10) 31.812, (11) 28.695, (12) 29.996, (13) 31.460, (14) 31.382, (15) 28.359, (16) 28.519. Please analyze the test results to see which factors have influence in reducing the completion time per lap.

3

Fractional Factorial Design 2_R^{K-P}

A fractional factorial design with two levels, denoted as 2_R^{K-P}, is often used in experimentation as an economical way to relate an objective function to influential factors. It can be used as a screening design to identify significant main effects, with the final objective to sequentially fit a higher-order relationship. A screening design is typically of resolution III (R = III), but an experimental design of resolution IV (R = IV) is also occasionally used. Sequential operations after screening designs or designs with high levels directly would lead to response surface models that can be used for conditional optimization in the operating ranges of the given design factors. Both main and interactive effects can be exploited using a design of resolution IV or higher, while quadratic and higher-order terms can only be unveiled using a (fractional) factorial design with at least three levels, such as an experimental design L^{K-P} (L ≥ 3) or composite design.

3.1. Fractional Factorial Design

Fractional factorial designs are cost-effective for screening important factors from a large pool of potential variables. The screening design aims to carry out an experimental plan that is intended to identify statistically significant factors from a list of potential factors. For example, suppose an engineer needs to investigate the effects of varying seven factors on a manufacturing process, with each factor taking two levels. An experiment based on a full factorial design, where the effects of every combination of levels of each factor are to be studied, would require 2^7 (= 128) experimental runs for this example. This is not economically correct nor is it feasible. In order to minimize the expense on experiments and timing on delivery, the experimenter might decide to neglect the investigation of higher-order interactions of the seven factors and pay attention to identifying their main effects and their low-order interactive effects that could be estimated from an experiment using a much smaller number of experimental treatments (runs). Generally speaking, the number of key factors that influence the outcome is usually no more than 11. Furthermore, contributions from three-factor interactions and above are usually minor or even negligible, as long as the working range of each factor is specified consciously.

3.1.1. Fractional Factorial Notation

The fractional factorial notation is $2_R{}^{K-P}$ [Box and Hunter 1961], as illustrated in Figure 3.1.1. Exponent K represents the number of factors, of which P factors are generated from the interactions of full factorial design 2K using alias equations. Since $2^{K-P} = 2^K/2^P$, the fractional factorial design 2^{K-P} allows an experimenter to use the design that has K factors but run only a 2^{-P} fraction of the corresponding full factorial design 2^K. In particular, the design matrix 2^{K-P} can be constructed using the following two steps [Bisgaard 1994]:

1. Construct a full factorial design matrix based on K-P factors.
2. Fill up the design matrix using alias equations for the remaining P factors.

A fractional factorial design allows many factors to be analyzed with only a few experiments, but it does so by confounding effects and interactions. The resolution of a design, i.e., subscript R in $2_R{}^{K-P}$, is the sum of the order of confounded effects [Fries and Hunter 1980]. Resolution is used to indicate the severity of confounding.

FIGURE 3.1.1 Fractional factorial notation for $2_R{}^{K-P}$.

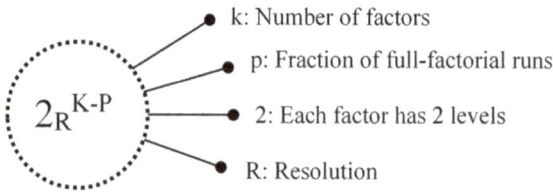

k: Number of factors

p: Fraction of full-factorial runs

$2_R{}^{K-P}$

2: Each factor has 2 levels

R: Resolution

Used with permission of Elsevier Science & Technology Journals, from "Minimum Aberration 2 K-P Designs," Technometrics, Fries , A. and Hunter , W. G., 1980; permission conveyed through Copyright Clearance Center, Inc.

The design with a higher resolution is considered better. A saturated design is one with resolution III, in which the number of design points (treatments) is equal to one more than the number of main and interactive effects to be estimated. Still, a saturated fractional factorial design allows an unbiased estimation of all main effects with the smallest possible variance and size, but the multifactor interactions cannot be exploited. Some examples of saturated designs are the Plackett–Burman two-level design and Taguchi's orthogonal arrays. Though these design matrices are convenient to use, it is hard to extend them to optimize for a robust design configuration. Even though effects could be confounded, and thus confused, with each other, a design with resolution V or above will minimize the impact of confounding, which allows an experimenter to estimate all main effects and two-factor interactions. Conducting fewer runs can save money and keep experiment costs low.

3.1.2. $2_{IV}{}^{4-1}$ Design

As $2^{4-1} = 2^4/2^1 = 2^3 = 8$, a 2^{4-1} design matrix yields a design that has eight runs, or eight treatments. There are three independent factors in this design, and they are identified as A, B, and C as shown in Table 3.1.1. There are eight and only eight distinct effects, including main effects and interactive effects, that can be generated by three factors: A, B, C, AB, AC, BC, ABC, and Mean. Factor D, the fourth factor, may be an independent variable in the nature of engineering, but it will be treated like a "dependent variable" in statistics using one of the following 14 possible aliasing opportunities:

$$R = II: \ D = A, D = -A, D = B, D = -B, D = C, D = -C;$$
$$R = III: \ D = A B, D = -A B, D = A C, D = -A C, D = B C, D = -B C;$$
and $\quad R = IV: D = A B C, D = -A B C$

Nevertheless, D = ABC and D = −ABC are the two best choices, as the probability for a three-factor interaction to occur is lower than that for a two-factor interaction, potentially.

Let D = ABC; the calculated effect in the "D = ABC column" in Table 3.1.1 is presumably the main effect of factor D. Thus, the alias equation, also called design generator or defining equation, for a $2_{IV}{}^{4-1}$ design matrix is:

$$D = A B C \tag{3.1.1}$$

or

$$D = -A B C \tag{3.1.2}$$

These alias equations are selected for generating the design matrix for $2_{IV}{}^{4-1}$. There are two operating rules for alias equations and they are:

1. A factor multiplied by itself is I (identity), e.g., I = A A = B B = C C = D D
2. A factor multiplied by I is itself, e.g., A I = A, B I = B, C I = C, D I = D

Thus, Equation (3.1.1) can be rewritten as AD = BC, BD = AC, CD = AB, A = BCD, B = ACD, C = ABD, and I = ABCD. Similarly, Equation (3.1.2) can be rewritten as AD = −BC, BD = −AC, CD = −AB, A = −BCD, B = −ACD, C = −ABD, and I = −ABCD.

Interactions between any two factors are so confounded that an interactive effect may be attributed to two possible causes, as follows:

$$\gamma_{AB} = \gamma_{AB} + \gamma_{CD} \tag{3.1.3}$$

$$\gamma_{AC} = \gamma_{AC} + \gamma_{BD} \tag{3.1.4}$$

and

$$\gamma_{AD} = \gamma_{AD} + \gamma_{BC} \tag{3.1.5}$$

Note that there are only two design generators for $2_{IV}{}^{4-1}$, i.e., Equations (3.1.1) and (3.1.2). The resolution of a design is given by the length of the shortest word in the defining relation. Since the shortest set of grouping these four factors (i.e., A, B, C, and D) to make I (identity) is four, the resolution of such a 2^{4-1} design is four (IV). In other words, no main effect is confounded with two-factor interaction in a $2_{IV}{}^{4-1}$ design. The design matrices generated using either one of the following two design generators: D = ABC and D = −ABC, are listed in Tables 3.1.1 and 3.1.2, respectively. When both design matrices are combined together, they constitute the design matrix of full factorial design 2^4.

TABLE 3.1.1 Fractional factorial design matrix of 2_{IV}^{4-1} based on D = ABC.

Run	a	b	c	d = abc	ab	ac	bc	Y	Y_p
1	−1	−1	−1	−1	1	1	1	37.02	36.17
2	1	−1	−1	1	−1	−1	1	76.72	77.84
3	−1	1	−1	1	−1	1	−1	68.20	70.92
4	1	1	−1	−1	1	−1	−1	46.08	43.10
5	−1	−1	1	1	1	−1	−1	96.54	92.67
6	1	−1	1	−1	−1	1	−1	63.02	64.85
7	−1	1	1	−1	−1	−1	1	55.92	57.93
8	1	1	1	1	1	1	1	99.56	99.60
Contrast	6.93	−0.89	21.76	34.75	3.84	−1.87	−1.16		
Effect	3.46	−0.44	10.88	17.37	1.92	−0.93	−0.58	67.88 (Average)	

© SAE International.

TABLE 3.1.2 Fractional factorial design matrix of 2^{4-1} based on D = −ABC.

Run	a	b	c	d = −abc	ab	ac	bc	Y	Y_p
1	−1	−1	−1	1	1	1	1	70.45	69.75
2	1	−1	−1	−1	−1	−1	1	44.72	42.20
3	−1	1	−1	−1	−1	1	−1	34.78	36.07
4	1	1	−1	1	1	−1	−1	73.94	75.87
5	−1	−1	1	−1	1	−1	−1	53.45	53.95
6	1	−1	1	1	−1	1	−1	93.90	93.75
7	−1	1	1	1	−1	−1	1	88.70	87.62
8	1	1	1	−1	1	1	1	59.32	60.07
Contrast	6.12	−1.45	17.87	33.68	−1.23	−0.59	1.78		
Effect	3.06	−0.72	8.94	16.84	−0.62	−0.30	0.89	64.91 (Average)	

© SAE International.

A group of college students carried out an experimental design on how the influences of the four factors listed in Table 3.1.3 would be on the cleanup of water on the car windshield. They begin with spreading 100 grams of water onto the windshield uniformly and apply the one-cycle wiping operation. The remaining water is then wiped and supposedly captured in the wiping paper that is dry initially. The leftover of water on the windshield, as absorbed by the wiping paper, is registered on the weight gain of the wiping paper. The water wiped off will be calculated as "100 grams – weight gain." The test results are listed in Tables 3.1.1 and 3.1.2.

TABLE 3.1.3 Factors and levels used for the study on characterizing wiper cleaning.

Factor	Level (−)	Level (+)
(A) Attack angle	Small	Large
(B) Moving speed	Low	Medium
(C) Pressing force	Low	High
(D) Wiping material	Hard (aged)	Soft (New)

© SAE International.

The calculation procedure for design contrasts and effects and statistical inferences based on the t-distribution for a fractional factorial design are the same as those for a full factorial design. Based on the physical test data given in Tables 3.1.1 and 3.1.2, the predictive equations for both design matrices and its corresponding full factorial design for this case study are given respectively as follows:

$$2^{4-1} (D = A B C): Y_p = 67.883 + 3.463 A + 10.878 C + 17.373 D (R_{pred} = 99.42\%)$$

$$2^{4-1} (D = - A B C): Y_p = 64.909 + 3.061 A + 8.936 C + 16.839 D (R_{pred} = 99.76\%)$$

and 2^4 (Full): $Y_p = 66.333 + 3.324 A + 9.844 C + 17.168 D$

Correlations between the predicted values and physical test data for both half-fractional factorial designs are 99.42% for 2^{4-1} (D = ABC) and 99.76% for 2^{4-1} (D = −ABC), respectively. Obviously, either of these two fractional factorial designs is a good approximation to full factorial design 2^4.

3.1.3. Foldover Pair

A 2^K factorial design can be broken into two 2^{K-1} design blocks [Montgomery and Runger 1996], such as Tables 3.1.1 and 3.1.2 for K = 4, that form a foldover pair. It means that a mirror-image foldover, i.e., folded over the dashed line in Table 3.1.4, of a 2^{4-1} design on factor D and reversing the sign of factor D will yield a 2^4 design as shown in Table 3.1.4 after both 2^{4-1} design matrices are combined together. Factor D is therefore the block factor that renders the blocking effect and leaves itself unconfounded with the main effects of the other K-1 factors. Note that a design matrix can be folded over on more than one block factors, as well as all variables, at the same time.

TABLE 3.1.4 2^4 Formed by folding 2_{IV}^{4-1} over on factor D.

Design	Run	A	B	C	D
2_{IV}^{4-1}	1	−1	−1	−1	−1
(D = ABC)	2	1	−1	−1	1
	3	−1	1	−1	1
	4	1	1	−1	−1
	5	−1	−1	1	1
	6	1	−1	1	−1
	7	−1	1	1	−1
	8	1	1	1	1
-------- Folded over ------					
Augmented runs	9	−1	−1	−1	1
	10	1	−1	−1	−1
	11	−1	1	−1	−1
	12	1	1	−1	1
	13	−1	−1	1	−1
	14	1	−1	1	1
	15	−1	1	1	1
	16	1	1	1	−1

3.2. Design Generators

Design generators are also called alias equations. Most of the frequently used 2_R^{K-P} are listed in Table 3.2.1. After conducting experimental tests following the design matrix, one can calculate the effects by following these steps:

1. Enter the average responses into the matrix for each treatment (run).
2. Work with the columns and determine the effect, which is the difference between the averaged response at the high level and the average response at the low level. Using the signs in the column in Excel, add up the responses and then divide by the number of pluses.

Note that the number of relatively important effects in a factorial experiment may be small. In general, low-order effects are more likely to be important than high-order effects, and the effects of the same order are equally likely to be important. It is called the effect of hierarchical ordering principle, also called heredity principle, thanks to a previous study [Li et al. 2006]. In order for an interaction to be significant, at least one of its parent factors should be significant. This follows the principle of marginality [McCullagh and Nelder 1989, Wu and Hamada 2009].

TABLE 3.2.1 Design generators (alias equations) for frequently used 2RK-P.

K	Fraction	No. of runs	Design generators
3	2_{III}^{3-1}	4	C = ± A B
4	2_{IV}^{4-1}	8	D = ± A B C
5	2_{V}^{5-1}	16	E = ± A B C D
	2_{III}^{5-2}	8	D = ± A B
			E = ± A C
6	2_{VI}^{6-1}	32	F = ± A B C D E
	2_{IV}^{6-2}	16	E = ± A B C
			F = ± B C D
	2_{III}^{6-3}	8	D = ± A B
			E = ± A C
			F = ± B C
7	2_{VII}^{7-1}	64	G = ± A B C D E F
	2_{IV}^{7-2}	32	F = ± A B C D
			G = ± A B D E
	2_{IV}^{7-3}	16	E = ± A B C
			F = ± B C D
			G = ± A C D
	2_{III}^{7-4}	8	D = ± A B
			E = ± A C
			F = ± B C
			G = ± A B C
8	2_{V}^{8-2}	64	G = ± A B C D
			H = ± A B E F

(Continued)

TABLE 3.2.1 (Continued) Design generators (alias equations) for frequently used 2RK-P.

K	Fraction	No. of runs	Design generators
	2_{IV}^{8-3}	32	F = ± A B C
			G = ± A B D
			H = ± B C D E
	2_{IV}^{8-4}	16	E = ± B C D
			F = ± A C D
			G = ± A B C
			H = ± A B D
9	2_{VI}^{9-2}	128	H = ± A C D F G
			J = ± B C E F G
	2_{IV}^{9-3}	64	G = ± A B C D
			H = ± A C E F
			J = ± C D E F
	2_{IV}^{9-4}	32	F = ± B C D E
			G = ± A C D E
			H = ± A B D E
			J = ± A B C E
	2_{III}^{9-5}	16	E = ± A B C
			F = ± A B D
			G = ± A C D
			H = ± B C D
			J = ± A B C D
10	2_{V}^{10-3}	128	H = ± A B C G
			J = ± A C D E
			K = ± A C D F
	2_{IV}^{10-4}	64	G = ± B C D F
			H = ± A C D F
			J = ± A B D E
			K = ± A B C E
	2_{IV}^{10-5}	32	F = ± A B C D
			G = ± A B C E
			H = ± A B D E
			J = ± A C D E
			K = ± B C D E
	2_{III}^{10-6}	16	E = ± A B C
			F = ± B C D
			G = ± A C D
			H = ± A B D
			J = ± A B C D
			K = ± A B
11	2_{IV}^{11-5}	64	G = ± C D E
			H = ± A B C D
			J = ± A B F
			K = ± B D E F
			L = ± A D E F

(Continued)

TABLE 3.2.1 (Continued) Design generators (alias equations) for frequently used 2RK-P.

K	Fraction	No. of runs	Design generators
	2_{IV}^{11-6}	32	F = ± A B C
			G = ± B C D
			H = ± C D E
			J = ± A C D
			K = ± A D E
			L = ± B D E
	2_{III}^{11-7}	16	E = ± A B C
			F = ± B C D
			G = ± A C D
			H = ± A B D
			J = ± A B C D
			K = ± A B
			L = ± A C
	L_{12} (R = III)	12	See Table 3.5.1 for design matrix
15	2_{III}^{15-11} (i.e., L_{16})	16	E = ± A B
			F = ± A C
			G = ± A D
			H = ± B C
			...
			(11 design generators)
31	2_{III}^{31-26}	32	F = ± A B
			G = ± A C
			H = ± A D
			I = ± A E
			J = ± B C
			K = ± B D
			...
			(26 design generators)

© SAE International.

3.3. Frequently Used Fractional Factorial Designs

Fractional factorial designs up to 21 factors (K = 21) are listed in Table 3.3.1. The analysis procedure for carrying out design matrix 2_{IV}^{4-1} (Section 3.1.2) can be extended to case studies on 2_V^{5-1}, 2_{III}^{6-3}, 2_{III}^{7-4}, and 2_{IV}^{8-4} designs, which are the design matrices applied frequently in the real-world applications. "Treatment" is the right terminology, but "run" is often used for convenience. In fact, there can be several replicated runs carried out within each treatment.

The common strategy to reduce experimentation cost is to run a fractional factorial design at resolution V (or IV) and confound higher-order interactions with main effects and two-factor interactions.

TABLE 3.3.1 List of fractional factorial design 2_R^{K-P}, up to K = 21.

#of Factors	# of Treatments (runs)							
	4	8	16	32	64	128	256	512
2	2^2							
3	2_{III}^{3-1}	2^3						
4	—	2_{IV}^{4-1}	2^4					
5	—	2_{III}^{5-2}	2_V^{5-1}	2^5				
6	—	2_{III}^{6-3}	2_{IV}^{6-2}	2_{VI}^{6-1}	2^6			
7	—	2_{III}^{7-4}	2_{IV}^{7-3}	2_{IV}^{7-2}	2_{VII}^{7-1}	2^7		
8	—	—	2_{IV}^{8-4}	2_{IV}^{8-3}	2_V^{8-2}	2_{VIII}^{8-1}	2^8	
9	—	—	2_{III}^{9-5}	2_{IV}^{9-4}	2_{IV}^{9-3}	2_{VI}^{9-2}	2_{IX}^{9-1}	2^9
10	—	—	2_{III}^{10-6}	2_{IV}^{10-5}	2_{IV}^{10-4}	2_{IV}^{10-3}	2_{VI}^{10-2}	2_X^{10-1}
11	—	—	2_{III}^{11-7}	2_{IV}^{11-6}	2_{IV}^{11-5}	2_{IV}^{11-4}	2_{VI}^{11-3}	2_{VII}^{11-2}
12	—	—	2_{III}^{12-8}	2_{IV}^{12-7}	2_{IV}^{12-6}	2_{IV}^{12-5}	2_{VI}^{12-4}	2_{VI}^{12-3}
13	—	—	2_{III}^{13-9}	2_{IV}^{13-8}	2_{IV}^{13-7}	2_{IV}^{13-6}	2_V^{13-5}	2_{VI}^{13-4}
14	—	—	2_{III}^{14-10}	2_{IV}^{14-9}	2_{IV}^{14-8}	2_{IV}^{14-7}	2_{IV}^{14-6}	2_{VI}^{14-5}
15	—	—	2_{III}^{15-11}	2_{IV}^{15-10}	2_{IV}^{15-9}	2_{IV}^{15-8}	2_V^{15-7}	2_{VI}^{15-6}
16	—	—	—	2_{IV}^{16-11}	2_{IV}^{16-10}	2_{IV}^{16-9}	2_V^{16-8}	2_{VI}^{16-7}
17	—	—	—	2_{III}^{17-12}	2_{IV}^{17-11}	2_{IV}^{17-10}	2_V^{17-9}	2_V^{17-8}
18	—	—	—	2_{III}^{18-13}	2_{IV}^{18-12}	2_{IV}^{18-11}	2_{IV}^{18-10}	2_V^{18-9}
19	—	—	—	2_{III}^{19-14}	2_{IV}^{19-13}	2_{IV}^{19-12}	2_{IV}^{19-11}	2_V^{19-10}
20	—	—	—	2_{III}^{20-15}	2_{IV}^{20-14}	2_{IV}^{20-13}	2_{IV}^{20-12}	2_V^{20-11}
21	—	—	—	2_{III}^{21-16}	2_{IV}^{21-15}	2_{IV}^{21-14}	2_{IV}^{21-13}	2_V^{21-12}

© SAE International.

3.3.1. 2_V^{5-1} Example—Time to Charge a Cell Phone

This is an excellent DOE for automotive engineering applications, because it has a resolution of V (five). It is very comfortable to make sense of the two-factor interactions by default on the assumption that three-factor interactions are generally insignificant. Five factors are a good size of independent variables whose effects are to be discovered in most applications. The design generator for a 2^{5-1} design matrix is either:

$$E = A\,B\,C\,D \tag{3.3.1}$$

or

$$E = -\,A\,B\,C\,D \tag{3.3.2}$$

An experiment on the battery-charging process for a hand-held cell phone was performed for the purpose of exploring the influences of five operating parameters (factors) on the time duration (minutes) required for charging the empty battery to in-full state by a charger. The five operating variables and their design levels considered in the study are given in Table 3.3.2. Note that there are two outlets in a battery charger; multi-tasking means that the other end is also loaded when one end of the charger is being used to charge the cell phone.

TABLE 3.3.2 Operating variables and their design levels for charging a cell phone.

Factor	Level (−1)	Level (+1)
A: Original charger wiring	No	Yes
B: Screen on or not	Off	On
C: Charger: initial state	Not in full	In full
D: Charger: multitasking or not	No	Yes
E: Charger: brand (having the same rating)	Supplier E⁻	Supplier E⁺

© SAE International.

Data resulting from the physical tests based on design matrix 2^{5-1} are denoted by Y and listed in Table 3.3.3. Since there is no replication (i.e., N = 1), the "not-so-significant" main effect (i.e., c) and two-factor interactions (i.e., "ab," "ad," and "cd"), as underlined in Table 3.3.3, are pooled together according to Equation (2.5.3) to formulate the sample random error as follows:

$$\text{Error} = [(\gamma_c^2 + \gamma_{ab}^2 + \gamma_{ad}^2 + \gamma_{cd}^2) / 4]^{1/2}$$

$$= \{[(-0.063)^2 + (-0.313)^2 + (-0.813)^2 + (-0.063)^2] / (1 + 1 + 1 + 1)\}^{1/2}$$

$$= 0.4375$$

Note that each effect (main or interactive) possesses one degree of freedom. The predictive equation based on statistic inference using $t_{4,95\%}$ is formulated as follows:

$$Y_p = 137.5625 - 1.0625\,a + 10.5625\,b - 1.1875\,d - 3.1875\,e$$

$$- 1.9375\,a\,c + 1.1875\,b\,c - 2.9375\,b\,d$$

$$- 0.9375\,a\,b\,c - 4.3125\,a\,b\,d - 1.4375\,a\,c\,d - 1.0625\,b\,c\,d$$

Variates a, b, c, d, and e are the five dimensionless variables derived from variables A, B, C, D, and E, respectively, based on Equation (2.4.5). The predicted values using the above equation are listed as column Y_p. After model checking, one can conclude that three two-factor and four three-factor interactions are significant in addition to four main effects. By far, factor B (screen on or not) has the greatest effect on the charge time.

TABLE 3.3.3 Fractional factorial design matrix 2_V^{5-1} on time to charge a cell phone.

(I) Main effects						
Run	a	b	c	d	e = abcd	Y (minutes)	Y_p
1	−1	−1	−1	−1	1	129	130.125
2	1	−1	−1	−1	−1	126	124.875
3	−1	1	−1	−1	−1	148	148.5
4	1	1	−1	−1	1	152	151.5
5	−1	−1	1	−1	−1	130	131.125
6	1	−1	1	−1	1	116	114.875
7	−1	1	1	−1	1	149	149.5
8	1	1	1	−1	−1	160	159.5
9	−1	−1	−1	1	−1	127	126.375
10	1	−1	−1	1	1	131	131.375
11	−1	1	−1	1	1	143	141.75
12	1	1	−1	1	−1	145	146
13	−1	−1	1	1	1	125	124.625
14	1	−1	1	1	−1	132	132.625
15	−1	1	1	1	−1	158	157
16	1	1	1	1	1	130	131.25

(Continued)

© SAE International.

TABLE 3.3.3 (Continued) Fractional factorial design matrix 2_V^{5-1} on time to charge a cell phone.

Contrast	−2.125	21.125	−0.125	−2.375	−6.375
Effect	−1.063	10.563	_−0.063_	−1.188	−3.188	137.5625 (average)
Error	0.4375	0.4375	0.4375	0.4375	0.4375
$|t_4|$	2.43	24.14	0.143	2.71	7.29
$t_{4,95\%}$	2.132	2.132	2.132	2.132	2.132
Significant?	Yes	Yes	No	Yes	Yes

(II) Two-factor and three-factor interactions

Run	ab	ac	ad	bc	bd	cd	abc	abd	acd	bcd
1	1	1	1	1	1	1	−1	−1	−1	−1
2	−1	−1	−1	1	1	1	1	1	1	−1
3	−1	1	1	−1	−1	1	1	1	−1	1
4	1	−1	−1	−1	−1	1	−1	−1	1	1
5	1	−1	1	−1	1	−1	1	−1	1	1
6	−1	1	−1	−1	1	−1	−1	1	−1	1
7	−1	−1	1	1	−1	−1	−1	1	1	−1
8	1	1	−1	1	−1	−1	1	−1	−1	−1
9	1	1	−1	1	−1	−1	−1	1	1	1
10	−1	−1	1	1	−1	−1	1	−1	−1	1
11	−1	1	−1	−1	1	−1	1	−1	1	−1
12	−1	−1	1	−1	1	−1	−1	1	−1	−1
13	1	−1	−1	−1	−1	1	1	1	−1	−1
14	−1	1	1	−1	−1	1	−1	−1	1	−1
15	−1	−1	−1	1	1	1	−1	−1	−1	1
16	1	1	1	1	1	1	1	1	1	1
Contrast	−0.625	−3.875	−1.625	2.375	−5.875	−0.125	−1.875	−8.625	−2.875	−2.125
Effect	_−0.313_	−1.938	_−0.813_	1.188	−2.938	_−0.063_	−0.938	−4.313	−1.438	−1.063
Error	0.4375	0.4375	0.4375	0.4375	0.4375	0.4375	0.4375	0.4375	0.4375	0.4375
$|T_4|$	0.71	4.43	1.86	2.71	6.71	0.143	2.143	9.86	3.29	2.43
$T_{4,95\%}$	2.132	2.132	2.132	2.132	2.132	2.132	2.132	2.132	2.132	2.132
Significant?	No	Yes	No	Yes	Yes	No	Yes	Yes	Yes	Yes

3.3.2. 2_{III}^{6-3} Example: Case Study on Shrink of Optical-Fiber Cable

Optical-fiber communication facilitates long-distance transmission at high speeds without electromagnetic interference. An optical-fiber cable, made of glass and plastics, is expected to achieve such optical conduction based on the total reflection theory. Nevertheless, the contraction of the outer coating of an optical fiber may deteriorate the communication capacity and shorten the reaching distance, though its primary functional requirement is to protect the optical fiber. A DOE was set up upon the six factors as listed in Table 3.3.4 to investigate the effect of these factors on the contraction rate of the outer coating [Jou et al. 2014].

TABLE 3.3.4 List of factors for study on shrink of optical-fiber cable [Jou et al. 2014].

Factor	Level (−)	Level (+)	Normalized variable
A: Die stretch ratio	1.5	2.1	$a = (A - 1.8)/0.6$
B: Operating temperature	162°C	175°C	$b = (B - 168.5)/13$
C: Coolant temperature	25°C	35°C	$c = (C - 30)/10$
D: Air cooling distance	100 mm	150 mm	$d = (D - 125)/50$
E: Take-up speed	20 m/min	40 m/min	$e = (E - 30)/20$
F: Set-out speed	20 m/min	40 m/min	$f = (F - 30)/20$

Reprinted from "Integrating the Taguchi Method and Response Surface Methodology for Process Parameter Optimization of the Injection Molding," Applied Mathematics and Information Sciences. © 2014 NSP.

When design matrix 2_{III}^{6-3} applies, three factors (namely A, B, and D for this case study) are required for setting up the full factorial design matrix as the 2^3 design does. As dictated by the Taguchi's L_8 design, the remaining design matrix for the other three factors, i.e., C, E, F, is obtained using the following alias equations:

$$C = - A B \tag{3.3.3}$$

$$E = - A D \tag{3.3.4}$$

and

$$F = - B D \tag{3.3.5}$$

The contrasts and effects calculated are listed in Table 3.3.5. Since replication = 1, the underlined small effects (four of them) are used to compute for the error using Equation (2.5.7) as:

$$S = \left(\frac{0.0019^2 + 0.0014^2 + 0.0003^2 + 0.0012^2}{4}\right)^{1/2} = 0.0013$$

It is concluded hereupon that factors B and D are the only two statistically significant factors when measured by the sample standard deviation. The predictive equation based on design 2_{III}^{6-3} is:

$$y = 0.0373 + 0.0187\, d - 0.0098\, b$$

which can be converted into:

$$Y = 0.0373 + 0.0187 \left(\frac{D - 125}{150 - 100}\right) - 0.0098 \left(\frac{B - 168.5}{175 - 162}\right)$$

$$= 0.0373 + 0.000374\, D - 0.04675 - 0.0007538\, B + 0.1270$$

$$= 0.1176 + 0.000374\, D - 0.0007538\, B$$

The predicted values and residuals for the eight runs (treatments) using the above predictive equation are listed in the last two columns, respectively in Table 3.3.5. The plot of residuals versus predicted values, Figure 3.3.1, shows the inadequacy of the linear regression model based on the two-level design. An experimental design based on the 3^2 design for these two influential factors is thus sought in order to explore the potential nonlinear modeling curvature (Chapter 4).

TABLE 3.3.5 Fractional factorial design matrix 2_{III}^{6-3} (also called Taguchi's L_8 design).

Run	D	B	A	C = −AB	E = −AD	F = −BD	Y (m/m)	Predicted	Residual
1	−1	−1	−1	−1	−1	−1	0.0352	0.0284	−0.0068
2	1	−1	−1	−1	1	1	0.0440	0.0658	0.0218
3	−1	1	−1	1	−1	1	0.0289	0.0088	−0.0201
4	1	1	−1	1	1	−1	0.0391	0.0462	0.0071
5	−1	−1	1	1	1	−1	0.0355	0.0284	−0.0071
6	1	−1	1	1	1	1	0.0442	0.0658	0.0216
7	−1	1	1	−1	1	1	0.0308	0.0088	−0.0220
8	1	1	1	−1	−1	−1	0.0405	0.0462	0.0057
Contrast	0.0374	−0.0196	0.0038	−0.0028	0.0006	−0.0024	—		
Effect	0.0187	−0.0098	_0.0019_	_−0.0014_	_0.0003_	_−0.0012_	0.0373 (Average)		
Error	0.0013	0.0013	0.0013	0.0013	0.0013	0.0013	0.0013		
$t_{4,1-\alpha}$	14.0	7.4	1.4	1.1	0.2	0.9	28.0		
$t_{4,95\%}$	2.132	2.132	2.132	2.132	2.132	2.132	2.132		
Significant	Yes	Yes	No	No	No	No	Yes		

FIGURE 3.3.1 Plot of residuals versus predicted values for the eight treatments.

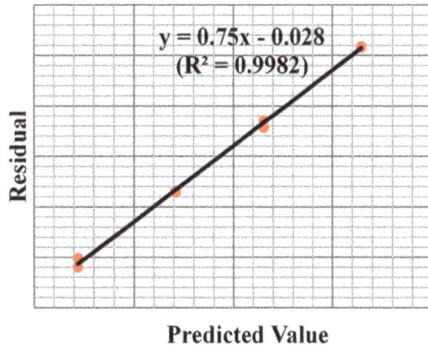

$$y = 0.75x - 0.028$$
$$(R^2 = 0.9982)$$

Residual

Predicted Value

3.4. **Foldover Design**

One theorem embedded in DOE is the foldover design, which is a follow-up step routinely after completing a two-level fractional factorial design with resolution III. The newly folded over design matrix in combination with the original design matrix will produce a design of resolution IV or higher. The folding takes the levels of all the factors in the original design and switches them in signs, i.e., (−1 ➔ 1) and (+1 ➔ −1), on some existing or new factors, which are called folding factors. Only the signs of the folding factors are fully switched in the follow-up design, and the other factors retain the same signs that they had in the initial design. Switching the sign of folding factor will de-confound that factor's main effect and all its associated two-factor interactions, when analyzing the two sets of fractional factorial design matrices together.

3.4.1. 2_{III}^{7-4} Design

A case study based on 2_{III}^{7-4} is taken here to illustrate the steps toward applying fractional factorial designs, because the total number of would-be-factors are usually no more than eight in most realistic automotive engineering applications. It is reasonable to have seven or eight predictive variables, but it is costive to run 128 tests (2^8 design) if not impossible to carry them out. An experimenter begins with a screening DOE based on the first seven factors using 2_{III}^{7-4}, which is a 1/16 replicate of the 2^7 full factorial design in eight runs having a resolution of III. A design matrix for such a design is given in Table 3.4.1. Assume that the seven factors are A, B, C, D, E, F, and G. The design matrix based on factors A, B, and C is first constructed using the 2^3 full factorial design. Then, the remaining design matrix for the other four factors is to be derived using the following alias equations:

$$D = A\,B \tag{3.4.1}$$

$$E = A\,C \tag{3.4.2}$$

$$F = B\,C \tag{3.4.3}$$

and

$$G = A\,B\,C \tag{3.4.4}$$

An alias indicates that two or more things are changed at the same time following their constraint equation(s), as shown above. Thus, the main effect of D is confounded with the interactive effect of A and B, the main effect of E is confounded with the interactive effect of A and C, the main effect of F is confounded with the interactive effect of B and C, and the main effect of G is confounded with the interactive effect of A, B, and C.

When the above two operating rules are applied to Equations (3.4.1)–(3.4.4), one has the design generator for 2_{III}^{7-4} as:

$$
\begin{aligned}
I &= A\,B\,D = AC\,E = B\,C\,F = D\,E\,F = CD\,G = B\,E\,G = A\,F\,G \\
&= B\,CD\,E = A\,CD\,F = A\,B\,E\,F = A\,B\,C\,G = A\,D\,E\,G = B\,D\,F\,G = C\,E\,F\,G \\
&= A\,B\,C\,D\,E\,F\,G
\end{aligned}
\tag{3.4.5}
$$

The shortest number of cascading factors leading to I is the resolution of the design matrix, i.e., III (three) for 2_{III}^{7-4}. When it is imperative to screen many factors to identify those that may be important (i.e., those that are related to the dependent variable of interest), a design matrix may be explored to let an experimenter test the largest number of main effects with the least number of runs (observations). It means to construct a resolution III design with as few runs as possible—8 runs for 2_{III}^{7-4}, which is also called L_8 in the automotive practice [Taguchi et al. 2005].

TABLE 3.4.1 Fractional factorial design matrix of 2_{III}^{7-4}.

Run	a	b	c	d = ab	e = ac	f = bc	g = abc
1	−1	−1	−1	1	1	1	−1
2	1	−1	−1	−1	−1	1	1
3	−1	1	−1	−1	1	−1	1
4	1	1	−1	1	−1	−1	−1
5	−1	−1	1	1	−1	−1	1
6	1	−1	1	−1	1	−1	−1
7	−1	1	1	−1	−1	1	−1
8	1	1	1	1	1	1	1

Assume that design 2_{III}^{7-4} was analyzed and some significant effects show up but there is not any degree of freedom for error. If one wants to look at another replicate of the original design, the alias equations associated with the next eight runs (9–16 runs) are then constructed as:

$$D = - A\,B \tag{3.4.6}$$

$$E = - A\,C \tag{3.4.7}$$

$$F = - B\,C \tag{3.4.8}$$

and

$$G = - A\,B\,C \tag{3.4.9}$$

It means that runs 9–16 are just another block of resolution III. The next eight more runs are expected to fulfill two missions. One is taking on another replicate and the other is moving to a higher resolution number. These can be done by applying foldover techniques as demonstrated in the next section.

3.4.2. 2_{IV}^{8-4} as Foldover Design from 2_{III}^{7-4} on an Additional Factor

In practice, the initial fraction is almost always sufficient by itself, but if the results turn out to be ambiguous, as a solution of last resort one can still modify the initial fraction and add a "folded fraction." In a resolution IV design, two-factor interactions are not confounded with any main effect, so this resolution IV design is a lot safer than a resolution III design, and it lends the experimenter to study up to eight factors in only 16 runs as quite a cost-effective solution.

A 2_{III}^{7-4} (seven-factor in eight runs with resolution III) shown in Table 3.4.1 can be folded over on one new factor (the 8th) to create a new design matrix 2_{IV}^{8-4} (eight-factor in 16 runs with resolution IV). When a resolution III design is folded over on an additional factor, it becomes a resolution IV design that has twice as many observations (runs) to be carried out. However, when a resolution IV design is folded over on an additional factor to double the number of observations (runs), it is still a resolution IV for most cases.

The 8th variable (factor H) is hereby employed for foldover to create a 2_{IV}^{8-4} that can resolve the ambiguities between main effects and their confounded two-factor interactions. The procedure for folding 2_{III}^{7-4} over on one additional factor is illustrated as follows:

1. Assign $(1, 1, 1, 1, 1, 1, 1, 1)^T$ to the additional column corresponding to the 8th variable.
   a. Copy the entire design matrix (runs 1–8) to create runs 9–16, which are identical to runs 1–8, i.e., run 1 to run 9, run 2 to run 10, run 3 to run 11, run 4 to run 12, run 5 to run 13, run 6 to run 14, run 7 to run 15, and run 8 to run 16.
   b. Another way is to really fold the entire design matrix (runs 1–8) over to create runs 9–16, i.e., copy run 8 to run 9, run 7 to run 10, run 6 to run 11, run 5 to run 12, run 4 to run 13, run 3 to run 14, run 2 to run 15, and run 1 to run 16.
2. Reverse all signs of runs 9–16.
3. The new design matrix (runs 1–16 in Table 3.4.2) is then a 2_{IV}^{8-4} design.

By folding over of a resolution III design on one additional factor, the experimenter gets more robust information on that factor (e.g., the 8th factor, H, given in Table 3.4.2) as all its two-factor interactions (i.e., AH, BH, CH, DH, EH, FH, and GH in 2_{III}^{7-4}) would be clear of all the other two-factor interactions (e.g., AB, AC, and BC). The newly formed 2_{IV}^{8-4} is of resolution IV. It serves this purpose well for upgrading resolution III designs, but not necessarily for resolution IV designs. By combining 2 (or more) fractional factorial designs, in which signs are switched, one can systematically isolate effects of potential interest. This procedure is called de-aliasing, which resolves the ambiguities between effects. Please see [Box et al. 2005] for detailed derivations and reasoning.

After the first eight runs (runs 1–8) are made, the design matrix (runs 1–8) is flipped over with signs changed and another variable, H, added to produce a new design matrix with resolution IV, i.e., 2^{8-4}_{IV}, embracing the 16 run conditions as exhibited in Table 3.4.2. In a design with resolution IV, main effects of individual factors, i.e., γ_A, γ_B, γ_C, γ_D, γ_E, γ_F, γ_G, and γ_H, are only confounded with three-factor interactions, which may be reasonably assumed to be remote and thus small. Thus, the main effects are outstanding factors if proven to be effective. Effects of two-factor interactions after foldover, denoted as γ_{AB}, γ_{AC}, γ_{AD}, γ_{AE}, γ_{AF}, γ_{AG}, and γ_{AH}, are confounded as [Box et al. 2005]:

$$\gamma_{AB} = \gamma_{AB} + \gamma_{CG} + \gamma_{DH} + \gamma_{EF} \qquad (A\,B = C\,G = D\,H = E\,F) \tag{3.4.10}$$

$$\gamma_{AC} = \gamma_{AC} + \gamma_{BG} + \gamma_{DF} + \gamma_{EH} \qquad (A\,C = B\,G = D\,F = E\,H) \tag{3.4.11}$$

$$\gamma_{AD} = \gamma_{AD} + \gamma_{BH} + \gamma_{CF} + \gamma_{EG} \qquad (A\,D = B\,H = C\,F = E\,G) \tag{3.4.12}$$

$$\gamma_{AE} = \gamma_{AE} + \gamma_{BF} + \gamma_{CH} + \gamma_{DG} \qquad (A\,E = B\,F = C\,H = D\,G) \tag{3.4.13}$$

$$\gamma_{AF} = \gamma_{AF} + \gamma_{BE} + \gamma_{CD} + \gamma_{GH} \qquad (A\,F = B\,E = C\,D = G\,H) \tag{3.4.14}$$

$$\gamma_{AG} = \gamma_{AG} + \gamma_{BC} + \gamma_{DE} + \gamma_{FH} \qquad (A\,G = B\,C = D\,E = F\,H) \tag{3.4.15}$$

and

$$\gamma_{AH} = \gamma_{AH} + \gamma_{BD} + \gamma_{CE} + \gamma_{FG} \qquad (A\,H = B\,D = C\,E = F\,G) \tag{3.4.16}$$

The above equations are given without derivations; detailed derivations can be found in [Box et al. 2005]. It is advantageous to do a screening design using 2_{III}^{7-4} first, then to continue with its follow-up foldover design. If no interaction is effective, one can conclude the DOE. Otherwise, by folding a fractional design that doubles the treatment size, one is able to separate each main effect from its confounded two-factor interactions. In summary, a design with resolution III may be folded over on all factors for obtaining a design with resolution IV, although only "foldover on the eighth factor" is discussed above. Note that the folding factor (factor H) can be a dummy one.

Furthermore, if it is desired to separate interactive effect γ_{AD} from interactive effect γ_{CF} in Equation (3.4.12), for example, one may fold the design on factor A (or alternatively on factor D, on C, or on E) hitting directly the target for breaking the "alias chain."

3.4.3. Foldover Design Example: Gasket Sealing

As an example, the engine coolant-sealing gasket for a heavy-duty diesel engine is used hereupon to demonstrate the application of the foldover design. The objective is to have a uniform pressure distribution (MPa) around the contact contour between the gasket and its contacting flanges. The eight design factors are identified in Table 3.4.2.

TABLE 3.4.2 List of factors and design levels for experimental tests on engine coolant sealing.

Variables	Level (−)	Level (+)
A: Corrugating profile (Figure 3.4.1)	Semi-open-stepped	Sine arc
B (mm): Corrugating height	0.4	0.3
C: Bolt pattern (Figure 3.4.2)	Diagonal	Opposite
D (mm): Gasket thickness	0.25	0.35
E (mm): Upper flange thickness	20	15
F: Upper flange material	HT250	YL112
G: Gasket material	SAE 1010	12Cr17Ni7
H: Lower flange material	HT250	YL112

Notes:
HT250: Grey iron 250
YL112: Die-cast aluminum alloy (Chinese specification)

FIGURE 3.4.1 Cross-sectional views of two corrugated caskets.

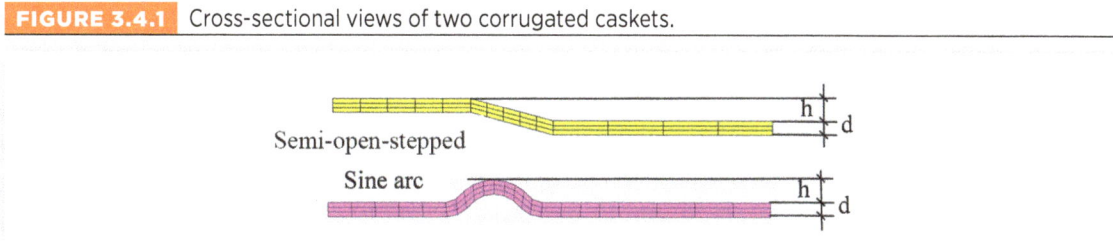

FIGURE 3.4.2 Bolt patterns: (a) diagonal—corner to corner and (b) opposite—side to side.

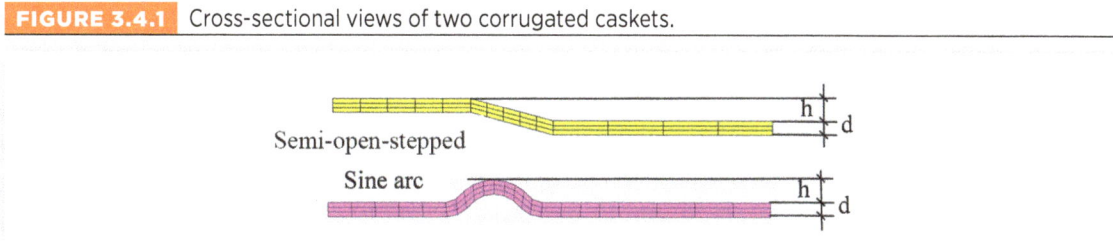

© SAE International.

FIGURE 3.4.3   Contact pressure distributions and sealing scores of the 16 DOE runs.

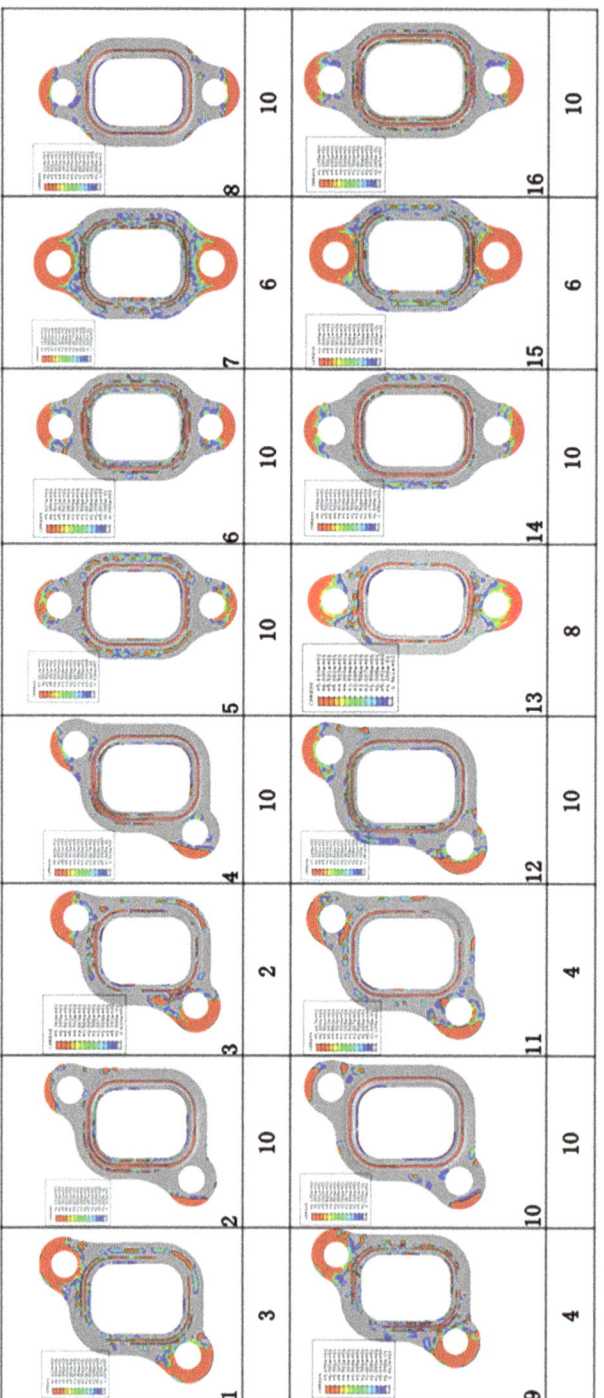

TABLE 3.4.3 Folding 2_{III}^{7-4} design over factor H to obtain 2_{IV}^{8-4} in two blocks (runs 1–8 and runs 9–18).

Run	a	b	c	d	e	f	g	h	Y	Y_p
1	−1	−1	−1	1	1	1	−1	1	13	18.4
2	1	−1	−1	−1	−1	1	1	1	81	84.7
3	−1	1	−1	−1	1	−1	1	1	20	19.4
4	1	1	−1	1	−1	−1	−1	1	88	89.2
5	−1	−1	1	1	−1	−1	1	1	93	83.1
6	1	−1	1	−1	1	−1	−1	1	78	76.6
7	−1	1	1	−1	−1	1	−1	1	40	45.1
8	1	1	1	1	1	1	1	1	87	10
9	1	1	1	−1	−1	−1	1	−1	15	9.6
10	−1	1	1	1	1	−1	−1	−1	83	79.3
11	1	−1	1	1	−1	1	−1	−1	10	10.6
12	−1	−1	1	−1	1	1	1	−1	85	83.8
13	1	1	−1	−1	1	1	−1	−1	48	57.9
14	−1	1	−1	1	−1	1	1	−1	86	87.4
15	1	−1	−1	1	1	−1	1	−1	25	19.9
16	−1	−1	−1	−1	−1	−1	−1	−1	91	94.4
Contrast	51.88	−6.38	19.13	3.38	−8.13	−5.38	10.38	7.13		
Effect	25.94	−3.19	9.56	*1.688*	−4.063	*−2.688*	5.188	3.563		
Error	2.14	2.14	2.14	2.14	2.14	2.14	2.14	2.14		
T_5	12.12	1.49	4.47	0.79	1.90	1.26	2.42	1.67		
$T_{5,90\%}$	1.476	1.476	1.476	1.476	1.476	1.476	1.476	1.476		
Significant	Yes	Yes	Yes	No	Yes	No	Yes	Yes		

(b) Interactions:

Run	ab	cd	ac	ad	bd	ae	bc
1	1	−1	1	−1	−1	−1	1
2	−1	1	−1	−1	1	−1	1
3	−1	1	1	1	−1	−1	−1
4	1	−1	−1	1	1	−1	−1
5	1	1	1	−1	−1	1	−1
6	−1	−1	−1	−1	1	1	−1
7	−1	−1	1	1	−1	1	1
8	1	1	1	1	1	1	1
9	1	1	−1	1	1	1	1
10	−1	−1	1	1	−1	1	1
11	−1	−1	−1	−1	1	1	−1
12	1	1	1	1	−1	1	−1
13	1	−1	−1	1	1	−1	−1
14	−1	1	1	1	−1	−1	−1
15	−1	1	−1	−1	1	−1	1
16	1	−1	1	−1	−1	−1	1
Contrast	12.13	5.13	−17.88	−1.13	−9.88	4.88	−9.13
Effect	6.063	*2.563*	−8.94	*−0.563*	−4.94	*2.44*	−4.563
Error	0.218	0.218	0.218	0.218	0.218	0.218	0.218
T_5	2.00	0.29	4.87	0.86	0.29	1.43	1.43
$t_{10,90\%}$	1.476	1.476	1.476	1.476	1.476	1.476	1.476
Significant	Yes	No	Yes	No	Yes	No	Yes

Ambiguities about two-factor interactions resolve the experimenter to take the sense of engineering context and to judge relative magnitudes of individual main effects. The test data originating from 16 finite element analyses and the resulting main effects of individual factors are listed in Table 3.4.3(a), while the interactive effects between two factors are shown in Table 3.4.3(b). Main and interactive effects on the uniformity of engine coolant sealing are plotted in Figure 3.4.4. Taking 0.384 (i.e., 5% of the average 7.688) as the wording of statistical significance, one can pool the 10 "insignificant effects" (underlined) to calculate the pooled sample standard deviation, and it is 0.218, which is assumed to be the random error. When measured by this amount of error and based on the $\alpha = 5\%$ criterion on t-statistic tail value (or $t_{10,95\%}$), the predictive equation for pressure uniformity is formulated as follows:

$$Y_p = 58.938 + 25.938 \, a - 3.188 \, b + 9.563 \, c - 4.063 \, e + 5.188 \, g - 3.563 \, h$$

$$+ \, 6.063 \, a \, b - 8.938 \, a \, c \quad - 4.938 \, b \, d - 4.563 \, b \, c$$

FIGURE 3.4.4   Main and interactive effects on engine coolant sealing.

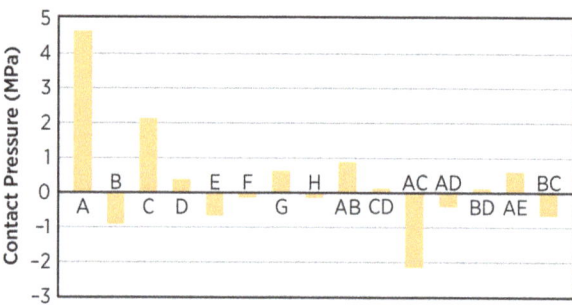

Since 2_{IV}^{8-4} has a resolution 4, the main effects (i.e., a, b, and c) given in the above equation are statistically significant as the probability for a three-factor interaction to be effective is assumed to be extremely rare. According to Equation (3.4.10) that $\gamma_{AB} = \gamma_{AB} + \gamma_{CG} + \gamma_{DH} + \gamma_{EF}$, the interactive effect of AB may come from CG, DH, and EF as well. However, factor G, D, H, E, or F has no significant main effect and thus the opportunity for any one of them to interact with any other factor effectively is here assumed to be statistically insignificant. Herein it lays the answer that AB is the sole contributor to η_{AB}. By the same token, it is understood that AC is the only contributor to η_{AC} in accordance with Equation (3.4.11). This general practice is called heredity principle or hierarchical ordering principle: When interactions are confounded with one another, the interaction that is most likely to be significant is the one that contains factors whose main effects are themselves significant.

Effectiveness of the predictive equation, i.e., Equation (3.4.16), can be validated using the regression of the test data against the predicted values based on the 16 run conditions—a straight trendline with a correlation of 97.1%, while adequacy of the predictive equation can be validated using the regression of the residuals against the predicted values—yielding a horizontal trendline with uniformly spotted fitting points. These diagnostic checking plots are presented in Figure 3.4.5. Note that if the correlation plot or residual plot is not satisfactorily persuading, one has to go back to those alias equations, i.e., Equations (3.4.10)–(3.4.16), doing more reasoning about the effectiveness of individual two-factor interactions or more experimental tests to resolve confounding.

FIGURE 3.4.5 Diagnostic checking: (a) correlation plot and (b) residual plot.

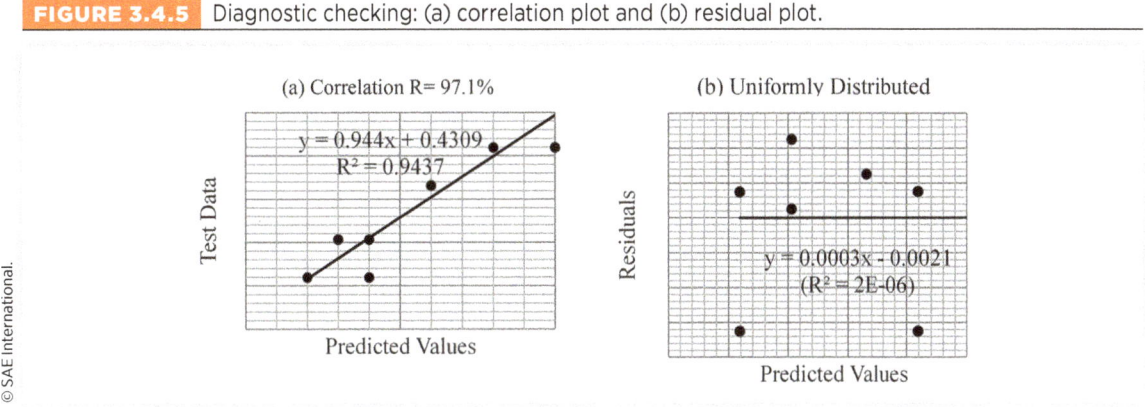

A normal probability plot, i.e., the q-q plot depicted in Figure 3.4.6, also supports the conclusions drawn above. Thus, the predictive equation as given is truly valid.

FIGURE 3.4.6 Normal probability plot of sealing gaskets for engine coolant.

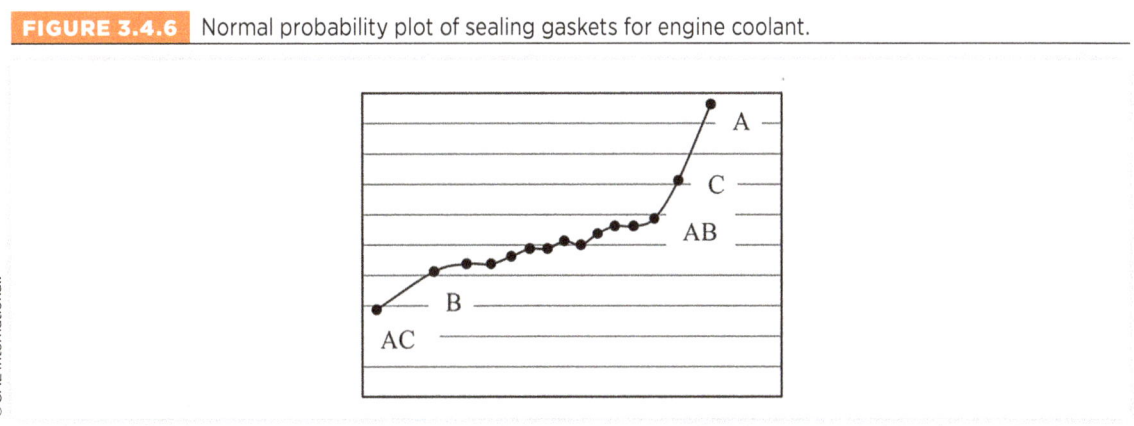

Two effective interactions are also observed in addition to the three individual main effects. One is the interaction between the sine-waved profile and a larger corrugating height that may hurt the uniformity of the contact pressure. The other is the interactive effect between the sine-waved profile and the "opposite" bolt pattern that may hurt works better to distribute the contact pressure uniformly, though both may help redistribute the contact pressure more uniformly as individual contributors.

Three main effects have significant influences on the sealing pressure distribution. The profile (or shape) of a corrugated metallic gasket has the greatest influence on the uniformity of sealing pressure. The corrugated gasket with sine-waved protrusion works better than the semi-open stepped gasket. The next influential factor is the bolt pattern—"opposite" work better than "diagonal." Both bolt patterns are demonstrated in Figure 3.4.2. The third effective factor is the corrugating height, i.e., h in Figure 3.4.1, as a larger corrugating height works better within the given work range between 0.3 mm and 0.4 mm.

3.4.4. Applying 2_{IV}^{8-4} Directly without Foldover

In fact, a similar design matrix that has 16 runs with resolution IV can be obtained for the eight design factors based on 2_{IV}^{8-4} using the following alias equations:

$$I = A B C G \qquad (3.4.17)$$

$$I = A\,B\,D\,H \tag{3.4.18}$$

$$I = A\,C\,D\,F \tag{3.4.19}$$

and

$$I = B\,C\,D\,E \tag{3.4.20}$$

If the experimental runs are not very time-consuming to carry out or it has been known that some two-factorial interactions are significant beforehand, one may jump to the above alias equations directly.

3.5. Fractional Factorial Design with Many Factors

When fractional factorial design with many parameters is checked against a specific goal of robustness, it may rely on a measure of its capacity to remain unaffected by deliberate variations in many potential parameters with minimum number of treatments, and provides an indication of its reliability during normal use [Xu 2009]. Though such a design matrix is mainly used as a screening design because of undesired confounding, it is very useful for economically detecting large main effects, assuming all interactions are negligible when compared with the few important main effects.

3.5.1. Placket-Burman Designs

Plackett–Burman designs are used to identify the viable factors early in the experimentation phase when most knowledge about the system is not available. Each of Plackett–Burman designs has two levels per continuous factor and is meant to screen linear terms. They allow practitioners to screen for the important factors that influence process output measures or product performance. General speaking, Plackett–Burman designs have the lowest number of runs in a single replicate for a given number of factors.

One of the Plackett–Burman designs is called L_{12}, which has been used in the automotive industry for identifying the main effects of 11 factors in 12 runs with resolution III, as shown in Table 3.5.1. This L_{12} is a two-level design matrix for 11 factors and it is so efficient that every main effect is confounded with two-factor interactions. The next one is the L_{20} whose design matrix is listed in Table 3.5.2. These factorial designs can be used as a screening design only. In DOE with many factors, a Plackett–Burman design is a starting point, and one should use it intending to follow up with more detailed experimentation. One may fold Plackett–Burman designs to estimate interactions that are confounded by the initial fraction of the design.

Specific Plackett and Burman design matrices, which have design levels higher than 2 for each factor were addressed in [Plackett and Burman 1946]. The general algorithm is to find experimental designs for investigating the dependence of a measured quantity on various independent potential factors, each taking 2, 3, 4, 5, or 7 levels, in such a way as to minimize the variance of the estimates of these dependencies using a very limited number of experiments. Plackett–Burman designs for two-levels that are mostly applied are L_{12}, L_{20}, L_{24}, L_{28}, and L_{32} [Ledolter and Swersey 2006]. Each of these Plackett–Burman designs is of resolution III, but it can be folded over to form a combined design with resolution IV for clarifying the contributions of individual factors and their interactions.

TABLE 3.5.1 L_{12} fractional factorial design matrix for 11 factors (Plackett–Burman design).

Run	a	b	c	d	e	f	g	h	i	j	k
1	1	1	1	1	1	1	1	1	1	1	1
2	1	1	1	1	1	-1	-1	-1	-1	-1	-1
3	1	1	-1	-1	-1	1	1	1	-1	-1	-1
4	1	-1	1	-1	-1	1	-1	-1	1	1	-1
5	1	-1	-1	1	-1	-1	1	-1	1	-1	1
6	1	-1	-1	-1	1	-1	-1	1	-1	1	1
7	-1	1	-1	-1	1	1	-1	-1	1	-1	1
8	-1	1	-1	1	-1	-1	-1	1	1	1	-1
9	-1	1	1	-1	-1	-1	1	-1	-1	1	1
10	-1	-1	-1	1	1	1	1	-1	-1	1	-1
11	-1	-1	1	-1	1	-1	1	1	1	-1	-1
12	-1	-1	1	1	-1	1	-1	1	-1	-1	1

TABLE 3.5.2 L_{20} fractional factorial design matrix for 19 factors (Plackett–Burman design).

Run	a	b	c	d	e	f	g	h	i	j	k	l	m	n	o	p	q	r	s
1	+	+	-	-	+	+	+	+	-	+	-	+	-	-	-	-	+	+	-
2	-	+	+	-	-	+	+	+	+	-	+	-	+	-	-	-	-	+	+
3	+	-	+	+	-	-	+	+	+	+	-	+	-	+	-	-	-	-	+
4	+	+	-	+	+	-	-	+	+	+	+	-	+	-	+	-	-	-	-
5	-	+	+	-	+	+	-	-	+	+	+	+	-	+	-	+	-	-	-
6	-	-	+	+	-	+	+	-	-	+	+	+	+	-	+	-	+	-	-
7	-	-	-	+	+	-	+	+	-	-	+	+	+	+	-	+	-	+	-
8	-	-	-	-	+	+	-	+	+	-	-	+	+	+	+	-	+	-	+
9	+	-	-	-	-	+	+	-	+	+	-	-	+	+	+	+	-	+	-
10	-	+	-	-	-	-	+	+	-	+	+	-	-	+	+	+	+	-	+
11	+	-	+	-	-	-	-	+	+	-	+	+	-	-	+	+	+	+	-
12	-	+	-	+	-	-	-	-	+	+	-	+	+	-	-	+	+	+	+
13	+	-	+	-	+	-	-	-	-	+	+	-	+	+	-	-	+	+	+
14	+	+	-	+	-	+	-	-	-	-	+	+	-	+	+	-	-	+	+
15	+	+	+	-	+	-	+	-	-	-	-	+	+	-	+	+	-	-	+
16	+	+	+	+	-	+	-	+	-	-	-	-	+	+	-	+	+	-	-
17	-	+	+	+	+	-	+	-	+	-	-	-	-	+	+	-	+	+	-
18	-	-	+	+	+	+	-	+	-	+	-	-	-	-	+	+	-	+	+
19	+	-	-	+	+	+	+	-	+	-	+	-	-	-	-	+	+	-	+
20	-	-	-	-	-	-	-	-	-	-	-	-	-	-	-	-	-	-	-

Note: "−" = −1 and "+" = +1

3.5.2. 2_{III}^{15-11} Example: Case Study on Angular Rigidity of Automotive Side Door Hinges

Working together, a hinge and a check function as the joint of a side door to a vehicle body. Hinges position the door relative to the body structure and pivot the door swing for passenger ingress and egress, and checks regulate the amount of door swing, while both are designed to withstand angular deflection. A hinge and a check can be individually attached to the vehicle body and work independently. When a check is integrated into a hinge such that they are attached to the same opening of the body pillar, the assembly becomes an integral hinge. Two hinges, namely upper and lower, are required to secure the door position. The distance between the upper and lower hinges is called the spread.

Potential failure modes of automotive hinge/check system can be classified into three categories: functional failure mode, structural failure mode, and noise-and-vibration failure mode. A functional failure means that the system fails to deliver the functional requirements of door operations, i.e., travels and efforts. A structural failure usually results in a safety problem. Among the potential structural problems, angular rigidity is of the greatest concern. Weak angular rigidity may result in impact between the door and vehicle body and/or cause the outer edge of the door to move downwards (mostly) or upwards permanently because of plastic deformation of the hinge, and thus the door is un-shut-able. An extensive study on the influences of 15 design parameters on the angular rigidity of automotive side door has been conducted by [Chiang et al. 2003] using both DOE and FEM.

Three parts are the minimum requirement for forming a door hinge assembly. They are door straps, body straps, and hinge pins, as depicted in Figure 3.5.1. Bolts are used to fasten the door strap to the door and body strap to the pillar; hinge pins, pivoting the door strap to the body strap and thus establish the swing axis, which is called hinge axis. M8 bolts are popular for regular passenger vehicles. Hinges and latches work together to secure the door closure [FVSS 1998]. The measurable obtained from the virtual angular rigidity test is to determine the angular deflection when a horizontal force perpendicular to the door plane is applied at the latch location. The body straps of both upper and lower hinges are fixed to the vehicle body and pivoted axis (between the door straps and body straps) are allowed to rotate. Relevant design parameters of a conventional hinge are shown in Figure 3.5.1. The same hinge and full-open angle are used for the upper and lower in order to identify the hinge performance unambiguously.

FIGURE 3.5.1   Conventional design configuration of an automotive side door hinge [Chiang et al. 2003].

(a) Door Hinge

(a) Body Strap

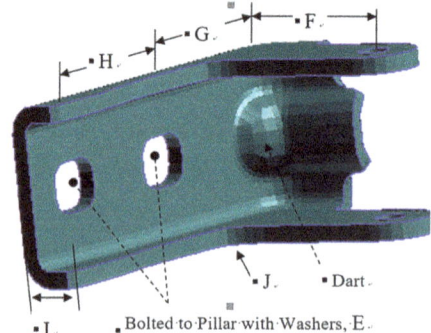

(b) Door Strap

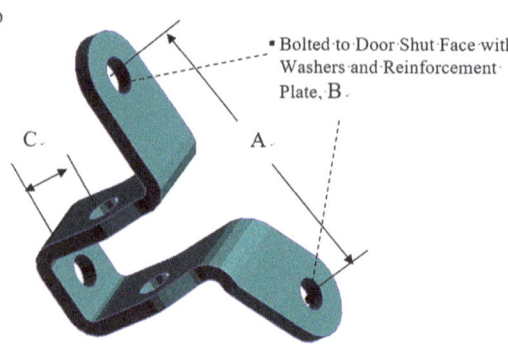

Influences of 15 design parameters and loading conditions on angular deflections of side door hinges subjected to over open forces were explored. Finite element analyses were employed to do angular rigidity tests and DOE was used to characterize the contributions from individual factors. Eight and six three-dimensional solid elements are employed to model the body and door straps. Beam elements are used to model bolts and locate the pivot axes. Both the vehicle body and door are modeled as rigid bodies in order not to let them complicate the analysis. Proper contact conditions are prescribed between the door strap and body strap. Both geometric and material nonlinearities are considered. The case study on this static nonlinear analysis is performed using the commercial code Abaqus. The load-deflection curve based on the finite element analysis is first calibrated by physical tests, as detailed in [Chiang et al. 2003]. Accuracy of finite element analyses was validated by load-deflection curves from physical testing. After then, DOE based on 2_{III}^{15-11} design matrix is set up for 15 factors, of which their designated design levels are given in Tables 3.5.3 and 3.5.4. The alias equations for defining the design matrix are given as follows:

$$E = A\,B$$

$$F = A\,C$$

$$G = A\,D$$

$$H = B\,C$$

$$I = B\,D$$

$$J = C\,D$$

$$K = A\,B\,C$$

$$L = A\,B\,D$$

$$M = A\,C\,D$$

$$N = B\,C\,D$$

and

$$P = A\,B\,C\,D$$

TABLE 3.5.3 Factors and design levels for experimental tests on side door hinge system.

Variables	Level (−)	Level (+)
A: Door strap—bolt spread (mm)	91.28	99.28
B: Door strap—washer and reinforcement	No	Yes
C: Door strap—contact offset (mm)	20	25
D: Body strap—dart	No	Yes
E: Body strap—washer	No	Yes
F: Body strap—free end height (mm)	25.38	30.38
G: Body strap—rear bolt offset (mm)	24.2	29.7
H: Body strap—bolt spacing (mm)	27	32.5
I: Body strap—flange height (mm)	11.1	14.3
J: Body strap—flange radius (mm)	5	7.68
K: Spread-upper and lower hinges (mm)	300	360
L: Material	Forged	HSLA 950A
M: Loading point—out of plane (mm)	−20	20
N: Loading point—span (mm)	900	1,300
P: Loading point—vertical offset (mm)	−100	100

Reprinted from "Robustness in Angular Rigidity of Automotive Side-Door Hinges," SAE 2003-01-1218. © SAE International.

TABLE 3.5.4 Design matrix for angular rigidity test of an automotive side door [Chiang et al. 2003].

Run	a	b	c	d	e	f	g	h	i	j	k	l	m	n	p	Y (%)	Y_p (%)
1	−1	−1	−1	−1	1	1	1	1	1	1	−1	−1	−1	−1	1	33.42	33.04
2	1	−1	−1	−1	−1	−1	−1	1	1	1	1	1	1	−1	−1	35.01	34.63
3	−1	1	−1	−1	−1	1	1	−1	−1	1	1	1	−1	1	−1	117.53	116.47
4	1	1	−1	−1	1	−1	−1	−1	−1	1	−1	−1	1	1	1	54.36	53.30
5	−1	−1	1	−1	1	−1	1	−1	1	−1	1	−1	1	1	−1	71.25	70.19
6	1	−1	1	−1	−1	1	−1	−1	1	−1	−1	1	−1	1	1	66.11	65.05
7	−1	1	1	−1	−1	−1	1	1	−1	−1	−1	1	1	−1	1	26.1	25.72
8	1	1	1	−1	1	1	−1	1	−1	−1	1	−1	−1	−1	−1	18.57	18.19
9	−1	−1	−1	1	1	1	−1	−1	−1	−1	1	1	1	1	−1	129.28	130.34
10	1	−1	−1	1	−1	−1	1	1	−1	−1	1	−1	−1	1	1	122.67	123.73
11	−1	1	−1	1	−1	1	−1	−1	1	−1	1	−1	1	−1	1	19.63	20.01
12	1	1	−1	1	1	−1	1	−1	1	−1	−1	1	−1	−1	−1	22.18	22.56
13	−1	−1	1	1	1	−1	−1	−1	−1	1	1	1	−1	−1	1	34.17	34.55
14	1	−1	1	1	−1	1	1	−1	−1	1	−1	−1	1	−1	−1	34.06	34.44
15	−1	1	1	1	−1	−1	−1	1	1	1	−1	−1	−1	1	−1	25.71	26.77
16	1	1	1	1	1	1	1	1	1	1	1	1	1	1	1	23.14	24.20
17	−1	1	1	1	1	−1	1	1	0	0	0	0	0	0	0	25.02	23.84

Notes:
Y: Angular rigidity test data obtained from finite element analysis
Y_p: Angular rigidity test data obtained from DOE predictor

Reprinted from "Robustness in Angular Rigidity of Automotive Side-Door Hinges," SAE 2003-01-1218. © SAE International.

The effects are calculated using the DOE-t procedure as shown in the previous sections. The relative size of effects is depicted in Figure 3.5.2. It shows that the impacts of loading point location (factor N), body-strap flange height (factor I), door strap-washer and reinforcement plate (factor B) are overwhelming. In addition to these four variables, influences of body-strap-flange radius (factor J), door strap bolt spread (factor A), vertical offset of loading point (factor P), material (factor L), rear offset of body strap (factor G), washer of body strap (factor E), spread between upper and lower hinges (factor K), free end height of contact wall of body strap (factor F), and out-of-plane distance of loading point (factor M) are moderate. It appears that the bolt spacing within the range of consideration has no impact on the angular rigidity.

FIGURE 3.5.2 Effects on the angular deflection when subjected to factorial variations.

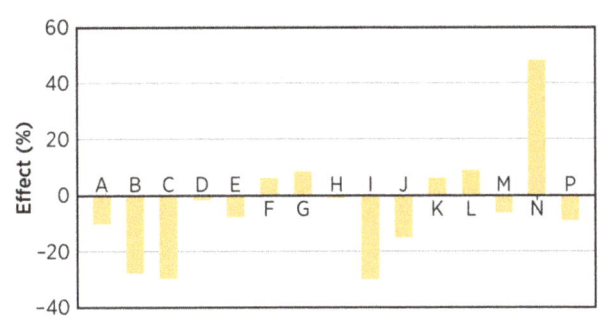

TABLE 3.5.5 Significance levels of design factors in the side door hinge system.

Variables	Effect	Standard error	t-Ratio	Significance level
Mean	52.0744	0.5618	92.71	<1%
A	−10.124	1.123375	9.012	<1%
B	−27.344	1.123375	24.34	<1%
C	−29.371	1.123375	26.15	<1%
D	−1.439*	1.123375	—	—
E	−7.556	1.123375	6.726	<1%
F	6.286	1.123375	5.596	<1%
G	8.5439	1.123375	7.512	<1%
H	−0.674*	1.123375	—	—
I	−30.036	1.123375	26.74	<1%
J	−14.799	1.123375	13.17	<1%
K	6.344	1.123375	5.647	<1%
L	9.231	1.123375	8.217	<1%
M	−5.941	1.123375	5.289	<1%
N	48.364	1.123375	43.05	<1%
P	−9.249	1.123375	8.235	<1%

* The standard error (sample) for the t-ratio is estimated by pooling these small effects.

The standard error (sample) for the t-ratio is estimated by pooling the two small effects, e.g., factors D and H, following Equation (2.5.3). Diagnostic checking of the DOE model is based on the t-ratio, which is calculated as (effect − hypothesized mean)/sample standard error. The hypothesized mean is zero as assumed from the null hypothesis being granted for this DOE. Significance levels for individual effects are shown in Table 3.5.5, of which the data obtained by checking against their t_2-distribution statistics. The higher the t-ratio, the higher the significance level of the design factor. The correlation between the test data (from FEA virtual tests) and the predictive values by DOE is 99.975%, which means that the DOE modeling is overwhelmingly truthful. Another diagnostic checking is based on the normal probability plot as shown in Figure 3.5.3.

FIGURE 3.5.3 Normal probability plot for angular rigidity of side door hinge system [Chiang et al. 2003].

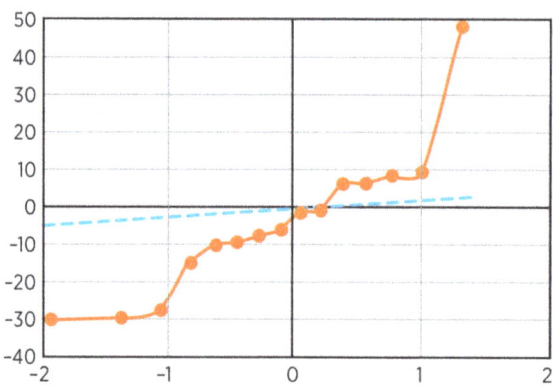

3.5.3. 2_{III}^{16-11} Example: Case Study on Sealing Tire Air

In order to improve tire durability and wheel fatigue design, finite element methods and analytic solutions have been applied to exploring the contact pressure between tires and their fitting rims. Three-dimensional finite element models are herein employed to characterize the behaviors of both inflated tires and deflected tires and DOE is utilized to examine the significance of main effects and their interactions [Chiang et al. 2000]. A Detailed modeling of orthotropic properties of elastomeric composites can be accessed in [Chiang and Tang 1995]. The tire used as the baseline for this study is P195/60R14, i.e., a passenger tire. Tire components included in the study are sidewall, apex, chafer, apex, liner, belts, carcass, and beads, of which the cross-sectional view of the finite element mesh is shown in Figure 3.5.4. Design parameters of a passenger car rim under investigation are bead seat angle, bead seta radius, rim radius, rim width, flange height, and flange radius, as presented in Figure 3.5.4. All the design parameters are summarized in Table 3.5.6.

FIGURE 3.5.4 Cross-sectional view of FEA mesh of tire and design parameters of rim [Chiang et al. 2000].

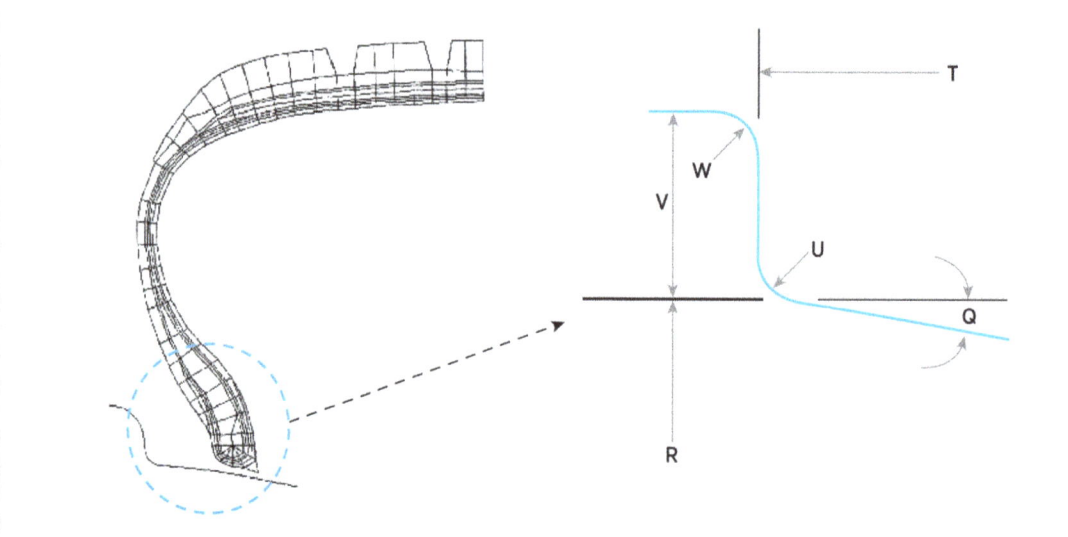

The finite element analysis is implemented using commercial codes Abaqus. Effectiveness of the finite element model is validated first as depicted in Figure 3.5.5, and then the model is applied to all the treatments to calculate the pressure distribution in the interfacial area between the tire and rim. A typical distribution of contact pressure at the taper along the peripheral direction is plotted in Figure 3.5.6, of which the peak pressure is defined as the sealing pressure. When the tire is deflected against the ground, the peak pressure around the toe increases and the peak pressure in the neighborhood of the flange gets reduced.

TABLE 3.5.6 Potential influential factors and their design levels investigated for studying sealing tire air [Chiang et al. 2000].

Component	Factor	Level (−)	Level (+)
Tire	A: Sidewall rubber	Soft	Stiff
	B: Apex rubber	Soft	Stiff
	C: Chafer rubber	Soft	Stiff
	D: Liner rubber	Soft	Stiff
	E: Belt material*	Soft	Stiff
	F: Carcass material*	Soft	Stiff
	G: Belt cord angle	14°	30°
	H: Carcass cord angle	85°	90°
	P: Bead size	Small	Large
Rim	Q: Bead seat angle	4°	6°
	R: Nominal radius (mm)	177.209	177.591
	T: Semi-rim radius (mm)	75.75	76.75
	U: Bead seat radius (mm)	5.5	6.5
	V: Flange height (mm)	16.7	19.5
	W: Flange radius (mm)	9.5	10.5
Ground-tire contact	L: Tire deflection (mm)	0	28

* The thickness and rubber material of belts are kept constant, so are those of carcass plies.

FIGURE 3.5.5 Validation of finite element accuracy as applied to a deflected tire [Chiang et al. 2000].

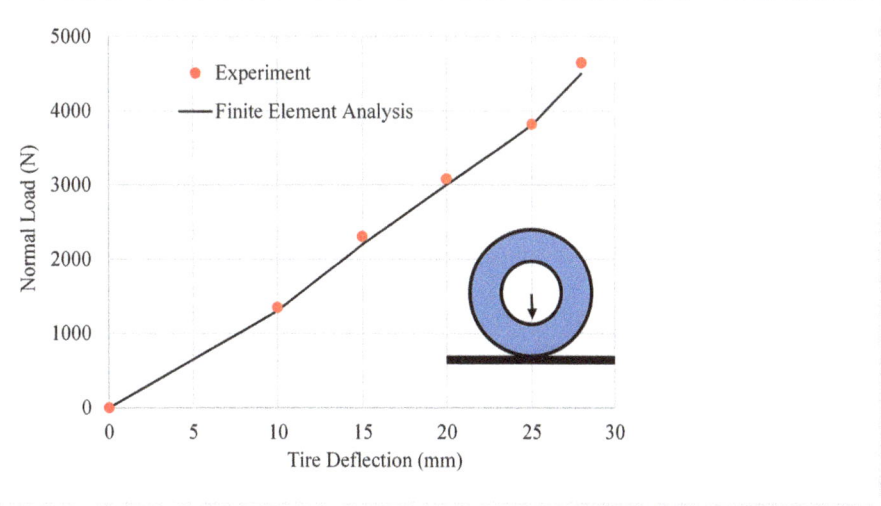

FIGURE 3.5.6   Seating pressure of P195/60R14 tire onto the rim along the peripheral direction [Chiang et al. 2000].

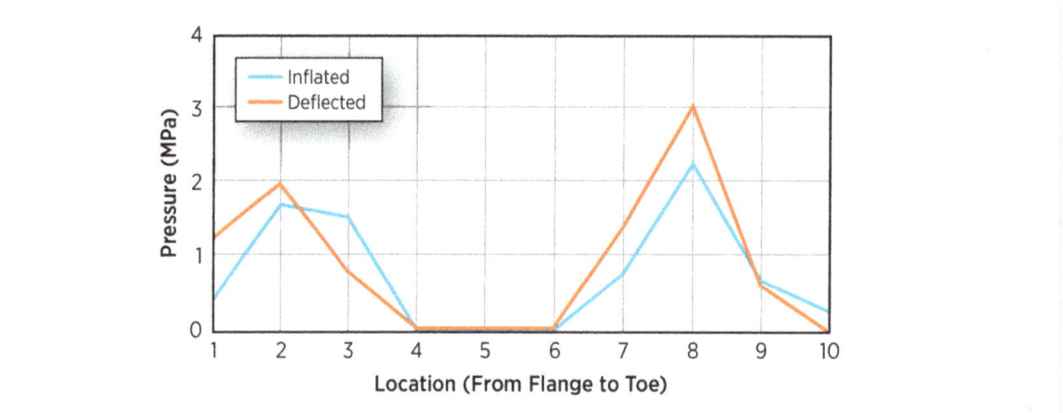

Used with permission of Inderscience Enterprises Limited (UK), from "Multi-Variable Effects on Sealing Pressure between Tires and Rims," Y J Chiang, 23, 2006; permission conveyed through Copyright Clearance Center, Inc.

Fractional factorial design $2_{\mathrm{III}}^{15\text{-}11}$ is employed to exploit the design matrix with the following alias equations:

$$
\begin{aligned}
E &= A\,B \\
F &= A\,C \\
G &= A\,D \\
H &= B\,C \\
P &= B\,D \\
Q &= C\,D \\
R &= A\,B\,C \\
T &= A\,B\,D \\
U &= A\,C\,D \\
V &= B\,C\,D \\
\text{and} \quad W &= A\,B\,C\,D
\end{aligned}
$$

The screening design matrix based on the above design generators are listed as the first 16 treatments in Table 3.5.7. These 16 run conditions are then folded over on a new factor (L), i.e., tire deflection against the ground as a typical passenger rests on the ground to form another fractional factorial design, i.e., $2_{\mathrm{IV}}^{16\text{-}11}$, of which the treatments are given as the first 32 treatments in Table 3.5.7.

TABLE 3.5.7 Test conditions of factorial design 2^{16-11} for investigating passenger tire sealing pressure [Chiang et al. 2000].

Used with permission of Inderscience Enterprises Limited (UK), from "Multi-Variable Effects on Sealing Pressure between Tires and Rims," Y J Chiang, 23, 2006; permission conveyed through Copyright Clearance Center, Inc.

Run	A	B	C	D	E	F	G	H	P	Q	R	T	U	V	W	L
1	−	−	−	−	+	+	+	+	+	+	−	−	−	−	+	−
2	+	−	−	−	−	−	−	+	+	+	+	+	+	−	−	−
3	−	+	−	−	−	+	+	−	−	+	+	+	−	+	−	−
4	+	+	−	−	+	−	−	−	−	+	−	−	+	+	+	−
5	−	−	+	−	+	−	+	−	+	−	+	−	+	+	−	−
6	+	−	+	−	−	+	−	+	−	−	+	−	+	+	−	−
7	−	+	+	−	−	−	+	+	−	−	+	+	+	−	+	−
8	+	+	+	−	+	+	−	+	−	−	+	−	−	−	−	−
9	−	−	−	+	+	+	+	−	+	−	−	+	+	+	−	−
10	+	−	−	+	−	−	+	+	−	−	+	−	−	+	+	−
11	−	+	−	+	−	+	−	−	+	−	+	−	+	−	+	−
12	+	+	−	+	+	−	−	+	−	−	+	−	−	−	−	−
13	−	−	+	+	+	−	−	−	−	+	+	+	−	−	+	−
14	+	−	+	+	+	+	−	−	−	+	−	−	+	−	−	−
15	−	+	+	+	−	−	−	+	+	+	+	−	−	+	−	−
16	+	+	+	+	+	+	+	+	+	+	+	+	+	+	+	−
17	+	+	+	+	−	−	−	−	−	+	+	+	+	−	−	+
18	−	+	+	+	+	+	+	−	−	−	−	−	+	+	+	+
19	+	−	+	+	+	−	+	+	−	−	−	−	+	−	+	+
20	−	−	+	+	−	+	+	+	+	−	+	+	−	−	−	+
21	+	+	−	+	−	+	−	+	−	+	−	+	−	−	+	+
22	−	+	−	+	+	+	−	+	+	−	+	+	−	+	−	+
23	+	−	−	+	+	+	−	−	+	+	+	+	−	−	+	+
24	−	−	−	+	−	−	+	−	+	+	−	+	+	+	+	+
25	+	+	+	−	−	+	−	+	+	+	+	−	−	−	+	+
26	−	+	+	−	+	+	−	−	+	+	−	+	+	−	−	+
27	+	−	+	−	+	−	+	+	−	+	−	+	−	+	+	+
28	−	−	+	−	−	+	−	+	−	+	+	−	+	+	+	+
29	+	+	−	−	−	+	+	+	+	−	−	−	+	+	−	+
30	−	+	−	−	+	−	−	+	+	−	+	+	−	+	+	+
31	+	−	−	−	+	+	+	−	−	−	+	+	−	+	+	+
32	−	−	−	−	−	−	−	−	−	−	−	−	−	−	−	+
33	+	+	+	+	−	−	−	−	−	+	+	+	+	−	−	−
34	−	+	+	+	+	+	+	−	−	−	−	−	+	+	+	−
35	+	−	+	+	+	−	−	+	+	−	−	−	+	−	+	−
36	−	−	+	+	−	+	+	+	+	−	+	+	−	−	−	−
37	+	+	−	+	−	+	−	+	−	+	−	+	−	−	+	−
38	−	+	−	+	+	+	−	+	+	−	+	+	−	+	−	−
39	+	−	−	+	+	+	−	−	+	+	+	+	−	−	+	−
40	−	−	−	+	−	−	+	−	+	+	−	+	+	+	+	−
41	+	+	+	−	−	+	−	+	+	+	+	−	−	−	+	−
42	−	+	+	−	+	+	−	−	+	+	−	+	+	−	−	−
43	+	−	+	−	+	−	+	+	−	+	−	+	−	+	+	−
44	−	−	+	−	−	+	−	+	−	+	+	−	+	+	+	−
45	+	+	−	−	+	+	+	+	+	−	−	−	+	+	−	−
46	−	+	−	−	+	−	−	+	+	−	+	+	−	+	+	−
47	+	−	−	−	+	+	+	−	−	−	+	+	+	−	+	−
48	−	−	−	−	−	−	−	−	−	−	−	−	−	−	−	−

Results from analyzing these 32 treatments show that some confounded two-factor interactions are so confounded that they cannot be separated. In order to resolve these statistical ambiguities, an additional 16 treatments were set up for the purpose of detaching factor L (deflection) from two-factor interactions. Main effects and interactions resulting from all the 48 treatments are plotted in Figure 3.5.7.

FIGURE 3.5.7   Pareto plot of design contrasts due to main effects and interactions [Chiang et al. 2000].

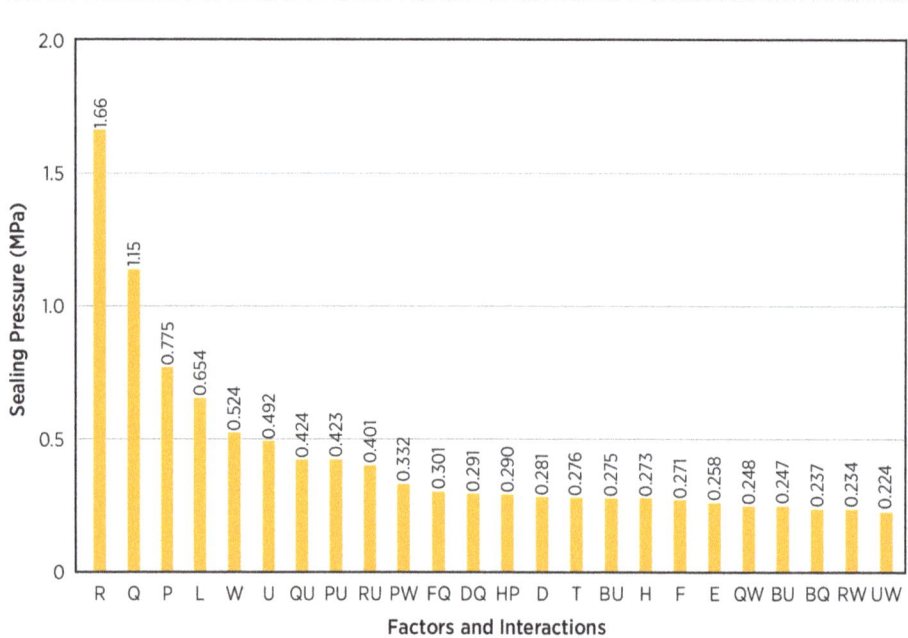

Confounding among two-factor interactions without involving with deflection were resolved by engineering senses, rules, and relative magnitudes of their corresponding individual main effects. Finally, 12 main effects and 12 two-factor interactions are shown to be statistically significant by checking t-ratios that are measured with a confidence interval of 95%, i.e., having a significance level (p-value) $\alpha = 5\%$. After all the insignificant main effects and interactions were discarded, the predictive equation for sealing pressure at the tire taper can be expressed as:

$$Y_p = 2.4345 + 0.124\,b - 0.14\,d + 0.129\,e - 0.135\,f - 0.137\,h + 0.327\,l$$

$$+ 0.388\,p - 0.575\,q + 0.829\,r - 0.138\,t + 0.246\,u + 0.262\,w$$

$$- 0.118\,b\,q - 0.137\,b\,u + 0.145\,d\,q + 0.15\,f\,q - 0.145\,h\,p + 0.212\,p\,u$$

$$+ 0.166\,p\,w - 0.212\,q\,u - 0.124\,q\,w + 0.201\,r\,u + 0.117\,r\,w - 0.112\,u\,w$$

The effectiveness and adequacy of the above predictive equation can be proven using the correlation plot and normal probability plot. During the calculation process, the first 32 treatments are regarded as the first block and the next 12 treatments as the second block. The averaged sealing pressure (i.e., 2.4345 MPa) is obtained from the first block only. Among all the factors investigated, the ranking of main effects is given as follows:

(a) The nominal rim (factor R) is the most significant factor. In other words, the dimensional tolerance for the interference fit between the tire and its mating rim has to be well controlled in order to seal the tire air.

(b) The second most significant factor is the bead seat angle (factor Q). A taper with a smaller slope tends to retain the tire air better.

(c) Bead size (factor P) is the third most significant factor. The stiffer the bed, the higher its capability to seal the tire air.

(d) Other eight main effects are also significant, but not as much as the top three mentioned above. In order to increase the sealing pressure, the following design guidelines are suggested: stiffer apex, softer liner, stiffer belt rubber, softer carcass, radial cord angle for carcass, narrow rim width, larger beat seat radius, and larger flange radius.

(e) There are 12 interactions whose impacts on sealing pressure are significant. Among them the top three are interactions between bead seat radius and bead angle, bead seat radius and bead size, and the bead seat radius and nominal rim radius. The all have positive influences on sealing pressure. The last nine two-factor interactions have minor contributions although influential.

In summary, the sealing pressure is dominantly controlled by rim geometry and bead stiffness (Tables 3.5.8 and 3.5.9).

TABLE 3.5.8 Contact pressure distribution from flange to toe [Chiang et al. 2000].

Test no.	\multicolumn{10}{c}{Location (from flange to toe)}									
	1	2	3	4	5	6	7	8	9	10
1	0.000	1.527	2.204	0.302	0	0	0	1.043	0.339	0.285
2	0.419	1.505	1.147	0	0	0	1.628	2.999	0.556	0.091
3	0.596	1.505	1.147	0	0	0	0	1.559	0.628	0.23
4	0.512	1.845	1.776	0	0	0	0	0.132	0	0.318
5	0.671	1.677	1.089	0	0	0	0.613	2.998	1.081	0.174
6	0.563	1.749	1.488	0	0	0	0	2.112	0.891	0.237
7	0	1.408	1.943	0	0	0	1.354	1.924	0.339	0.334
8	0.385	1.77	1.554	0	0	0	0.062	2.951	1.088	0.072
9	0.694	1.708	1.321	0	0	0	0	1.885	0.858	0.234
10	0.556	1.721	1.414	0	0	0	0.021	2.689	1.212	0.153
11	0.455	1.313	0.602	0	0	0	4.458	5.413	0.718	0.022
12	0	1.669	2.048	0	0	0	0	2.066	0.931	0.235
13	0.174	1.611	1.949	0	0	0	0.094	2.058	0.579	0.13
14	0	1.75	2.365	0	0	0.022	0.018	0.404	0.034	0.376
15	0.678	1.788	1.347	0	0	0	0	0.42	0.063	0.331
16	0.615	1.645	0.915	0	0	0	1.145	3.184	0.858	0.133
17	1683	2.092	0.000	0	0	0	1.341	3.615	0.859	0
18	1.355	2.029	0.976	0	0	0	0.000	2.256	0.769	0
19	1.073	1.831	0.984	0	0	0	2.073	3.475	0.488	0
20	1.553	2.146	0.536	0	0	0	0	1.450	0.345	0
21	0.986	2.018	1.642	0	0	0	0	1.356	0.243	0
22	1.203	1.85	0.845	0	0	0	1.389	2.601	0.306	0
23	1.706	2.208	0.146	0	0	0	1.053	2.911	0.66	0
24	1.174	1.954	1.264	0	0	0	0.276	1.847	0.537	0.063
25	1.085	1.881	0.997	0	0	0	2.043	3.714	0.507	0
26	0.974	1.994	1.563	0	0	0	0.175	1.784	0.357	0
27	1.453	2.105	0.764	0	0	0	0.000	1.202	0.281	0.097
28	1.569	1.983	0.112	0	0	0	1.071	2.626	0.321	0
29	1.531	2.043	0.583	0	0	0	0.611	2.828	0.928	0
30	1.487	1.819	0.042	0	0	0	3.126	4.938	0.917	0
31	1.117	1.780	0.68	0	0	0	3.076	4.059	0.462	0
32	1.117	1.959	1.362	0	0	0	0	2.126	0.704	0
33	0.835	1.849	0.864	0	0	0	0.431	2.932	1.102	0.157
34	0.361	1.688	1.835	0	0	0	0	1.281	0.353	0.353
35	0.175	1.572	1.616	0	0	0	1.431	2.817	0.57	0.179
36	0.514	1.844	1.773	0	0	0	0	0.149	0.000	0.32
37	0	1.683	2.208	0.263	0	0	0	0.427	0.028	0.300
38	0.176	1.573	1.728	0	0	0	0.655	1.944	0.449	0.172
39	0.95	1.843	0.892	0	0	0	0	2.160	0.849	0.123
40	0.290	1.655	1.914	0	0	0	0	0.824	0.306	0.331
41	0.135	1.567	1.613	0	0	0	1.034	3.174	0.788	0.077
42	0.085	1.678	2.136	0	0	0	0	0.759	0.097	0.287
43	0.381	1.843	1.786	0	0	0	0	0.112	0.017	0.419
44	0.725	1.701	1.315	0	0	0	0.093	1.847	0.531	0.168
45	0.581	1.797	1.397	0	0	0	0	1.930	0.917	0.251
46	0.792	1.571	0.715	0	0	0	2.074	4.395	1.233	0.047
47	0.078	1.434	1.35	0	0	0	2.354	3.519	0.685	0.153
48	0.133	1.709	2.041	0	0	0	0	1.266	0.463	0.259

Used with permission of Inderscience Enterprises Limited (UK), from "Multi-Variable Effects on Sealing Pressure between Tires and Rims," Y J Chiang, 23, 2006; permission conveyed through Copyright Clearance Center, Inc.

TABLE 3.5.9 Significance levels of effects on peak sealing pressure [Chiang et al. 2000].

Variable	Effect	Standard Error[†]	t-Ratio*	Confidence Level
Mean	2.4345	0.0485	50.20	
Main effects				
A	0.093			
$\overline{B}$	0.247	0.097	2.55	99.1%
$\overline{C}$	−0.094			
$\overline{D}$	−0.281	0.097	2.90	99.6%
$\overline{E}$	0.258	0.097	2.66	99.3%
$\overline{F}$	−0.271	0.097	2.79	99.5%
$\overline{G}$	−0.137			
$\overline{H}$	−0.273	0.0915	2.98	99.7%
$\overline{P}$	0.775	0.097	7.99	>99.9%
$\overline{Q}$	−1.150	0.097	11.86	>99.9%
$\overline{R}$	1.657	0.097	17.01	>99.9 %
$\overline{T}$	−0.276	0.0915	3.02	99.7%
$\overline{U}$	0.492	0.0915	5.38	>99.9%
$\overline{V}$	−0.059			
$\overline{W}$	0.524	0.097	5.40	>99.9%
$\overline{L}$	0.654	0.097	6.74	>99.9%
Two-factor interactions				
$AB+CR+DT+FH+GP+\underline{QW}+UV$	−0.248	0.097	2.56	99.1%
$AC+BR+DU+EH+GQ+\underline{\overline{PW}}+TV$	0.332	0.097	3.42	99.9%
$AD+BT+CU+EP+\underline{FQ}+\overline{HW}+RV$	0.301	0.0915	3.28	99.8%
$AE+CH+DP+FR+\overline{GT}+QV+UW$	−0.224	0.097	2.31	98.5%
$AF+BH+\underline{DQ}+ER+GU+PV+\overline{TW}$	0.291	0.0915	3.18	99.8%
$AG+BP+\overline{CQ}+ET+FU+HV+\underline{RW}$	0.234	0.097	2.41	98.8%
$AH+BF+CE+DW+GV+\underline{PU}+\overline{QT}$	0.423	0.097	4.36	>99.9%
$AP+BG+CW+DE+FV+\overline{HU}+QR$	0.141			
$AQ+BWP+CG+DF+EV+HT+PR$	0.107			
$AR+BC+DV+EF+GW+PQ+TU$	−0.008			
$AT+BD+CV+EG+FW+HQ+\underline{RU}$	0.401	0.097	4.14	>99.9%
$AU+BV+CD+EW+FG+HP+\overline{RT}$	−0.290	0.097	2.99	99.7%
$AV+BU+CT+DR+EQ+\overline{FP}+GH$	−0.275	0.097	2.84	99.5%
$AW+\underline{\overline{BQ}}+CP+DH+EU+FT+GR$	−0.237	0.0915	2.59	99.2%
AL	−0.067			
$BE+CF+DG+HR+PT+\underline{QU}+VW$	−0.424	0.0915	4.63	>99.9%
BL	−0.047			
CL	0.053			
DL	0.044			
EL	−0.048			
FL	0.0715			
GL	0.050			
HL	0.029			
PL	0.014			
QL	0.021			
RL	−0.104			
TL	0.0635			
UL	−0.0454			
VL	0.0145			
OL	−0.080			

* The standard error for the t-ratio is estimated by pooling small effects and interactions, which are not underlined in this table. The effects of significant main effects and 2-factor interactions (underlined) were revised after the insignificant ones were discarded.

† Having a significance level of 5% on one side of the t-distribution density function.

References

Bisgaard, S. (1994), "A Note on the Definition of Resolution for Blocked 2^{K-P} Designs," *Technometrics*, 36, pp. 308-311.

Box, G. E. P. and Hunter, J. S. (1961), "The 2^{K-P} Fractional Factorial Designs," *Technometrics*, 3, pp. 449-458.

Box, G. E. P., Hunter, J. S., and Hunter, W. G. (2005), *Statistics for Experimenters: Design, Innovation, and Discovery*, 2nd Edition, 672 pages, John Wiley & Sons; ISBN: 978-0-471-71813-0.

Chiang, Y. J. and Tang, C. (1995), "Accuracy Assessment to Applying 20-Node Solid Elements to Pressurized Composite Shells," *Finite Elements in Analysis and Design*, 20, pp. 219-231.

Chiang, Y. J., Shih, C. D., Lin, C. C., Lee, H. K., Tseng, Y. Y., Tsai, K. C., and Cheng, Y. H. (2000), "Multi-Variable Effects on Sealing Pressure between Tires and Rims," *International Journal of Vehicle Design*, 23(1/2), pp. 78-93.

Chiang, Y. J., Akhter, A., Wang, Y., and Winfree, R. (2003), "Robustness in Angular Rigidity of Automotive Side-Door Hinges," *SAE Technical Paper 2003-01-1218*, https://doi.org/10.4271/2003-01-1218.

Fang, J. et al. (2018), "Reliability Improvement of Diamond Drill Bits Using Design of Experiments," *Quality Engineering*, 30(2), pp. 339-350.

Federal Vehicle Safety Standard (1998), "Door Locks and Door Retention Components—Passenger Cars, Multiple Passenger Vehicles, and Trucks," FMVSS Part 571 S 206, DOT, Washington, DC.

Fries, A. and Hunter, W. G. (1980), "Minimum Aberration 2^{K-P} Designs," *Technometrics*, 22, pp. 601-608.

Jou, Y. et al. (2014), "Integrating the Taguchi Method and Response Surface Methodology for Process Parameter Optimization of the Injection Molding," *Applied Mathematics and Information Sciences*, 8(3), pp. 1277-1285.

Ledolter, J. and Swersey, A. J. (2006), "Using a Fractional-Factorial Design to Increase Direct Mail Response at Mother Jones Magazine" *Quality Engineering*, 18, pp. 469-475.

Li, X., Sudarsanam, N., and Frey, D. D. (2006), "Regularities in Data from Factorial Experiments," *Complexity*, 11(5), pp. 32-45.

McCullagh, P. and Nelder, J. (1989), *Generalized Linear Models*, 2nd Edition, Chapman & Hall, Boca Raton, FL, 532 pages; ISBN: 9780412317606.

Montgomery, D. C. and Runger, G. C. (1996). "Foldovers of 2^{K-P} Resolution IV Designs," *Journal of Quality Technology*, 28, pp. 446-450.

Plackett, R. L. and Burman, J. P. (1946), "The Design of Optimum Multifactorial Experiments," *Biometrika*, 33(4), pp. 305-325.

Sousa, A. M. F. et al. (2013), "Design of Experimental Design as a Tool for the Processing and Characterization of HDPE Composites with Sponge-Gourds (Luffa-Cylindrica) Agrofiber Residue," *Journal of Sustainable Development* 6(4) pp. 106-117.

Taguchi, G., Chowdhur, S., and Wu, Y. (2005), *Taguchi's Quality Engineering Handbook*, John Wiley & Sons, Inc., Hoboken, NJ.

Wu, C. F. J. and Hamada, M. S. (2009), *Experiments: Planning, Analysis and Parameter Design Optimization*, 2nd Edition, Wiley, New York; ISBN: 978-0471699460.

Xu, H. (2009), "Algorithmic Construction of Efficient Fractional Factorial Designs with Large Run Sizes," *Technometrics*, 51(3), pp. 262-277.

Problems

P3.1: This is an exercise to illustrate how to identify the influential factors on enlengthening the MTBF (mean time between failures) for a nonrepairable component, or MTTR (mean time to repair) for a repairable component, of a product with regard to the overall availability? The following life data (MTBF in hours) are obtained from a DOE concerning the product's MTBF:

Run #	Temperature (°C)	Humidity (%)	Speed (km/hr)	Weight (kg)	MTBF (hr)
1	4	30	50	1000	67
2	4	30	80	2000	60
3	4	60	50	2000	50
4	4	60	80	1000	56
5	27	30	50	2000	77
6	27	30	80	1000	70
7	27	60	50	1000	67
8	27	60	80	2000	75

P3.2: An experiment was performed (NIST Ceramics Division: Material Science and Engineering Laboratory [NIST]) to study the effect of grinding process on the strength (MPa) of a high-performance silicon nitride ceramic material. The grinding variables considered in the study and their design levels are given as follows:

Variable	Level (−1)	Level (+1)	Description
E: Batch	1st	2nd	Two material batches
D: Direction	Longitudinal	Transverse	Grinding direction
C: Grit	140–170	80–100	Grinding material coarseness
B: Feed (mm/min)	0.05	0.125	Material sample feed
A: Speed (m/s)	0.025	0.125	Grinding table speed

1. Calculate design contrasts based on 2^{5-1}
2. Calculate main and interactive effects
3. Pareto chart for main and interactive effects
4. Formulate the predictive equation
5. Checking statistical significance of main and interactive effects based on t-test
6. Do diagnostic checking—predicted values versus test results
7. Do diagnostic checking—residual versus predicted values
8. Do diagnostic checking—normal probability plots

The experimental trials were randomized over all 32 runs of the full factorial design (2^5). The test sequence has been randomized. The experimental data obtained are shown as follows:

Run #	Test#	E	D	C	B	A	Y (Strength: MPa)
1	17	−	−	−	−	−	680
2	30	−	−	−	−	+	722
3	14	−	−	−	+	−	702
4	8	−	−	−	+	+	667
5	32	−	−	+	−	−	704
6	20	−	−	+	−	+	642
7	26	−	−	+	+	−	693
8	24	−	−	+	+	+	669
9	10	−	+	−	−	−	492
10	16	−	+	−	−	+	476
11	27	−	+	−	+	−	479
12	18	−	+	−	+	+	568
13	3	−	+	+	−	−	445
14	19	−	+	+	−	+	410
15	31	−	+	+	+	−	429
16	15	−	+	+	+	+	491
17	12	+	−	−	−	−	607
18	1	+	−	−	−	+	621
19	4	+	−	−	+	−	611
20	23	+	−	−	+	+	638
21	2	+	−	+	−	−	585
22	28	+	−	+	−	+	586
23	11	+	−	+	+	−	602
24	9	+	−	+	+	+	608
25	25	+	+	−	−	−	443
26	21	+	+	−	−	+	434
27	6	+	+	−	+	−	418
28	7	+	+	−	+	+	511
29	5	+	+	+	−	−	392
30	13	+	+	+	−	+	343
31	22	+	+	+	+	−	386
32	29	+	+	+	+	+	447

The exercise given here is to encourage students to do an independent analysis based on fractional factorial design that could lead to the same result as what obtained from full factorial design (conducted by NIST).

P3.3: A Plackett–Burman DOE has been employed to predict the life expectancy of diamond drill bits in terms of endured length in minutes. The experimental treatments are given in Table P3.3 and the design levels of the 11 factors are listed as follows [Fang et al. 2018]:

Factor	Description	Level (−)	Level (+)
A	Solute A concentration	57 g/L	68 g/L
B	Solute B concentration	4.22 g/L	5.22 g/L
C	Catalyst concentration	31 g/L	39 g/L
D	Ph value	4.1	4.7
E	Ni-diamond composite electro plating time	100 minutes	140 minutes
F	Thickness-raised Ni electroplating time	85 minutes	95 minutes
G	Supersonic wave thickness-raised Ni electroplating time	5 minutes	9 minutes
H	Solution temperature	46°C	49°C
I	Current density of Ni-diamond composite electroplating	0.3 A/dm	0.7 A/dm
J	Current density of thickness-raised Ni electroplating	0.6 A/dm	1.4 A/dm
K	Current density of supersonic wave thickness-raised Ni electroplating	15 A/dm	19 A/dm

TABLE P3.3 L_{12} fractional factorial design matrix for diamond drill bits experiment.

Run	a	b	d	d	e	f	g	h	i	j	k	Life
1	−1	−1	−1	−1	−1	−1	−1	−1	−1	−1	−1	450
2	−1	−1	1	1	1	−1	1	1	−1	1	−1	40
3	−1	1	1	−1	1	−1	−1	−1	1	1	1	420
4	1	−1	−1	−1	1	1	1	−1	1	1	−1	270
5	1	−1	1	1	−1	1	−1	−1	−1	1	1	840
6	−1	1	1	1	−1	1	1	−1	1	−1	−1	300
7	1	1	−1	1	1	−1	1	−1	−1	−1	1	990
8	−1	−1	−1	1	1	1	−1	1	1	−1	1	280
9	1	1	−1	1	−1	−1	−1	1	1	1	−1	990
10	−1	1	−1	−1	−1	1	1	1	−1	1	1	60
11	1	1	1	−1	1	1	−1	1	−1	−1	−1	10
12	1	−1	1	−1	−1	−1	1	1	1	−1	1	20

P3.4: Factors and their design levels for identifying the mechanical properties of extruded plastics reinforced with natural fibers (HDPE/sponge-gourds) are given as follows [Sousa et al. 2013]:

Factor	Level (−)	Level (O)	Level (−)
A: Fiber content (weight)	10%	20%	30%
B: Fiber diameter (mm)	0.63	0.39	0.15
C: Screw speed (RPM)	300	375	450
D: Curing temperatures (°C)	Die: 180	Die: 200	Die: 220
	Zone 2–5: 140	Zone 6–9: 160	Zone 6–9: 180
	Zone 6–9: 160	Zone 6–9: 160	Zone 6–9: 200

The test results of ultimate tensile strength (σ_{uts}), Young's modulus (E), and Izod impact energy per unit length (I_{zod}) based on the Box-Behnken design are listed as follows:

Treatment	A	B	C	D	σ_{uts} (MPa)	E (GPa)	I_{zod} (J/m)
1	−1	−1	−1	−1	17.89	0.809	57.00
2	1	−1	−1	−1	17.83	1.331	52.87
3	−1	1	−1	−1	17.91	0.821	63.09
4	1	1	−1	−1	17.29	1.410	53.46
5	−1	−1	1	−1	17.72	0.820	73.03
6	1	−1	1	−1	17.99	1.421	59.67
7	−1	1	1	−1	17.85	0.836	59.10
8	1	1	1	−1	17.40	1.439	52.00
9	−1	−1	−1	1	14.22	0.510	92.78
10	1	−1	−1	1	18.10	1.474	64.77
11	−1	1	−1	1	17.81	0.839	63.05
12	1	1	−1	1	16.94	1.496	46.97
13	−1	−1	1	1	17.54	0.802	75.31
14	1	−1	1	1	18.26	1.489	65.92
15	−1	1	1	1	17.78	0.755	59.34
16	1	1	1	1	17.64	1.369	56.77

The experimental procedure is detailed in [Sousa et al. 2013]. Please formulate the predictive equation for each mechanical property. The exercise given here is to let students do analysis of variance (that is not presented in the original paper) as an independent analysis.

4

General Factorial Design L^{K-P}

esign of experiments based on F-distribution (DOE-F) is a statistical technique that systematically determines which inputs have a significant impact on the output in a cost-effective and timely manner. The design renders a set of experimental tests, i.e., model runs of different treatments, of which each treatment may yield replicated responses as a function of multiple variables at arbitrary levels nonlinearly. It may cast a more complete response surface model, including quadratic terms, as it is exactly a response surface method (RSM) for an experimental design of two factors.

4.1. RSM and F-Distribution

Response surface methodology is a collection of mathematical and statistical techniques based on the fit of a nonlinear polynomial equation to the experimental data [Myers et al. 2009]. The resulting method is intended to describe the behavior of the data set to make statistical inferences. It is well suitable when a response or a set of responses of interest are influenced by several factors at higher than two levels. The methodology is to identify potential influential factors and consequently yield a nonlinear predictive equation, including mostly linear terms, quadratic terms, and interactive terms, which leads to a reliable design configuration for further processing such as optimization. Since the DOE method is based on F-distribution, it is called DOE-F. In general, the objective of applying DOE-F is usually to simultaneously optimize the levels of these variables to attain the best system performance.

4.1.1. High-Level Factorial Design

If a factor is not significant when applying DOE-t, drop it from the metamodel and reanalyze the data before going to a higher-order model. When all DOE-t metamodels are proven to be inadequate, a DOE with higher levels for the significant factors as verified by DOE-t are to be taken, and then DOE-F may prevail. Nevertheless, a three- or higher-level factorial design is sometimes demanded arbitrarily, such as

(a) For a variable factor, it is common to study the effect of the factor on the response at its current design setting and two other settings around it (i.e., $Y \pm \Delta Y$) to figure out the response gradients in both directions of variation. For example, a dimension with bilateral tolerances.

(b) For an attribute factor, it may be set at three levels or above. For example, at least three suppliers are often required for justifying a buy-into deal in a bidding process.

High-level factorial design matrices based on DOE-F are employed if nonlinear curvilinear terms are to be exploited in the fitting equation. Equations derived from DOE-F are applicable to factors of any level, although they are demonstrated using examples with three levels. The three levels are referred to as low, intermediate, and high levels and they are numerically coded as $(-1, 0, 1)$, $(1, 2, 3)$, or $(0, 1, 2)$. A L^K design means that there are K factors involved in the design and each factor has L levels. L can be different for each individual factor. For example, $2^4 \times 3^2$ means there are four factors, of which each has two levels and two factors of which each has three levels.

There are two ways to explore the nonlinearity embedded in a DOE. One is a factorial design 3^{K-P} and the other is a composite design. A general three-level design (3^{K-P}) is useful for investigating quadratic effects, but it may be required to do a good size of treatments (runs) and sometimes it is often prohibitive to carry it out. Alternatively, a two-level design with center points, called composite design, is much less complicated while it is still an easy and good way to establish the presence or absence of curvature.

4.1.2. F-Distribution

The F-statistic can be used for analyzing a factorial design with factors having three levels and above. The F-distribution is a statistical distribution of the ratios of two independent estimators of the population variances. There are two degrees of freedom associated with each factor of design 3^K in an F-distribution. The probability density function and cumulative distribution function of F-distribution in statistics are, respectively [Ross 1987]:

$$f(F) = \frac{\Gamma[\tfrac{1}{2}(u+v)](u/v)^{u/2} F^{u/2-1}}{\Gamma[\tfrac{1}{2}u]\,\Gamma[\tfrac{1}{2}v]\,[(u/v)F+1]^{(u+v)/2}}, \quad 0 < F < \infty \qquad (4.1.1)$$

and

$$F(F) = 2\,u^{(u-2)/2}\left(\frac{F}{v}\right)^{u/2} \frac{2\,H_g[\tfrac{1}{2}(u+v),\ \tfrac{1}{2}u;\ 1+\tfrac{1}{2}u;\ -u\,x/v]}{B_e(\tfrac{1}{2}u,\ \tfrac{1}{2}v)} \qquad (4.1.2)$$

where
   F is the F variate
   $\Gamma[]$ is the gamma function
   u and v are degrees of freedom
   H_g is the hypergeometric function
   B_e is the beta function

Parameters u and v are degrees of freedom for the prediction sample variance and residual sample variance, respectively, when the F-test is applied in the DOE. The probability density function of F-distribution is plotted in Figure 4.1.1. There is a distinct curve for each set of degrees of freedom (u, v). Each probability density curve is not symmetrical but skewed to the right side. As the degrees of freedom for the numerator and denominator increase, the curve reshapes to approximate the normal distribution. As an example, the plot of $F_{4,10}$ is shown in Figure 4.1.2 and its p-value (i.e., α) is depicted as the shaded area. According to Equation (4.1.2), percentage points of the accumulative F-distribution are tabulated in Table 4.1.1.

FIGURE 4.1.1 Typical probability density functions of $F_{u,v}$ distribution according to Equation (4.1.1).

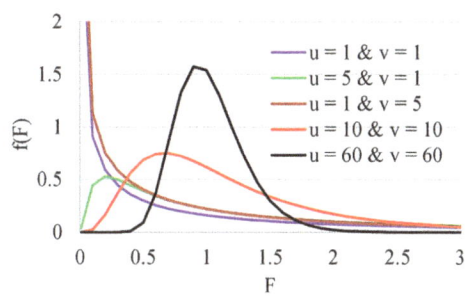

FIGURE 4.1.2 Tailing area: p-value (α = 10%) at $F_{4,10,\alpha}$ = 2.605 [Table 4.1.1(B)].

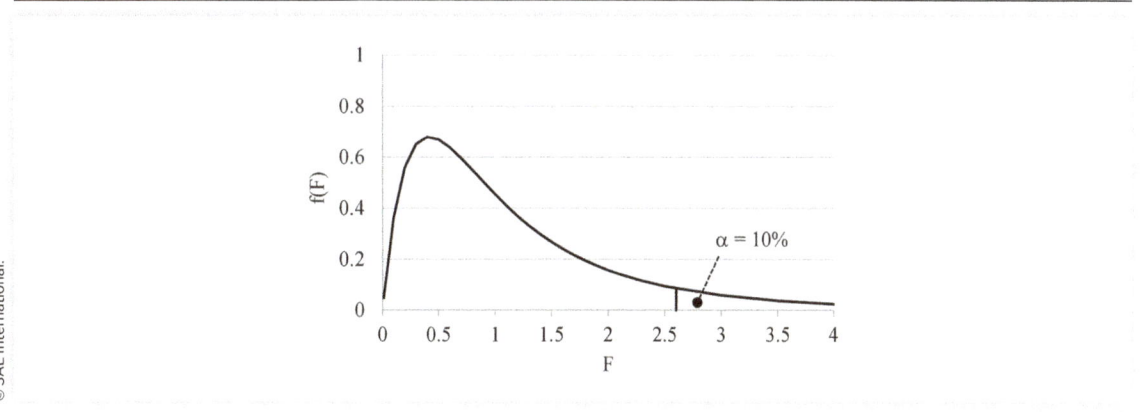

For the DOE, subscript u of $F_{u,v,\alpha}$ stands for the variation of the design factor of interest or its interaction with other design factors, while subscript v refers to the degrees of freedom for the random errors available. As demonstrated in Figure 4.1.3, three frequently used series of $F_{u,v,\alpha}$ values in Table 4.1.1 are given as follows:

1. For 2^{k-p}: u = 1, at a significance level α = 5% (p-value)
2. For main effects of 3^{k-p}: u = 2, at a significance level α = 5% (p-value)
3. For two-factor interactions of 3^{k-p}: u = 4, at a significance level α = 5% (p-value)

FIGURE 4.1.3 $F_{1,v,\alpha}$, $F_{2,v,\alpha}$, and $F_{4,v,\alpha}$ with p-value α = 5%.

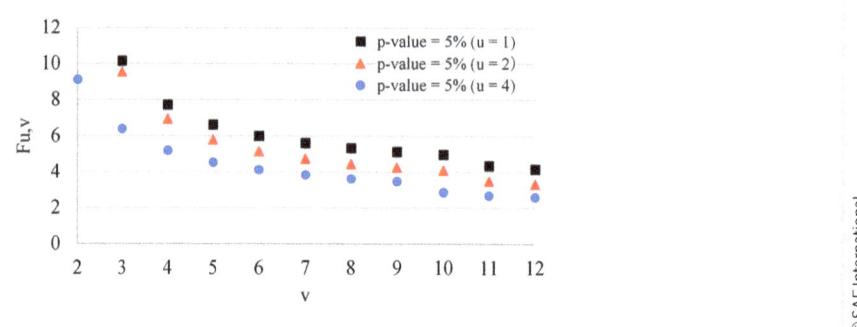

© SAE International.

TABLE 4.1.1 Percentage points of the accumulative F-distribution according to Equation (4.1.2).

(A) α = 5%

$F_{u,v,\,5\%}$	$u = 1$	$u = 2$	$u = 3$	$u = 4$	$u = 5$	$u = 6$	$u = 7$	$u = 8$	$u = 9$	$u = 10$
$v = 1$	161.5	199.5	215.7	224.6	230.2	234.0	236.8	238.9	240.5	241.9
$v = 2$	18.51	19.00	19.16	19.25	19.30	19.33	19.35	19.37	19.39	19.40
$v = 3$	10.13	9.55	9.28	9.12	9.01	8.94	8.89	8.85	8.81	8.79
$v = 4$	7.71	6.94	6.59	6.39	6.26	6.16	6.09	6.04	6.00	5.96
$v = 5$	6.61	5.79	5.41	5.19	5.05	4.95	4.88	4.82	4.77	4.74
$v = 6$	5.99	5.14	4.76	4.53	4.39	4.28	4.21	4.15	4.10	4.06
$v = 7$	5.59	4.74	4.35	4.12	3.97	3.87	3.79	3.73	3.68	3.64
$v = 8$	5.32	4.46	4.07	3.84	3.69	3.58	3.50	3.44	3.39	3.35
$v = 9$	5.12	4.26	3.86	3.63	3.48	3.37	3.29	3.23	3.18	3.14
$v = 10$	4.97	4.10	3.71	3.48	3.33	3.22	3.14	3.07	3.02	2.98
$v = 20$	4.35	3.49	3.10	2.87	2.71	2.60	2.51	2.45	2.39	2.35
$v = 30$	4.17	3.32	2.92	2.69	2.53	2.42	2.33	2.27	2.21	2.17
$v = 40$	4.09	3.23	2.84	2.61	2.45	2.34	2.25	2.18	2.12	2.08
$v = 60$	4.00	3.15	2.76	2.53	2.37	2.25	2.17	2.10	2.04	1.99
$v = 100$	3.94	3.09	2.70	2.46	2.31	2.19	2.10	2.03	1.98	1.93

(B) α = 10%

$F_{u,v,\,10\%}$	$u = 1$	$u = 2$	$u = 3$	$u = 4$	$u = 5$	$u = 6$	$u = 7$	$u = 8$	$u = 9$	$u = 10$
$v = 1$	39.86	49.50	53.59	55.83	57.24	58.20	58.91	59.44	59.86	60.19
$v = 2$	8.526	9.000	9.162	9.243	9.293	9.326	9.349	9.367	9.381	9.392
$v = 3$	5.538	5.462	5.391	5.343	5.309	5.285	5.266	5.252	5.240	5.230
$v = 4$	4.545	4.325	4.191	4.107	4.051	4.010	3.979	3.955	3.936	3.920
$v = 5$	4.060	3.780	3.619	3.520	3.453	3.405	3.368	3.339	3.316	3.297
$v = 6$	3.776	3.463	3.289	3.181	3.108	3.055	3.014	2.983	2.958	2.937
$v = 7$	3.589	3.257	3.074	2.961	2.883	2.827	2.785	2.752	2.725	2.703
$v = 8$	3.458	3.113	2.924	2.806	2.726	2.668	2.624	2.589	2.561	2.538
$v = 9$	3.360	3.006	2.813	2.693	2.611	2.551	2.505	2.469	2.440	2.416
$v = 10$	3.285	2.924	2.728	2.605	2.522	2.461	2.414	2.377	2.347	2.327
$v = 20$	2.975	2.589	2.380	2.249	2.158	2.091	2.039	1.999	1.965	1.937
$v = 30$	2.881	2.489	2.276	2.142	2.049	1.980	1.927	1.884	1.849	1.819
$v = 40$	2.835	2.440	2.226	2.091	1.997	1.927	1.873	1.829	1.793	1.763
$v = 60$	2.791	2.393	2.177	2.041	1.946	1.875	1.819	1.775	1.738	1.707
$v = 120$	2.748	2.347	2.130	1.992	1.896	1.824	1.767	1.722	1.684	1.652
$v = \infty$	2.706	2.303	2.084	1.945	1.847	1.774	1.717	1.670	1.632	1.599

© SAE International.

It can be seen that more treatments and/or replications, i.e., with increasing v, will reduce the required value of $F_{u,v,\alpha}$ to meet the required significance level (5%, taken for granted usually).

4.2. ANOVA for L^K Design

Analysis of variance (ANOVA) has been applied to L^K design, of which superscript K is the number of factors and L is the arbitrary number of levels. Traditionally, it is also called L-way factorial design. The L values are referred to as the levels of a factor, and the number of levels can vary among factors. A mixed design, such as 2^23^1, is also feasible, of which two factors take two levels each and the third factor takes three levels. A three- or higher-level factorial design is required when there is a nonlinear relationship between the response and any factor, since it is not possible to detect such a curved effect with two levels. Nevertheless, 3^K design is more general in the real-world practice.

ANOVA is used to detect significant factors in a multilevel factorial design model, of which each factor takes on a certain number of levels that can be equal or greater than two. To apply ANOVA to an experimental design, the following assumptions must be met:

(a) The resulting residuals are independent
(b) The resulting residuals have the same variance consistently
(c) The resulting residuals are normally distributed
(d) All factors are orthogonal to (independent of) each other

4.2.1. I^{1}J^{1}M^1 Design and 3^3 Design

For simplicity, a full factorial design I^{1}J^{1}M^1, which is a multiple-level design of three factors, will be used here to interpret the procedure for carrying out the general case L^K. When I = J = M = 3, it reduces to a full factorial design 3^3. Analytic calculations based on the design matrix of I^{1}J^{1}M^1 described here are used for illustrating how to carry out the analysis of variances and the procedure can be generalized for the following three different design configurations:

1. 3^K: Full factorial design
2. 3^{K-P}: Fractional factorial design
3. Central composite design

First, consider full factorial design 3^3. Each experimental outcome obtained from full factorial design 3^3 with no replication (only one outcome for each treatment) can be expressed as [Box and Behnken 1960]:

$$Y_{ijm} = \mu + A_i + B_j + C_m + A_i B_j + A_i C_m + B_j C_m + A_i B_j C_m + \epsilon_{ijm} \qquad (4.2.1)$$

where
  μ is the mean (grand average)
  A_i, B_j, and C_m are the factors A, B, and C at levels i, j, and m, respectively
  i, j, and m are indices for levels
  ϵ_{ijm} is the residual from the treatment at (A_i, B_j, C_m) level

As an example, the design matrix for design matrix 3^3 is given in Table 4.2.1 and its 27 runs (treatment combinations) are depicted in Figure 4.2.1. The levels can be represented by "–1, 0, 1," "0, 1, 2," or "1, 2, 3." It can be realized that there are 27 runs (treatments) in total for the 3^3 design with no replication. In such a case,

1. Each main effect, i.e., A, B, or C, has 2 DOFs each
2. Each two-factor interaction, i.e., AB, AC, or BC has 4 (= 2^2) DOFs each, and
3. Each three-factor interaction, i.e., ABC, has 8 (= 2^3) DOFs each

TABLE 4.2.1 27 runs (treatments) for the 3^3 full factorial design in coded variables.

Run	a	b	c	or	a	b	c	or	a	b	c
1	−1	−1	−1		1	1	1		0	0	0
2	−1	−1	0		1	1	2		0	0	1
3	−1	−1	1		1	1	3		0	0	2
4	−1	0	−1		1	2	1		0	1	0
5	−1	0	0		1	2	2		0	1	1
6	−1	0	1		1	2	3		0	1	2
7	−1	1	−1		1	3	1		0	2	0
8	−1	1	0		1	3	2		0	2	1
9	−1	1	1		1	3	3		0	2	2
10	0	−1	−1		2	1	1		1	0	0
11	0	−1	0		2	1	2		1	0	1
12	0	−1	1		2	1	3		1	0	2
13	0	0	−1		2	2	1		1	1	0
14	0	0	0		2	2	2		1	1	1
15	0	0	1		2	2	3		1	1	2
16	0	1	−1		2	3	1		1	2	0
17	0	1	0		2	3	2		1	2	1
18	0	1	1		2	3	3		1	2	2
19	1	−1	−1		3	1	1		2	0	0
20	1	−1	0		3	1	2		2	0	1
21	1	−1	1		3	1	3		2	0	2
22	1	0	−1		3	2	1		2	1	0
23	1	0	0		3	2	2		2	1	1
24	1	0	1		3	2	3		2	1	2
25	1	1	−1		3	3	1		2	2	0
26	1	1	0		3	3	2		2	2	1
27	1	1	1		3	3	3		2	2	2

FIGURE 4.2.1 3^3 Design—27 runs.

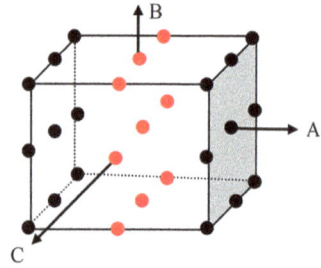

In summary, a K-factor interaction has 2^K DOFs. Full factorial design 3^3 contains 26 (= 2 + 2 + 2 + 4 + 4 + 4 + 8) DOFs that are carried by its associated main effects and interactions. When the mean is taken into consideration, there are 27 unknowns, which could be solved using 27 data obtained from 27 (3^3) experimental runs based on the F-test.

Carrying out experimental tests with replications is always encouraged. Replications will help the experimenter to calculate the residuals, which can be used for assuring diagnostic checking of the metamodel. However, when there is no replication, the fit is exact and there is no further information for the error term, i.e., ϵ_{ijm} in Equation (4.2.1). In case there is no replication, the small effects and even three-factor interactions can be assumed to be remote and statistically meaningless, and their associated DOFs can be then utilized for estimating the error.

Based on the ANOVA, the sum of squares originated from different sources of variances (i.e., main effects A, B, and C, and interactions AB, AC, BC, and ABC), as well as the residual sum of squares, can be calculated.

The ANOVA demonstrated in Table 4.2.2 is used herein to summarize the analysis of a general design matrix involved with three factors while each factor has its own distinct design levels, denoted as I, J, and M levels. Inserting n (n = 1, 2, …, N) into Equation (4.2.1) when implemented in the test with N replications for each factor, one may rewrite a new equation for each experimental outcome (Y_{ijmn}) into the following components:

$$Y_{ijmn} = \bar{Y} + (\bar{Y}_{A,i} - \bar{Y}) + (\bar{Y}_{B,j} - \bar{Y}) + (\bar{Y}_{C,m} - \bar{Y})$$
$$+ (\bar{Y}_{AB,ij} - \bar{Y}_{A,i} - \bar{Y}_{B,j} + \bar{Y}) + (\bar{Y}_{AC,im} - \bar{Y}_{A,i} - \bar{Y}_{C,m} + \bar{Y}) + (\bar{Y}_{BC,jm} - \bar{Y}_{B,j} - \bar{Y}_{C,m} + \bar{Y})$$
$$+ (\bar{Y}_{ABC,ijm} - \bar{Y}_{AB,ij} - \bar{Y}_{AC,im} - \bar{Y}_{BC,jm} + \bar{Y}_{A,i} + \bar{Y}_{B,j} + \bar{Y}_{C,k} - \bar{Y}) \qquad (4.2.2)$$
$$+ (Y_{ijmn} - \bar{Y}_{ABC,ijm})$$

while

$$\varepsilon_{ijmn} = Y_{ijmn} - \bar{Y}_{ABC,ijm} \qquad (4.2.3)$$

where
$\bar{Y}$ is the grand average, i.e., the average of all the data acquired
$\bar{Y}_{A,i}$ is the averaged effect of factor A
$\bar{Y}_{B,j}$ is the averaged effect of factor B
$\bar{Y}_{C,m}$ is the averaged effect of factor C
$\bar{Y}_{AB,ij}$ is the averaged effect of the interaction between factors A and B
$\bar{Y}_{AC,im}$ is the averaged effect of the interaction between factors A and C
$\bar{Y}_{BC,jm}$ is the averaged effect of the interaction between factors B and C
$\bar{Y}_{ABC,ijm}$ is the averaged effect of the interaction among factors A, B, and C
n is the replication, n = 1, 2, 3, …, N
N is the total number of replications, no replication applied when N = 1
ε is the sample random error, also called white noise

If the null hypothesis is true, all the treatments should yield the results that respond proportionally to the assigned factorial levels numerically. The null hypothesis states that $\bar{Y}$ (grand average) is a fair estimate of every experimental outcome, and all the effects are attributed to the measured random variation. The averages (sample means) of different combinations of factors A, B, and C are calculated as follows:

$$\bar{Y} = [\sum_{i=1}^{I} \sum_{j=1}^{J} \sum_{m=1}^{M} \sum_{n=1}^{N} Y_{ijmn}] / (I\,J\,M\,N) \qquad (4.2.4)$$

$$\bar{Y}_{ABC,ijm} = [\sum_{n=1}^{N} Y_{ijmn}] / N \qquad (4.2.5)$$

$$\bar{Y}_{AB,ij} = [\sum_{m=1}^{M} \sum_{n=1}^{N} Y_{ijmn}] / (M\,N) \qquad (4.2.6)$$

$$\bar{Y}_{AC,im} = [\sum_{j=1}^{J} \sum_{n=1}^{N} Y_{ijmn}] / (J\,N) \qquad (4.2.7)$$

$$\bar{Y}_{BC,jm} = [\sum_{i=1}^{I} \sum_{n=1}^{N} Y_{ijmn}] / (I\,N) \qquad (4.2.8)$$

$$\bar{Y}_{A,i} = [\sum_{j=1}^{J} \sum_{m=1}^{M} \sum_{n=1}^{N} Y_{ijmn}] / (J\,M\,N) \qquad (4.2.9)$$

$$\bar{Y}_{B,j} = [\sum_{i=1}^{I} \sum_{m=1}^{M} \sum_{n=1}^{N} Y_{ijmn}] / (I\,M\,N) \qquad (4.2.10)$$

and

$$\bar{Y}_{C,m} = [\sum_{i=1}^{I} \sum_{j=1}^{J} \sum_{n=1}^{N} Y_{ijmn}] / (I\,J\,N) \qquad (4.2.11)$$

where

    I is the total number of levels of variable A
    J is the total number of levels of variable B
    M is the total number of levels of variable C
    N is the total number of replications

Relocating the grand average to the left side of "=" sign in Equation (4.2.2) and having both sides squared, one has the variance of an individual piece of data from the grand average that consists of eight distinct orthogonal components, i.e.,

$$(Y_{ijmn} - \bar{Y})^2 = [(\bar{Y}_{A,i} - \bar{Y}) + (\bar{Y}_{B,j} - \bar{Y}) + (\bar{Y}_{C,m} - \bar{Y})$$

$$+ (\bar{Y}_{AB,ij} - \bar{Y}_{A,i} - \bar{Y}_{B,j} + \bar{Y})$$

$$+ (\bar{Y}_{AC,im} - \bar{Y}_{A,i} - \bar{Y}_{C,m} + \bar{Y})$$

$$+ (\bar{Y}_{BC,jm} - \bar{Y}_{B,j} - \bar{Y}_{C,m} + \bar{Y}) \qquad (4.2.12)$$

$$+ (\bar{Y}_{ABC,ijm} - \bar{Y}_{AB,ij} - \bar{Y}_{AC,im} - \bar{Y}_{BC,jm} + \bar{Y}_{A,i} + \bar{Y}_{B,j} + \bar{Y}_{C,m} - \bar{Y})$$

$$+ (Y_{ijmn} - \bar{Y}_{ABC,ijm})]^2$$

where

$\bar{Y}_{A,i} - \bar{Y}$ is the main effect of factor A at level i

$\bar{Y}_{B,j} - \bar{Y}$ is the main effect of factor B at level j

$\bar{Y}_{C,m} - \bar{Y}$ is the main effect of factor C at level m

$\bar{Y}_{AB,ij} - \bar{Y}_{A,i} - \bar{Y}_{B,j} + \bar{Y}$ is the interactive effect of factor A at level i and factor B at level j

$\bar{Y}_{AC,im} - \bar{Y}_{A,i} - \bar{Y}_{C,m} + \bar{Y}$ is the interactive effect of factor A at level i and factor C at level m

$\bar{Y}_{BC,jm} - \bar{Y}_{B,j} - \bar{Y}_{C,m} + \bar{Y}$ is the interactive effect of factor B at level j and factor C at level m

$\bar{Y}_{ABC,ijm} - \bar{Y}_{AB,ij} - \bar{Y}_{AC,im} - \bar{Y}_{BC,jm} + \bar{Y}_{A,i} + \bar{Y}_{B,j} + \bar{Y}_{C,m} - \bar{Y}$ is the interactive effect of

- Factor A at level i,

- Factor B at level j, and

- Factor C at level m

The inherent condition of orthogonality imposed on the eight components means that the total variation in Y from the mean can be decomposed into the contributions associated with the following eight effects and interactions: A, B, C, AB, AC, BC, ABC, and the residuals. Orthogonality means that they are independent of each other and that would allow one to use statistics such as the arithmetic average or the sum of squares. Since the additivity of the sums of squares (SS) in a random block is true and factors A, B, and C are orthogonal to each other like the (x, y, z) coordinate system in the three-dimensional space, the following equation can be established using Equation (4.2.12):

$$SS_T = SS_A + SS_B + SS_C + SS_{AB} + SS_{AC} + SS_{BC} + SS_{ABC} + SS_E \qquad (4.2.13)$$

where

$$SS_{ave} = I \, J \, M \, N \, \bar{Y}^2 \qquad \text{(Grand Average)} \qquad (4.2.14)$$

$$SS_T = \sum_{i=1}^{I} \sum_{j=1}^{J} \sum_{m=1}^{M} \sum_{n=1}^{N} (Y_{ijmn} - \bar{Y})^2 \qquad \text{(Subtotal)} \qquad (4.2.15)$$

$$SS_A = J \, M \, N \sum_{i=1}^{I} (\bar{Y}_{A,i} - \bar{Y})^2 \qquad \text{(Factor A)} \qquad (4.2.16)$$

$$SS_B = I \, M \, N \sum_{j=1}^{J} (\bar{Y}_{B,j} - \bar{Y})^2 \qquad \text{(Factor B)} \qquad (4.2.17)$$

$$SS_C = I \, J \, N \sum_{m=1}^{M} (\bar{Y}_{C,m} - \bar{Y})^2 \qquad \text{(Factor C)} \qquad (4.2.18)$$

$$SS_{AB} = M \, N \sum_{i=1}^{I} \sum_{j=1}^{J} (\bar{Y}_{AB,ij} - \bar{Y}_{A,i} - \bar{Y}_{B,j} + \bar{Y})^2 \qquad \text{(Interaction AB)} \qquad (4.2.19)$$

$$SS_{AC} = J N \sum_{i=1}^{I} \sum_{m=1}^{M} (\bar{Y}_{AC,im} - \bar{Y}_{A,i} - \bar{Y}_{C,m} + \bar{Y})^2 \quad \text{(Interaction AC)}$$ (4.2.20)

$$SS_{BC} = I N \sum_{j=1}^{J} \sum_{m=1}^{M} (\bar{Y}_{BC,jm} - \bar{Y}_{B,j} - \bar{Y}_{C,m} + \bar{Y})^2 \quad \text{(Interaction BC)}$$ (4.2.21)

$$SS_{ABC} = N \sum_{i=1}^{I} \sum_{j=1}^{J} \sum_{m=1}^{M} (\bar{Y}_{ABC,ijm} - \bar{Y}_{AB,ij} - \bar{Y}_{AC,im} - \bar{Y}_{BC,jm} + \bar{Y}_{A,i} + \bar{Y}_{B,j} + \bar{Y}_{C,m} - \bar{Y})^2$$

$$\text{(Interaction ABC)}$$ (4.2.22)

and

$$SS_E = \sum_{i=1}^{I} \sum_{j=1}^{J} \sum_{m=1}^{M} \sum_{n=1}^{N} (Y_{ijmn} - \bar{Y}_{ABC,ijm})^2 \quad \text{(Error)}$$ (4.2.23)

note that

$$SS - SS_{ave} = SS_T$$ (4.2.24)

where

$$SS = \sum_{i=1}^{I} \sum_{j=1}^{J} \sum_{m=1}^{M} \sum_{n=1}^{N} Y_{ijmn}^2 \quad \text{(Total Sum of Squares)}$$ (4.2.25)

As shown above, the ANOVA summarizes how much of the variance in the data (sum of squares) is accounted for by the main effects (sums of squares of factors), the interactive effects (sums of squares of interactive terms), and how much is due to random error (residual sum of squares). The ANOVA table (Table 4.2.2) provides a formal F-test for the factorial main effects and interactions. Each treatment in the DOE can be regarded as a group. In an ANOVA model, the total variation (total SS) is partitioned into variation between groups (between SS) and variation within groups (within SS). This methodology is to be illustrated using the DOE of three factors in the following subsection.

The F-ratio in ANOVA is defined as the ratio of variability between groups (between treatments in DOE) divided by the variability within groups (within individual treatments in DOE), as:

$$\text{F-ratio} = \frac{\text{Variability between Treatments}}{\text{Variability within Treatments}}$$ (4.2.26)

The denominator in the above equation is supposedly function as the experimental error (sample random error). The variability within treatments is measured by the mean square of error (MS_E) as demonstrated in Table 4.2.2, instead of the sum of squares, which appears to depend on both the error and sample size. ANOVA is a method for testing differences among treatment means by analyzing variance [Sahai and Ageel 2000].

TABLE 4.2.2 ANOVA table of multiple-level design of three factors based on Equations (4.2.14)–(4.2.25).

Variance	SS	DOF	MS (mean square)	$F_{u,v}$
A	SS_A	$I-1$	$MS_A = SS_A/(I-1)$	$F_{(I-1),(N-1)IJM} = MS_A/MS_E$
B	SS_B	$J-1$	$MS_B = SS_B/(J-1)$	$F_{(J-1),(N-1)IJM} = MS_B/MS_E$
C	SS_C	$M-1$	$MS_C = SS_C/(M-1)$	$F_{(M-1),(N-1)IJM} = MS_C/MS_E$
AB	SS_{AB}	$(I-1)(J-1)$	$MS_{AB} = SS_{AB}/[(I-1)(J-1)]$	$F_{(I-1)(J-1),(N-1)IJM} = MS_{AB}/MS_E$
AC	SS_{AC}	$(I-1)(M-1)$	$MS_{AC} = SS_{AC}/[(I-1)(M-1)]$	$F_{(I-1)(M-1),(N-1)IJM} = MS_{AC}/MS_E$
BC	SS_{BC}	$(J-1)(M-1)$	$MS_{BC} = SS_{BC}/[(J-1)(M-1)]$	$F_{(J-1)(M-1),(N-1)IJM} = MS_{BC}/MS_E$
ABC	SS_{ABC}	$(I-1)(J-1)(M-1)$	$MS_{ABC} = SS_{ABC}/[(I-1)(J-1)(M-1)]$	$F_{(I-1)(J-1)(M-1),(N-1)IJM} = MS_{ABC}/MS_E$
Error	SS_E	$(N-1)IJM$	$MS_E = SS_E/[(N-1)IJM]$	–
Subtotal	SS_T	$IJMN-1$	–	–
Grand average	SS_{ave}	1	–	–
Grand total	SS	$IJMN$	–	–

Notes:
I, J, and M: Levels of factors A, B, and C, respectively; for 3^3, I = J = M = 3
N: Number of replications for each run to produce the error message
DOF: Degrees of freedom
SS: Sum of squares
MS: Mean square
$F_{u,v}$: F-distribution statistic of (u, v) DOFs
Subtotal: Total excluding the grand average ($\bar{Y}$)

4.2.2. Null Hypothesis and F-Test

ANOVA avoids the Type-I error problem by the simple device of pooling all observations and making one global test of differences among treatments. The F-test based on the ANOVA is conducted to assure the adequacy of fit for a DOE with factors of high levels, i.e., three and above. The adequacy of fit is to be decided using the following null hypothesis:

$$H_0: \mu_q = \mu \tag{4.2.27a}$$

$$H_1: \mu_q \neq \mu \tag{4.2.27b}$$

where
μ is the common population mean
μ_q is the treatment mean, q = 1, 2, 3, …, Q, of which Q is the total number of treatments
i, j, and m are the indices for levels, up to I, J, and M levels, respectively

As exhibited above, the mean square (variance) is computed as to how much each individual "*treatment mean*" differs from the "*overall mean*" of overall observations in the experimental design. If the null hypothesis is true, the sample variance for each treatment (run) must be consistently of the same size statistically. Since each of the main and interactive effects, i.e., A, B, C, AB, AC, BC, and ABC, has its own DOFs, its corresponding mean square (i.e., variance or sum of squares/DOF) is a fair representation of variability. In other words, the mean square of an individual main or interactive effect, i.e., A, B, C, AB, AC, BC, or ABC, shall estimate the same common population variance of each of the I × J × M combinations if the null prevails. The ability of a statistical F-test to resolve a difference between treatments increases with increasing sample size, including the number of replicated runs N.

As long as the relationship between the response and design factors are normally distributed and its variance is consistently constant, the mean square of each main effect or interaction as measured by the mean square of residual errors fits the F-distribution. The numerical value for F-statistic is obtained by the ratio of these two sample variances [Ross 1987]:

$$\frac{S_p^2}{S_r^2} = F_{u,v} \tag{4.2.28}$$

where

S_p^2 is the prediction sample variance of a main or interactive effect

S_r^2 is the residual sample variance, i.e., experimental error

A lower p-value (probability values), as yielded by a higher $F_{u,v}$ (F-statistic) value, provides stronger evidence against the null hypothesis. If the null hypothesis is rejected, it can be concluded that at least one of the population means is different from at least one other population mean.

Note that Student's t-distribution used for 2^K and 2^{K-P} designs is actually a special case of the F-distribution, as they are related to each other by:

$$t^2_{v,\alpha} = F_{1,v,2\alpha} \tag{4.2.29}$$

of which α and 2α indicate the right tail areas of the t and F distributions, respectively. For example, $t^2_{10,\,0.025} = F_{1,10,\,0.05}$ or $t^2_{10,\,0.05} = F_{1,10,\,0.1}$. Nevertheless, the t-distribution is less sensitive to the normality assumption.

It was pointed out that in a completely randomized design, with additivity between treatments (runs) and the experimental variation, ANOVA with replications can be tested out to be distribution-free by means of a permutation test and that the common F-test is a good approximation [Hinkelmann and Kempthorne 2008].

However, the additivity between treatments and the experimental variation is crucial, and inequality in variances resulting from different treatments can be an indication of no additivity in this sense. In other words, if there are several test data in each treatment and the variances of any two treatments are not of the same variance pattern, special statistical inference algorithms like the maximum likelihood method (Chapters 7, 8, and 9, Volume II) must be utilized instead of performing statistical inference based on the F-distribution.

4.2.3. Coded Variables for Three-Level Factorial Designs

There are three distinct sets of coded variables in use for three-level factorial designs (3^K). They are illustrated using factorial design 3^3 as follows:

1. System (–1, 0, –1): Coded variables a, b, and c are calculated as a two-level factorial according to Equations (2.4.5)–(2.4.7).

2. System (0, 1, 2): Numerical numbers 0, 1, and 2 are frequently used for coded variables a, b, and c that represent factors A, B, and C in factorial design 3^3 as displayed in Table 4.2.4. It is set up for the convenience of applying the modulus 3 calculus—any multiple of 3 is equal to 0. Thus, these three dimensionless variables a, b, and c can be calculated as:

$$a = (A - A_{average} / A_{range/2}) + 1 \tag{4.2.30}$$

$$b = (B - B_{average} / B_{range/2}) + 1 \tag{4.2.31}$$

and

$$c = (C - C_{average} / C_{range/2}) + 1 \qquad (4.2.32)$$

3. System (1, 2, 3): Coded variables a, b, and c can be calculated as:

$$a = (A - A_{average} / A_{range/2}) + 2 \qquad (4.2.33)$$

$$b = (B - B_{average} / B_{range/2}) + 2 \qquad (4.2.34)$$

and

$$c = (C - C_{average} / C_{range/2}) + 2 \qquad (4.2.35)$$

System (0, 1, 2) is preferred for mathematical operations based on modulus 3 calculus (To be detailed in Section 4.3.3).

4.2.4. 3^3 Example—Pull Strength of Seat-Belt Knuckle

This is a case study on the strength of automotive seat-belt knuckle based on ANOVA using the data obtained from [Wu and Hamada 2009]. Experimental tests are set up to study the main and interactive effects on the pull strength of truck seat belts of the following three factors: (A) pressure, (B) die flat, and (C) crimp length, as given in Table 4.2.3. Three design levels are assigned to each design factor, and the dimensions of die flat and crimp length are identified in Figure 4.2.2. The design matrix of the full factorial design 3^3 is rephrased in Table 4.2.4.

TABLE 4.2.3 Factors and their design levels for fastening a seat belt.

Factor	Level (0)	Level (1)	Level (2)
A: Pressure (MPa)	7.58	9.65	11.72
B: Die flat (mm)	10	10.2	10.4
C: Crimp length (mm)	18	22.5	27

Reprinted from "Experiments: Planning, Analysis, and Optimization," Wu, Chien-Fu., Hamada,Michael © 2009 John Wiley & Sons - Books.

FIGURE 4.2.2   Dimensions of die flat (factor B) and crimp length (factor C) of a belt knuckle.

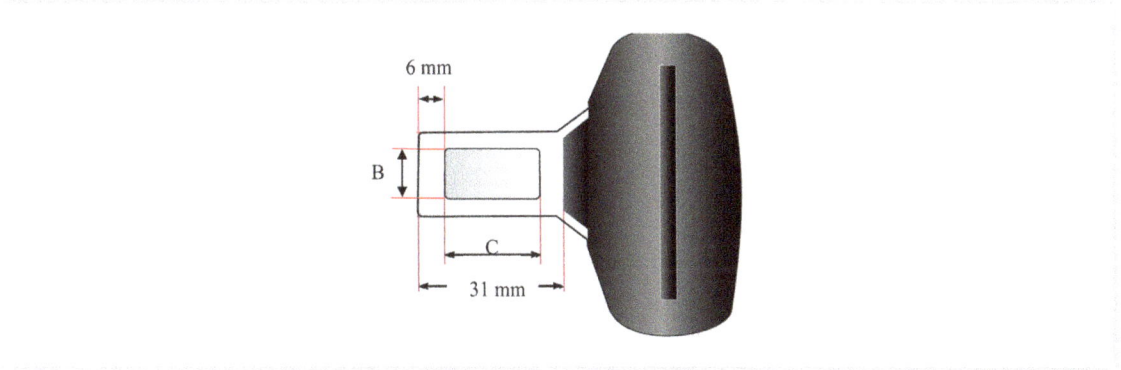

Here is the requirement, originating from the US Federal Motor Vehicle Safety Standards, that the pull strength of the crimp for this vehicle is no less than 13.5 kN [FMVSS 1994]. The objective function (pull strength) is expected to exceed the legal requirement, and it is the more the better. Three replicated strength measurements have been conducted for each treatment (run condition). Data obtained from physical tests are listed in Table 4.2.4.

Following Table 4.2.2, one is able to obtain the ANOVA table as shown in Table 4.2.5. Given that the significance level is set at p-value = 5%, factor A (pressure) has the greatest impact on the tensile strength, factor C (crimp length) is the next influential factor, and their interactive effect (AC) is also important. Influences by interactions AB and ABC are also statistically significant.

Calculations for the decomposed orthogonal components of interactions, i.e., AB ($A \times B$, $A \times B^2$), AC ($A \times C$, $A \times C^2$), BC ($B \times C$, $B \times C^2$), and ABC ($A \times B \times C$, $A \times B \times C^2$, $A \times B^2 \times C$, $A \times B^2 \times C^2$) are not detailed here. This is to be addressed further in Section 4.3.3.

TABLE 4.2.4 Factorial design 3^3 for evaluating the strength of car seat belt [Wu and Hamada 2009].

Run	a	b	c	Pull strength (kN)		
1	0	0	0	23.0	29.0	26.5
2	0	0	1	23.8	27.2	23.2
3	0	0	2	13.7	16.8	18.9
4	0	1	0	24.7	29.2	28.1
5	0	1	1	21.1	19.6	24.2
6	0	1	2	24.6	18.0	20.1
7	0	2	0	25.3	27.8	27.6
8	0	2	1	25.5	27.9	26.0
9	0	2	2	25.6	21.3	24.1
10	1	0	0	30.4	30.7	30.9
11	1	0	1	29.1	28.1	21.3
12	1	0	2	27.4	25.9	26.5
13	1	1	0	30.5	30.3	30.7
14	1	1	1	30.2	29.8	30.2
15	1	1	2	29.0	28.9	29.2
16	1	2	0	28.8	31.0	29.9
17	1	2	1	30.4	31.3	22.5
18	1	2	2	22.1	25.3	25.6
19	2	0	0	31.8	30.8	27.7
20	2	0	1	30.7	31.4	31.8
21	2	0	2	30.8	32.0	29.7
22	2	1	0	32.1	31.9	31.2
23	2	1	1	31.2	31.3	32.0
24	2	1	2	27.6	27.8	28.9
25	2	2	0	31.8	30.6	31.0
26	2	2	1	31.9	32.3	30.9
27	2	2	2	31.4	31.4	30.9

TABLE 4.2.5 ANOVA for evaluating car seat-belt strength.

Source	SS	DOF	MS	$F_{u,v}$	F (α = 5%)
A	SS_A = 688.8	2	MS_A = 344.4	$F_{2,54}$ = 87.30	3.168
B	SS_B = 19.39	2	MS_B = 9.697	$F_{2,54}$ = 2.458	3.168
C	SS_C = 187.7	2	MS_C = 93.86	$F_{2,54}$ = 23.79	3.168
AB	SS_{AB} = 65.71	4	MS_{AB} = 16.43	$F_{4,54}$ = 4.164	2.543
$A \times B$	$SS_{A \times B}$ = 54.00	2	$MS_{A \times B}$ = 26.98	$F_{2,54}$ = 6.839	3.168
$A \times B^2$	$SS_{A \times B^2}$ = 11.29	2	$MS_{A \times B^2}$ = 5.647	$F_{2,54}$ = 1.431	3.168
AC	SS_{AC} = 75.06	4	MS_{AC} = 18.76	$F_{4,54}$ = 4.756	2.543
$A \times C$	$SS_{A \times C}$ = 59.07	2	$MS_{A \times C}$ = 29.54	$F_{2,54}$ = 7.488	3.168
$A \times C^2$	$SS_{A \times C^2}$ = 17.54	2	$MS_{A \times C^2}$ = 8.771	$F_{2,54}$ = 2.223	3.168
BC	SS_{BC} = 8.619	4	MS_{BC} = 2.155	$F_{4,54}$ = 0.546	2.543
$B \times C$	$SS_{B \times C}$ = 8.453	2	$MS_{B \times C}$ = 4.227	$F_{2,54}$ = 1.071	3.168
$B \times C^2$	$SS_{B \times C^2}$ = 0.418	2	$MS_{B \times C^2}$ = 0.209	$F_{2,54}$ = 0.053	3.168
ABC	SS_{ABC} = 102.2	8	MS_{ABC} = 12.78	$F_{8,54}$ = 3.239	2.115
$A \times B \times C$	$SS_{A \times B \times C}$ = 88.9	2	$MS_{A \times B \times C}$ = 44.45	$F_{2,54}$ = 11.27	3.168
$A \times B \times C^2$	$SS_{A \times B \times C^2}$ = 5.20	2	$MS_{A \times B \times C^2}$ = 2.60	$F_{2,54}$ = 0.659	3.168
$A \times B^2 \times C$	$SS_{A \times B^2 \times C}$ = 4.07	2	$MS_{A \times B^2 \times C}$ = 2.03	$F_{2,54}$ = 0.515	3.168
$A \times B^2 \times C^2$	$SS_{A \times B^2 \times C^2}$ = 4.86	2	$MS_{A \times B^2 \times C^2}$ = 2.43	$F_{2,54}$ = 0.616	3.168
Error	SS_E = 213.03	54	MS_E = 3.945	—	—
Subtotal	SS_T = 1,360.5	80	—	—	—
Grand ave.	—	1	—	—	—
Total	—	81	—	—	—

4.2.5. 3^K Design

The corresponding variance of outcome y that can be sourced from the variances of individual independent variables up to the second order is:

$$
\begin{aligned}
\sigma_y^2 = \ & (\partial y / \partial x_1)^2 \, \sigma_{x1}^2 + (\partial y / \partial x_2)^2 \, \sigma_{x2}^2 + (\partial y / \partial x_3)^2 \, \sigma_{x3}^2 + \ldots + (\partial y / \partial x_k)^2 \, \sigma_{xK}^2 \\
& + 2 \, (\partial y / \partial x_1) \, (\partial y / \partial x_2) \, \rho_{x1,x2} \, \sigma_{x1} \, \sigma_{x2} + 2 \, (\partial y / \partial x_1) \, (\partial y / \partial x_2) \, \rho_{x1,x3} \, \sigma_{x1} \, \sigma_{x3} + \ldots \\
& + 2 \, (\partial y / \partial x_2) \, (\partial y / \partial x_3) \, \rho_{x2,x3} \, \sigma_{x2} \, \sigma_{x3} + 2 \, (\partial y / \partial x_2) \, (\partial y / \partial x_4) \, \rho_{x2,x4} \, \sigma_{x2} \, \sigma_{x4} + \ldots \\
& + 2 \, (\partial y / \partial x_3) \, (\partial y / \partial x_4) \, \rho_{x3,x4} \, \sigma_{x3} \, \sigma_{x4} + 2 \, (\partial y / \partial x_3) \, (\partial y / \partial x_5) \, \rho_{x3,x5} \, \sigma_{x3} \, \sigma_{x5} + \ldots \\
& + \ldots + 2 \, (\partial y / \partial x_{k-1}) \, (\partial y / \partial x_k) \, \rho_{xK-1,xK} \, \sigma_{xK-1} \, \sigma_{xK}
\end{aligned}
\tag{4.2.36}
$$

where
σ_y is the standard deviation of outcome y
σ_{xi} is the standard deviation of variable x_i, i = 1, 2, …, k
$\rho_{xi,xj}$ is the correlation between variables x_i and x_j, where i = 1, 2, …, K and j = 1, 2, …, K
K is the number of independent variables

As a special case, the implementation of the above equation for the 3^3 design can be obtained from the sum of squares for each concerned orthogonal component as demonstrated in the last section. By the same token, sample variances corresponding to a general 3^K factorial design can be derived. For example, for the

3^4 design, the sum of squares for the outcome and each contributing component can be obtained by applying the least-squares method to the following equation:

$$Y_{ijmpn} - \bar{Y} = (\bar{Y}_{A,i} - \bar{Y}) + (\bar{Y}_{B,j} - \bar{Y}) + (\bar{Y}_{C,m} - \bar{Y}) + (\bar{Y}_{D,p} - \bar{Y})$$

$$+ (\bar{Y}_{AB,ij} - \bar{Y}_{A,i} - \bar{Y}_{B,j} + \bar{Y}) + (\bar{Y}_{AC,im} - \bar{Y}_{A,i} - \bar{Y}_{C,m} + \bar{Y}) + (\bar{Y}_{AD,jp} - \bar{Y}_{B,j} - \bar{Y}_{D,p} + \bar{Y})$$

$$+ (\bar{Y}_{BC,ij} - \bar{Y}_{B,j} - \bar{Y}_{C,m} + \bar{Y}) + (\bar{Y}_{BD,jp} - \bar{Y}_{B,j} - \bar{Y}_{D,p} + \bar{Y}) + (\bar{Y}_{CD,mp} - \bar{Y}_{C,m} - \bar{Y}_{D,p} + \bar{Y})$$

$$+ (\bar{Y}_{ABC,ijm} - \bar{Y}_{AB,ij} - \bar{Y}_{AC,im} - \bar{Y}_{BC,jm} + \bar{Y}_{A,i} + \bar{Y}_{B,j} + \bar{Y}_{C,m} - \bar{Y}) \qquad (4.2.37)$$

$$+ (\bar{Y}_{ACD,imp} - \bar{Y}_{AC,im} - \bar{Y}_{AD,ip} - \bar{Y}_{CD,mp} + \bar{Y}_{A,i} + \bar{Y}_{C,m} + \bar{Y}_{D,p} - \bar{Y})$$

$$+ (\bar{Y}_{BCD,jmp} - \bar{Y}_{BC,jm} - \bar{Y}_{BD,jp} - \bar{Y}_{CD,mp} + \bar{Y}_{B,j} + \bar{Y}_{C,m} + \bar{Y}_{D,p} - \bar{Y})$$

$$+ \ldots + (Y_{ijmpn} - \bar{Y}_{ABCD,ijmp})$$

4.3. Predictive Equation by Multifactorial DOE Regression

A regression model can be used to describe the relationship between a set of independent variables and the dependent variable. It is applicable to both continuous and categorical variables. Regression analysis produces a predictive equation where the coefficients represent the relationship between these independent and dependent variables. A predictive equation is useful to accomplish what-if scenarios that take the correlation to the transparent level. Although data cannot be collected at all levels and factors, the predictive equation formulated using these data can be utilized to estimate the response to any desired factorial variations within the test ranges. Nevertheless, any conclusion drawn outside the test ranges should be taken cautiously.

4.3.1. ANOVA and Regression

Regression and ANOVA are identical approaches except for the nature of the explanatory variables. Regression yields a statistical model that can be used to predict a continuous outcome on the basis of one or more "continuous predictor variables," while ANOVA is another statistical model that can be used to predict a continuous outcome on the basis of one or more "categorical predictor variables" such as $(-1, 0, 1)$. If the experimenter has a single categorical variable, and it only has two levels (e.g., a binary category variable such as -1 and 1), then the approach would be described as a DOE-t. A single categorical predictor variable with three or more levels or multiple categorical predictor variables with any number of levels would be considered an ANOVA model or DOE-F.

Regression and ANOVA may be combined on experiments by fitting a series of regression lines, this is called ANCOVA (Analysis of Covariance). The concept of partitioning variation into sums of squares (SS) in an ANOVA model provides a robust way to examine a complex regression model.

4.3.2. Least-Squares Regression for ANOVA with 3^3 Factor Design

Consider Equation (4.2.1) for a multifactorial regression for an experimental design of three factors. The corresponding predictive equation for main and interactive effects in response to three independent factors can be rewritten as [Rhinehart 2016]:

$$Y = \mu + e_A\,A + e_B\,B + e_C\,C + \lambda_{AB}\,e_A\,e_B\,A\,B + \lambda_{AC}\,e_A\,e_C\,A\,C + \lambda_{BC}\,e_B\,e_C\,B\,C$$

$$+ \lambda_{ABC}\,e_A\,e_B\,e_C\,A\,B\,C \tag{4.3.1a}$$

Geometrically, one may think of this as fitting a hypersurface with four variates, as $Y = f(A, B, C)$, based on a set of IJM points in that space via the least-squares criterion. Each piece of data collected from the experimental design can then be expressed as:

$$Y_{ijm} = \mu + e_A\,A_i + e_B\,B_j + e_C\,C_m + \lambda_{AB}\,e_A\,e_B\,A_i\,B_j + \lambda_{AC}\,e_A\,e_C\,A_i\,C_m$$

$$+ \lambda_{BC}\,e_B\,e_C\,B_j\,C_m + \lambda_{ABC}\,e_A\,e_B\,e_C\,A_i\,B_j\,C_m + \varepsilon_{ijm} \tag{4.3.1b}$$

or

$$\varepsilon_{ijm} = Y_{ijm} - [\mu + e_A\,A_i + e_B\,B_j + e_C\,C_m + \lambda_{AB}\,e_A\,e_B\,A_i\,B_j + \lambda_{AC}\,e_A\,e_C\,A_i\,C_m$$

$$+ \lambda_{BC}\,e_B\,e_C\,B_j\,C_m + \lambda_{ABC}\,e_A\,e_B\,e_C\,A_i\,B_j\,C_m] \tag{4.3.1c}$$

where
   μ is the mean
   e_A, e_B, and e_C are the main effects of factors A, B, and C, respectively
   λ_{AB} is the coefficient of correlation between factors A and B
   λ_{AC} is the coefficient of correlation between factors A and C
   λ_{BC} is the coefficient of correlation between factors B and C
   λ_{ABC} is the coefficient of correlation among factors A, B, and C
   ε_{ijm} is the sample random error
   i, j, and m are the applied levels of factors A, B, and C, respectively

Note that $1 \le i \le I$, $1 \le j \le J$, and $1 \le m \le M$, where I, J, and M are the total levels taken by factors A, B, and C, respectively. The goal of multifactorial regression is to find out the values of μ, e_A, e_B, e_C, λ_{AB}, λ_{AC}, λ_{BC} and λ_{ABC} that minimize error ε.

The standard deviation due to residual ε. i.e., random error, characterizes the variability around the regression curve, and thus, the smaller the ε value, the better the fit. The coefficients and related summary outputs in Equation (4.3.1) explain the dependence of the design factors being tested on the assumption that data used in the ANOVA model meet the following three prerequisites: normally distributed, having constant variances, and independent of the response value. The variance of the errors should be consistent for all observations. In other words, the variance does not change for each observation or for the specified range of observations. This recognition is known as homoscedasticity, which means the same scattering pattern. If the variance differs from an observation to another, it is a case study on heteroscedasticity and then ANOVA based on the F-distribution is not applicable.

The least-squares method is frequently employed to fulfill the mathematical regression analysis that finds the "geometry of the best fit" for a set of data by "least-squares" as shown in Figure 4.3.1. The concept of least-squares is based on two independent variables (factors) that the distance between the data points to the fitted plane is to be minimized. The linear square plane turns into a 3-D surface if higher-order terms (quadratic or above) are used. Each point of data is representative of the relationship between known independent variables and the unknown dependent variable. Taking the sum of squares of the residuals, i.e., Equation (4.3.1), leads to the following equation:

$$L = \sum_{i=1}^{I}\ \sum_{j=1}^{J}\ \sum_{m=1}^{M}\ \varepsilon_{ijm}^{2} \tag{4.3.2}$$

The least-squares method is then used to minimize the above equation, in which the ε_{ijm} consists of random disturbances with zero mean, i.e., $E[\varepsilon_{ijm}] = 0$.

FIGURE 4.3.1   Geometric interpretation of least-squares properties of for $y = f(x_1, x_2)$.

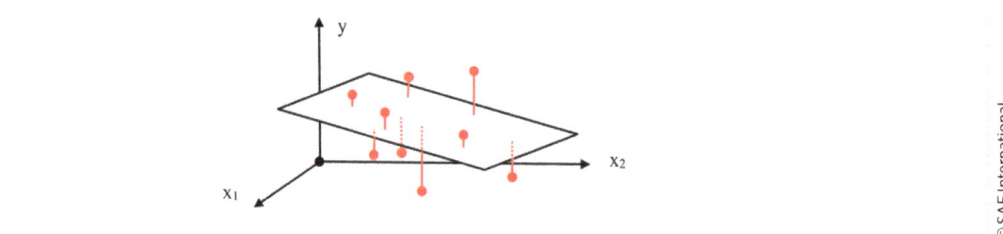

Considering Equations (4.3.1)–(4.3.2), the following equation is proposed to approximate the above equation based on the test data,

$$L = \sum_{i=1}^{I} \sum_{j=1}^{J} \sum_{m=1}^{M} [Y_{ijm} - (\mu + e_A A_i + e_B B_j + e_C C_m + \lambda_{AB} e_A e_B A_i B_j + \lambda_{AC} e_A e_C A_i C_m + \lambda_{BC} e_B e_C B_j C_m + \lambda_{ABC} e_A e_B e_C A_i B_j C_m)]^2 \tag{4.3.3}$$

The algorithm is used to minimize the sum of squares, i.e., L, in search of the following unknown parameters and sample mean,

1. Sample mean μ
2. Three main effects: e_A, e_B, and e_C
3. Three coefficients of correlation for two-factor interactions: λ_{AB}, λ_{AC}, and λ_{BC}
4. One coefficient of correlation for the three-factor interaction: λ_{ABC}

Taking partial differentiations with respect to these eight unknown variables and setting each of them to zero lead to the following eight equations:

$$\partial L / \partial \mu = 0 \tag{4.3.4}$$

$$\partial L / \partial e_A = 0 \tag{4.3.5}$$

$$\partial L / \partial e_B = 0 \tag{4.3.6}$$

$$\partial L / \partial e_C = 0 \tag{4.3.7}$$

$$\partial L / \partial \lambda_{AB} = 0 \tag{4.3.8}$$

$$\partial L / \partial \lambda_{AC} = 0 \tag{4.3.9}$$

$$\partial L / \partial \lambda_{BC} = 0 \tag{4.3.10}$$

© SAE International.

and

$$\partial L \, / \, \partial \lambda_{ABC} = 0 \tag{4.3.11}$$

The above eight simultaneous equations can be solved for the eight unknown mean and coefficients as follows:

$$\mu = \bar{Y} \tag{4.3.12}$$

$$e_A = \frac{1}{I} \sum_{i=1}^{I} (\bar{Y}_{A,i} - \bar{Y}) \tag{4.3.13}$$

$$e_B = \frac{1}{J} \sum_{j=1}^{J} (\bar{Y}_{B,j} - \bar{Y}) \tag{4.3.14}$$

$$e_C = \frac{1}{M} \sum_{m=1}^{M} (\bar{Y}_{C,i} - \bar{Y}) \tag{4.3.15}$$

$$\lambda_{AB} = \frac{\displaystyle\sum_{i=1}^{I} \sum_{j=1}^{J} \sum_{m=1}^{M} [(\bar{Y}_{A,i} - \bar{Y})(\bar{Y}_{B,j} - \bar{Y})\, \bar{Y}_{ABC,ijm}]}{M \displaystyle\sum_{i=1}^{I} \sum_{j=1}^{J} [(\bar{Y}_{A,i} - \bar{Y})^2 (\bar{Y}_{B,j} - \bar{Y})^2]} \tag{4.3.16}$$

$$\lambda_{AC} = \frac{\displaystyle\sum_{i=1}^{I} \sum_{j=1}^{J} \sum_{m=1}^{M} [(\bar{Y}_{A,i} - \bar{Y})(\bar{Y}_{C,m} - \bar{Y})\, \bar{Y}_{ABC,ijm}]}{J \displaystyle\sum_{i=1}^{I} \sum_{m=1}^{M} [(\bar{Y}_{A,i} - \bar{Y})^2 (\bar{Y}_{C,m} - \bar{Y})^2]} \tag{4.3.17}$$

$$\lambda_{BC} = \frac{\displaystyle\sum_{i=1}^{I} \sum_{j=1}^{J} \sum_{m=1}^{M} [(\bar{Y}_{B,j} - \bar{Y})(\bar{Y}_{C,m} - \bar{Y})\, \bar{Y}_{ABC,ijm}]}{I \displaystyle\sum_{j=1}^{J} \sum_{m=1}^{M} [(\bar{Y}_{B,j} - \bar{Y})^2 (\bar{Y}_{C,m} - \bar{Y})^2]} \tag{4.3.18}$$

and

$$\lambda_{ABC} = \frac{\displaystyle\sum_{i=1}^{I} \sum_{j=1}^{J} \sum_{m=1}^{M} [(\bar{Y}_{A,i} - \bar{Y})(\bar{Y}_{B,j} - \bar{Y})(\bar{Y}_{C,m} - \bar{Y})\, \bar{Y}_{ABC,ijm}]}{\displaystyle\sum_{i=1}^{I} \sum_{j=1}^{J} \sum_{m=1}^{M} [(\bar{Y}_{A,i} - \bar{Y})^2 (\bar{Y}_{B,j} - \bar{Y})^2 (\bar{Y}_{C,m} - \bar{Y})^2]} \tag{4.3.19}$$

For example, considering the design matrix 3^3 with no replication, one will have 27 DOFs (from 27 observations) to account for the above eight unknowns plus the residual. If replications are considered as usually encouraged, the residuals resulting from Equation (4.3.1) have to be revised as $\varepsilon_{ijmn} = Y_{ijmn} - \bar{Y}_{ABC,ijm}$. Equations (4.3.12)–(4.3.19) given above can be applied to any DOE with three factors, of which each may have an arbitrary number of levels.

4.3.3. Decomposition of Interactions in Factorial Design 3^3

Note that SS_{AB} given by Equation (4.2.19) can be further divided into two components [Wu and Hamada 2009]. One is the sum of squares of the direct interaction between factor and factor B (i.e., A×B) and the other is the interaction between factor A and pseudo-factor B^2 (i.e., A×B^2):

$$SS_{AB} = SS_{AXB} + SS_{AXB^2} \tag{4.3.20}$$

Similarly,

$$SS_{AC} = SS_{AXC} + SS_{AXC^2} \tag{4.3.21}$$

and

$$SS_{BC} = SS_{BXC} + SS_{BXC^2} \tag{4.3.22}$$

The above three equations are given without mathematical derivations. In other words, each of the two-factor interactions of factorial design 3^3 can be further decomposed into two parts as:

$$AB \rightarrow AxB \text{ and } AxB^2 \tag{4.3.23}$$

$$AC \rightarrow AxC \text{ and } AxC^2 \tag{4.3.24}$$

and

$$BC \rightarrow BxC \text{ and } BxC^2 \tag{4.3.25}$$

where B^2 and C^2 are quadratic terms of factor B and C, respectively. It means that each of the two-factor interactions consists of two components, of which each has two DOFs.

Nevertheless, interaction A×B (AC or BC) is usually more statistically significant than A×B^2 (A×C^2 or B×C^2) for general cases, since the opportunity for A×B^2 to occur is less than A×B. It is evidenced by the data exhibited in Table 4.2.5. This is theorized according to the hierarchical ordering principle [Li et al. 2006], also called the sparsity effect principle that given numerous variables in a factorial experiment, only a few will likely be significant—typically main effects and two-way interactions. Note that A×B^2, A×C^2, and B×C^2 are regarded as three-way interactions, although they are two-factor interactions.

Furthermore, the three-factor interaction can be decomposed into four components:

$$ABC \rightarrow A \times B \times C,\ A \times B^2 \times C,\ A \times B \times C^2,\ and\ A \times B^2 \times C^2 \qquad (4.3.26)$$

It means that the three-factor interaction consists of four components, each of which has two DOFs. Again, the sparsity effect principal reigns. The occurrence of interaction $A \times B^2 \times C$, $A \times B \times C^2$, or $A \times B^2 \times C^2$ is less possible than $A \times B \times C$, as evidenced by the data exhibited in Table 4.2.5.

The decomposition of the interactive terms for full factorial design 3^3, derived from modulus 3 calculus based on (0, 1, 2) system using Equations (4.3.23)–(4.3.26), without individual main effects (i.e., A, B, and C) and quadratic terms (i.e., A^2, B^2, and C^2), is exhibited in Table 4.3.1. Modulus 3 calculus will be further addressed in Section 4.7.

TABLE 4.3.1 Decomposition of interactions for full factorial design 3^3 in (0, 1, 2) system.

Run	AB		AC		BC		ABC			
	A×B	A×B²	A×C	A×C²	B×C	B×C²	A×B×C	A×B×C²	A×B²×C	A×B²×C²
1	0	0	0	0	0	0	0	0	0	0
2	0	0	1	2	1	2	1	2	1	2
3	0	0	2	1	2	1	2	1	2	1
4	1	2	0	0	1	1	1	1	2	2
5	1	2	1	2	2	3	2	0	3	1
6	1	2	2	1	0	2	0	2	1	0
7	2	1	0	0	2	2	2	2	1	1
8	2	1	1	2	0	4	0	1	2	0
9	2	1	2	1	1	0	1	0	0	2
10	1	1	1	1	0	0	1	1	1	1
11	1	1	2	0	1	2	2	0	2	0
12	1	1	0	2	2	1	0	2	0	2
13	2	0	1	1	1	1	2	2	0	0
14	2	0	2	0	2	0	0	1	1	2
15	2	0	0	2	0	2	1	0	2	1
16	0	2	1	1	2	2	0	0	2	2
17	0	2	2	0	0	1	1	2	0	1
18	0	2	3	2	1	0	2	1	1	0
19	2	2	2	2	0	0	2	2	2	2
20	2	2	0	1	1	2	0	1	3	1
21	2	2	1	0	2	1	1	0	1	0
22	0	1	2	2	1	1	0	3	1	1
23	0	1	0	1	2	0	1	2	2	0
24	0	1	1	0	0	2	2	1	0	2
25	1	0	2	2	2	2	1	1	0	0
26	1	0	0	1	0	1	2	0	1	2
27	1	0	1	0	1	0	0	2	2	1

4.4. Generic Concept of Multiple Linear Regression

Multiple linear regression is the most useful form of regression analysis for the DOE. It is often employed to explain the relationship between one continuous output (dependent) variable and numerous input (independent) variables. The independent variables, which can be continuous or categorical, are also referred to as predictors or regressors.

The procedure for the least-squares formulation can be applied to the design of the experiment with an arbitrary number of factors. There are closed-form solutions to the unknown coefficients for a linear regression model, but it is not true for a nonlinear regression model. In contrast to a linear problem, a nonlinear least-squares problem may not have a closed-form solution and it is usually solved by iteration. However, a generic concept of linear regression for nonlinear models can be sought in a pseudo-linear way and it will be explained here using an example. Assume that it is sought to fit a quadratic function of two variables,

$$Y(A, B) = c_1 + c_2\, A + c_3\, B + c_4\, A^2 + c_5\, A\, B + c_6\, B^2 \qquad (4.4.1)$$

This is a set of trivariate data (i.e., Y, A, and B) by regression in a two-factor design. $Y(A, B)$ is a nonlinear polynomial with regard to independent variables A and B. Equation (4.4.1) can be solved using the Newton–Raphson method. However, it can also be simply solved in a linear way. Instead of thinking of Y as a quadratic function of variables A and B, one may conceptually rewrite Equation (4.4.1) as:

$$Y(A, B, A^2, AB, B^2) = c_1 + c_2\, A + c_3\, B + c_4\, A^2 + c_5\, A\, B + c_6\, B^2 \qquad (4.4.2)$$

This means that Y consists of a linear function of five pseudo-variables, i.e., A, B, A^2, AB, B^2. It turns out that the six unknown constants and coefficients, i.e., c_1, c_2, c_3, c_4, c_5, and c_6, may result from fitting a multivariate linear equation using the trivariate data (Y, A, B), as long as there are at least six data. The same notion obviously extends to a higher-degree polynomial that fits in with a set of multivariate data. For example, assuming that the experimenter is only interested in fitting a second-order polynomial to a 3^3 design, then:

$$Y(A, B, C, A^2, B^2, C^2, AB, AC, BC) = c_1 + c_2\, A + c_3\, B + c_4\, C$$
$$+ c_5\, A^2 + c_6\, B^2 + c_7\, C^2 + c_8\, A\, B + c_9\, A\, C + c_{10}\, B\, C \qquad (4.4.3)$$

Again, the 10 unknown constant and coefficients, i.e., $c_1, c_2, c_3, c_4, c_5, c_6, c_7, c_8, c_9$, and c_{10}, may result from fitting a multivariate linear equation using the available data (Y, A, B, C), as long as there are at least 10 data. The fulfillment of multiple linear regression is illustrated in Example 4.4.1 using full factorial design 2^2, whose predictive equation can also be derived from the DOE-t for 2^K.

EXAMPLE 4.4.1

Data resulting from an experimental design on "time to failure" are listed below:

Treatment	a	b	Y (days)
q = 1	−1	−1	27
q = 2	1	−1	25
q = 3	−1	1	50
q = 4	1	1	55

Find out the regression model using the least-squares method for the experimental design given above.

Solution:

There are two independent variables, i.e., A and B, and one dependent variable, Y. Their relationship can be formulated using the following relationship:

$$Y = c_1 + c_2\,A + c_3\,B + c_4\,A\,B$$

For the given data set, the curve-fitting by the generic concept of multiple linear regression can be obtained by minimizing the following sum of squares of sample errors:

$$E_r = \sum_{q=1}^{4} [Y_q - (c_1 + c_2\,A_q + c_3\,B_q + c_4\,A_q\,B_q)]^2$$

The first derivatives of E_r with respect to the four knowns are set to zeroes as:

$$\frac{\partial E_r}{\partial c_1} = 2 \sum_{q=1}^{4} [Y_q - (c_1 + c_2\,A_q + c_3\,B_q + c_4\,A_q\,B_q)] = 0$$

$$\frac{\partial E_r}{\partial c_2} = 2 \sum_{q=1}^{4} A_q\,[Y_q - (c_1 + c_2\,A_q + c_3\,B_q + c_4\,A_q\,B_q)] = 0$$

$$\frac{\partial E_r}{\partial c_3} = 2 \sum_{q=1}^{4} B_q\,[Y_q - (c_1 + c_2\,A_q + c_3\,B_q + c_4\,A_q\,B_q)] = 0$$

and $$\frac{\partial E_r}{\partial c_4} = 2 \sum_{q=1}^{4} A_q\,B_q\,[Y_q - (c_1 + c_2\,A_q + c_3\,B_q + c_4\,A_q\,B_q)] = 0$$

The above four equations can be written as:

$$4\,c_1 + (\textstyle\sum A_q)\,c_2 + (\textstyle\sum B_q)\,c_3 + (\textstyle\sum A_q\,B_q)\,c_4 = \textstyle\sum Y_q$$

$$(\textstyle\sum A_q)\,c_1 + (\textstyle\sum A_q{}^2)\,c_2 + (\textstyle\sum A_q\,B_q)\,c_3 + (\textstyle\sum A_q{}^2\,B_q)\,c_4 = \textstyle\sum A_q\,Y_q$$

$$(\textstyle\sum B_q)\,c_1 + (\textstyle\sum A_q\,B_q)\,c_2 + (\textstyle\sum B_q{}^2)\,c_3 + (\textstyle\sum A_q\,B_q{}^2)\,c_4 = \textstyle\sum B_q\,Y_q$$

and $$(\textstyle\sum A_q\,B_q)\,c_1 + (\textstyle\sum A_q{}^2\,B_q)\,c_2 + (\textstyle\sum A_q\,B_q{}^2)\,c_3 + (\textstyle\sum A_q{}^2\,B_q{}^2)\,c_4 = \textstyle\sum A_q\,B_q\,Y_q$$

These four equations can be rearranged in matrix notation as:

$$
\begin{bmatrix}
4 & \sum A_q & \sum B_q & \sum A_q B_q) \\
\sum A_q & \sum A_q^2 & \sum A_q B_q & \sum A_q^2 B_q \\
\sum B_q & \sum A_q B_q & \sum B_q^2 & \sum A_q B_q^2 \\
\sum A_q B_q & \sum A_q^2 B_q & \sum A_q B_q^2 & \sum A_q^2 B_q^2
\end{bmatrix}
\begin{Bmatrix} c_1 \\ c_2 \\ c_3 \\ c_4 \end{Bmatrix}
=
\begin{Bmatrix} \sum Y_q \\ \sum A_q Y_q \\ \sum B_q Y_q \\ \sum A_q B_q Y_q \end{Bmatrix}
$$

Substituting the test data that:

$$(A_1, B_1, Y_1) = (-1, -1, 27)$$
$$(A_2, B_2, Y_2) = (1, -1, 25)$$
$$(A_3, B_3, Y_3) = (-1, 1, 50)$$
$$(A_4, B_4, Y_4) = (1, 1, 55)$$

into the above equations, one has four simultaneous equations to be solved for four unknowns, as:

$$c_1 = 39.25$$
$$c_2 = 0.75$$
$$c_3 = 13.25$$
$$\text{and} \quad c_4 = 1.75$$

Thus, the predictive equation is:

$$Y_p = 39.25 + 0.75\,A + 13.25\,B + 1.75\,A\,B$$

The same predictive equation can be obtained using DOE-t method (Chapter 2). Note that if there are five replicates (N = 5) for each treatment, there will be 20 equations to be solved for these four unknowns.

The solution procedure for multiple linear regression based on the full factorial design 2^2 presented above can be extended to a general case. Consider a case that there are Q experimental treatments (runs) with no replication. Once a nonlinear regression is translated into a multiple linear regression, the equation can be expressed in matrix notation as follows:

$$\{Y\}_{Q\times1} = [u]_{Q\times V}\,\{c\}_{V\times1} + \{\varepsilon\}_{Q\times1} \tag{4.4.4}$$

i.e.
$$
\begin{Bmatrix} Y_1 \\ \cdots \\ Y_q \\ \cdots \\ Y_Q \end{Bmatrix}
=
\begin{bmatrix}
u_{11} & u_{12} & \cdots & u_{1V} \\
\cdots & \cdots & \cdots & \cdots \\
u_{q1} & u_{q2} & \cdots & u_{qV} \\
\cdots & \cdots & \cdots & \cdots \\
u_{Q1} & u_{Q2} & \cdots & u_{QV}
\end{bmatrix}
\begin{Bmatrix} c_1 \\ \cdots \\ c_q \\ \cdots \\ c_V \end{Bmatrix}
+
\begin{Bmatrix} \varepsilon_1 \\ \cdots \\ \varepsilon_q \\ \cdots \\ \varepsilon_Q \end{Bmatrix}
\tag{4.4.5}
$$

where
$\{Y\}$ is the response vector, i.e., $\{Y_1, Y_2, \ldots, Y_Q\}^T$, data obtained from experimental tests
$\{c\}$ is the vector of formed unknowns, e.g., $\{c_1, c_2, c_3, c_4, c_5, c_6\}^T$ in Equation (4.4.4)
$\{\varepsilon\}$ is the vector of residuals, of which expectation $E(\{\varepsilon\}) = \{0\}$, i.e., sample random error
$[u]$ is the matrix that formed using test data, which relates $\{Y\}$ to $\{c\}$
Q is the number of runs with no replication in the DOE applied, for simplicity; $q = 1, 2, \ldots, Q$
V is the number of unknowns in the multiple linear equation; $v = 1, 2, \ldots, V$

All elements of matrix $[u]_{Q\times V}$ containing the observations on the explanatory variables in vector $\{c\}$ are nonstochastic. By default, $Q \geq V+1$, that will suffice to provide the equations to be solved for all the "V variables" and the constant (i.e., c_1). As the regression proceeds using least-squares method, the mission all comes down to minimize $\{\varepsilon\}^T\{\varepsilon\}$, which can be expressed as follows:

$$\{\varepsilon\}^T \{\varepsilon\} = \{\varepsilon\}^T_{1\times Q} \{\varepsilon\}_{Q\times 1} = \sum_{q=1}^{Q} \varepsilon_q^2 \tag{4.4.6}$$

Note that the following two stochastic properties must be met for using the least square estimate: (a) The diagonal terms of expectation $E[\{\varepsilon\}\{\varepsilon\}^T]$ must be consistent with σ_ε^2 and every one of its off-diagonal terms must be zero. Substituting Equation (4.4.4) into the above equation, one has:

$$\begin{aligned}\{\varepsilon\}^T \{\varepsilon\} &= [\{Y\}_{Q\times 1} - [u]_{Q\times V} \{c\}_{V\times 1}]^T [\{Y\}_{Q\times 1} - [u]_{Q\times V} \{c\}_{V\times 1}] \\ &= [\{Y\}^T_{1\times Q} - \{c\}^T_{1\times V} [u]^T_{V\times Q}] [\{Y\}_{Q\times 1} - [u]_{Q\times V} \{c\}_{V\times 1}] \\ &= \{Y\}^T_{1\times Q} \{Y\}_{Q\times 1} - 2 \{c\}^T_{1\times V} [u]^T_{V\times Q} \{Y\}_{Q\times 1} \\ &\quad + \{c\}^T_{1\times V} [u]^T_{V\times Q} [u]_{Q\times V} \{c\}_{V\times 1} \end{aligned} \tag{4.4.7}$$

Taking a partial differentiation of $\{\varepsilon\}^T\{\varepsilon\}$ with respect to the unknown coefficients $\{c\}$ and setting the result to zero yield:

$$\frac{\partial(\{\varepsilon\}^T \{\varepsilon\})}{\partial\{c\}} = -2 [u]^T_{V\times Q} \{Y\}_{Q\times 1} + 2 [u]^T_{V\times Q} [u]_{Q\times V} \{c\}_{V\times 1} = \{0\}_{V\times 1} \tag{4.4.8}$$

i.e.

$$([u]^T_{V\times Q} [u]_{Q\times V}) \{c\}_{V\times 1} = [u]^T_{V\times Q} \{Y\}_{Q\times 1} \tag{4.4.9}$$

As long as $([u]^T_{V\times Q} [u]_{Q\times V})$ is not singular, the above equation can be solved for the unknown constant and coefficients,

$$\{c\}_{V\times 1} = ([u]^T_{V\times Q} [u]_{Q\times V})^{-1} ([u]^T_{V\times Q} \{Y\}_{Q\times 1}) \tag{4.4.10}$$

Equation (4.4.9) can also be solved by inverting $([u]^T_{V\times Q} [u]_{Q\times V})$ directly or some other methods. Equation (4.4.10) is the classical formula for the least-squares estimate in matrix notation, while Equations (4.3.12)–(4.3.19) are the explicit solutions to the special case of eight distinct scalar unknowns. The next step is to do the proof of minimum by taking the particle differentiation of $\{\varepsilon\}^T\{\varepsilon\}$ twice,

$$\frac{\partial^2(\{\varepsilon\}^T \{\varepsilon\})}{\partial\{c\} \, \partial\{c\}^T} = 2 [u]^T_{V\times Q} [u]_{Q\times V} \tag{4.4.11}$$

The matrix derived above on the right side of the equal sign in the above equation is a positive definite matrix and thus the solution is the minimum of Equation (4.4.7). The least-squares estimate is an unbiased estimation method and the assumption of normality is not needed.

4.4.1. Response Surface Regression by Linear and Quadratic Terms

Test results from an experimental design are supposedly to be analyzed using response surface regression for the follow-up optimization [Box and Hunter 1957]. Without taking block effects into consideration, the relationship between the response and input variables is obtained by fitting them into the second-order polynomial equation as a general practice [Khuri and Cornell 2019]:

$$
\begin{aligned}
Y = {} & \gamma_0 + \gamma_1 x_1 + \gamma_2 x_2 + \gamma_3 x_3 + \gamma_{11} x_1^2 + \gamma_{22} x_2^2 + \gamma_{33} x_3^2 + \ldots + \beta_{KK} x_K^2 \\
& + \gamma_{12} x_1 x_2 + \gamma_{13} x_1 x_3 + \gamma_{23} x_2 x_3 + \ldots + \gamma_{1K} x_1 x_K \\
& + \gamma_{23} x_2 x_3 + \gamma_{24} x_2 x_4 + \ldots + \gamma_{2K} x_2 x_K \\
& + \text{higher-order terms} + \varepsilon
\end{aligned}
\tag{4.4.12}
$$

where subscript K is the number of factors. When $K = 2$ and $K = 3$, the above equation reduces to the following two equations, respectively:

$$
Y \approx \gamma_0 + \gamma_1 x_1 + \gamma_2 x_2 + \gamma_{11} x_1^2 + \gamma_{22} x_2^2 + \gamma_{12} x_1 x_2 + \varepsilon \quad (K = 2)
\tag{4.4.13}
$$

and

$$
\begin{aligned}
Y \approx {} & \gamma_0 + \gamma_1 x_1 + \gamma_2 x_2 + \gamma_3 x_3 + \gamma_{11} x_1^2 + \gamma_{22} x_2^2 + \gamma_{33} x_3^2 \\
& + \gamma_{12} x_1 x_2 + \gamma_{13} x_1 x_3 + \gamma_{23} x_2 x_3 + \varepsilon \quad (K = 3)
\end{aligned}
\tag{4.4.14}
$$

4.4.2. Example: Case Study on Shrinkage of Optical-Fiber Cable (Continued from Section 3.3.2)

The study of the influences of six potential factors on the contraction rate of optical-fiber communication cables using 2_{III}^{6-3} design, as exhibited in Tables 3.3.4 and 3.3.5, has shown that there are only two statistically significant factors. Their experimental conditions are rephrased below:

Factor	Low (−)	High (+)	Dimensionless variable
B (Operating temperature)	162°C	175°C	$b = (B - 168.5)/13$
D (Air-cooling distance)	100 mm	150 mm	$d = (D - 125)/50$

It is proposed to conduct a study using the 3^2 design to further explore nonlinear effects with the design matrix given in Table 4.4.1 [Jou et al. 2014]. Since there are two significant factors only, one may conduct multiple linear regressions based on Equation (4.4.13),

$$
Y \approx \gamma_0 + \gamma_1 B + \gamma_2 D + \gamma_{11} B^2 + \gamma_{22} D^2 + \gamma_{12} B D
$$

The regression model up to quadratic terms was done using the generic concept of linear regression for nonlinear models. How to interpret regression output? To answer questions using regression analysis, one needs to sequentially fit and verify that it is the best model by looking through the regression coefficients and related p-values.

TABLE 4.4.1 Optimization of injection molding of optical fibers using the 3^2 design [Jou et al. 2014].

Run	b	d	B	D	Y	Predicted	Residual
1	−1	−1	167	5	0.036	0.035957	0.000043
2	0	−1	173.5	5	0.032	0.033285	−0.001286
3	1	−1	180	5	0.041	0.042275	−0.001275
4	−1	0	167	17.5	0.028	0.026738	0.001262
5	0	0	173.5	17.5	0.026	0.024067	−0.001067
6	1	0	180	17.5	0.030	0.033056	−0.003056
7	−1	1	167	30	0.028	0.029082	−0.001082
8	0	1	173.5	30	0.026	0.02641	−0.000415
9	1	1	180	30	0.040	0.0354	0.0046

Notes:
B: Operating temperature
D: Air-cooling distance
Y: Contraction ratio

The coefficients represent the average change in the dependent variable given a one-unit change in the independent variable while controlling the other independent variables. When there is a low p-value, which is typically assumed to be less than 5% or 10%, the corresponding independent variable is statistically significant. In the first run using Minitab, it is found out that one two-factor interaction and one quadratic term are not effective since each of them has a p-value that is larger than the preset parting value of 10% in this case. If there are several insignificant effects, one shall remove one effect at a time; and rerun the model until the accountable effects in the model are significant. In this case, interaction BD has the worst p-value, one may drop the interactive term BD and rerun the analysis. Eventually, the predictive equation turns out to be:

$$y = 4.11 - 0.0474\ B - 0.001507\ D + 0.000138\ B^2 + 0.000037\ D^2 \qquad (4.4.15)$$

of which the correlation is 93.2% (R = 93.2%). The significance of each individual effect is shown in Table 4.4.2.

TABLE 4.4.2 Optimization of injection molding of optical fibers using the 3^2 design.

Variance	SS	DOF	MS	$F_{u,v}$	P-value
Regression	$SS_R = 234 \times 10^{-6}$	4	58×10^{-6}	$F_{4,4} = 6.66$	4.7%
B	67×10^{-6}	1	67×10^{-6}	$F_{1,4} = 7.60$	5.1%
D	87×10^{-6}	1	87×10^{-6}	$F_{1,4} = 9.89$	3.5%
B^2	68×10^{-6}	1	68×10^{-6}	$F_{1,4} = 7.75$	5.0%
D^2	68×10^{-6}	1	68×10^{-6}	$F_{1,4} = 7.75$	5.0%
Error	$SS_E = 35 \times 10^{-6}$	4	9×10^{-6}	—	—
Subtotal	$SS_T = 269 \times 10^{-6}$	8			
Grand ave.	SS_{ave}	1			
Total	SS	9			

Predicted values for individual treatments (run conditions) are calculated and listed in Table 4.4.1. The contraction rate as a function of the operating temperature and air-cooling distance is plotted in Figure 4.4.2. The model adequacy is proved by the plot of residuals versus the predicted values as given in Figure 4.4.3, which exhibits no correlation between the residuals and predicted values.

FIGURE 4.4.2   Response surface of concentration as a function of temperature and air-cooling distance.

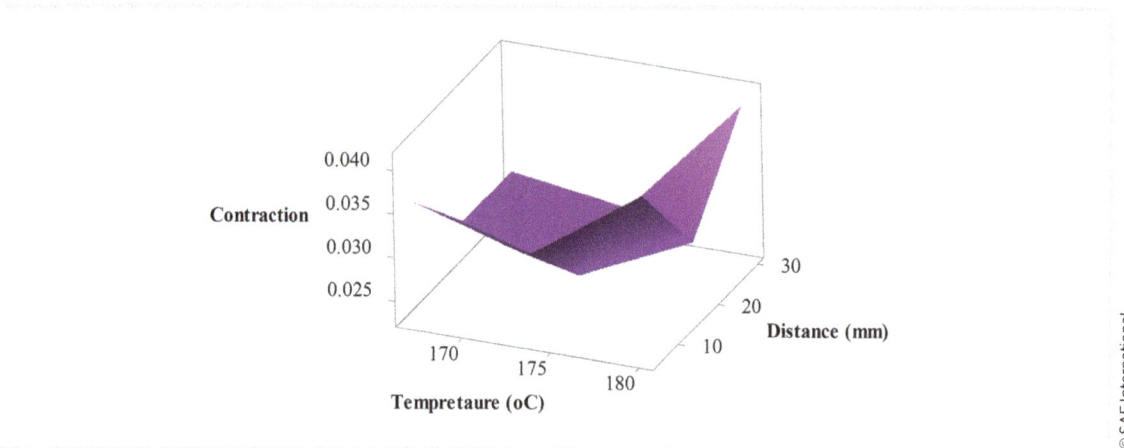

FIGURE 4.4.3   Diagnostic checking using residual plot—residuals versus predicted values.

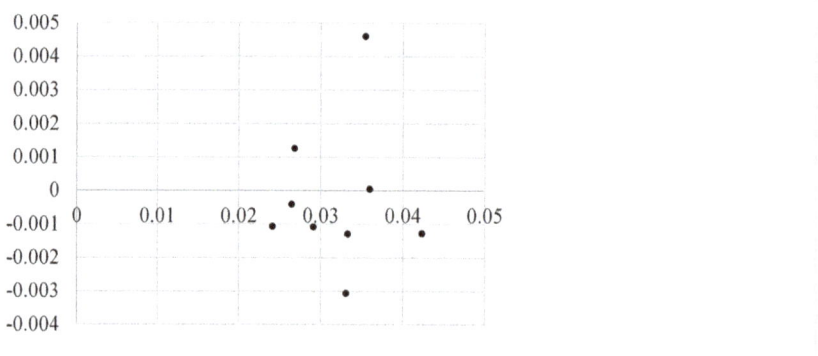

4.5. Checking Adequacy of Predictive Equations by Regression

Data resulting from all the treatments (run combinations) can be keyed into statistical software (e.g., Minitab) for calculating the coefficients given in Equation (4.4.12). The total sum of squares is the summation of the regression sum of squares and the residual sum of squares (also called error),

$$SS_T = SS_R + SS_E$$

$$\text{or} \quad SS_R = SS_T - SS_E$$

(4.5.1)

where

SS_R is the regression sum of squares, subscript R means regression

SS_E is the residual sum of squares, subscript E means error

4.5.1. Correlation and Adjusted Correlation

The significance of regression is demonstrated in Table 4.5.1. The residual sum of squares, i.e., SS_E, is a measurement of the amount of modeling accuracy of regression. The smaller the SS_E, the better the regression model. When both sides of Equation (4.5.1) are divided by SS_T, it is made dimensionless on both sides, as:

$$100\% = SS_R / SS_T + SS_E / SS_T = R^2 + (100\% - R^2) \tag{4.5.2}$$

where

$$R^2 \equiv SS_R / SS_T = 1 - SS_E / SS_T \tag{4.5.3}$$

of which R^2 is called the square of correlation R. R^2 can be used for checking the lack of adequacy. It is derived from the sample and is a positively biased estimate of the proportion of the variance of the dependent variable accounted for by the regression model. Another correlation factor is R_{adj}^2, called adjusted R^2, which corrects positive bias to provide a value that would be expected in the population. R_{adj}^2 is more reliable for checking the lack of adequacy in regression. Let N_R stand for the DOFs associated with regression. R_{adj}^2 is defined as the ratio of the residual mean square to the total mean square,

$$R_{adj}^2 \equiv 1 - (MS_E / MS_T) \tag{4.5.4}$$

Thus,

$$R_{adj}^2 = 100\% - \frac{SS_E / (N - N_R - 1)}{SS_T / (N - 1)} = 100\% - \frac{SS_E / (N_T - N_R)}{SS_T / N_T} \tag{4.5.5}$$

where
  MS_E is the residual mean square, also called mean square of error
  SS_T is the total mean square (corrected)
  N_T is the corrected total DOFs and $N_T = N - 1$
  N is the total DOFs
  N_R is the DOFs associated with regression

The corrected means that the "one" DOF corresponding to the "grand average" is excluded. In general, both $R^2 \geq 90\%$ and $R_{adj}^2 \geq 90\%$ are expected for justifying the regression accuracy. In light of Equations (4.5.3) and (4.5.5), R_{adj}^2 is related R^2 as:

$$R_{adj}^2 = 1 - (1 - R^2) \left(\frac{N_T - 1}{N_T - N_R - 1}\right) \tag{4.5.5a}$$

TABLE 4.5.1 Analysis of variance for significance of regression.

Source	SS	DOF	MS	$F_{u,v}$
Regression	SS_R	N_R	$MS_R = SS_R/N_R$	$F_{k, N-K-1} = MS_R/MS_E$
Error	SS_E	$N - N_R - 1$	$MS_E = SS_E/(N - N_R - 1)$	
Subtotal	SS_T	$N - 1$	$MS_T = SS_T/(N - 1)$	
Average	SS_{ave}	1		
Total	SS	N		

© SAE International.

Adjusted model correlation R_{adj}^2 is preferred, because model correlation R^2 always increases as more terms are added to the model.

4.5.2. Checking Lack of Fit

The accuracy of a DOE model is traditionally checked using the negative thinking such as "lack of fit." The lack of fit (LOF) of a multiple linear model, as well as a nonlinear model, occurs when the model does not adequately represent the mean response as a function of the factor level, especially when there is a complex curvature. The residual sum of squares SS_E can be partitioned into two components,

$$SS_E = SS_{PE} + SS_{LOF} \tag{4.5.6}$$

where

SS_{PE} is the sum of squares owing to the pure error
SS_{LOF} is the sum of squares subjected to the LOF

The LOF of a multilevel DOE model can be checked using replicated runs at the origin ($a = 0, b = 0, c = 0, \ldots$), which can be used to calculate the pure error. When the N_o replicated runs at the origin are to be used to estimate sample average at the origin, the sum of squares, and mean squares for the pure error are, respectively:

$$\bar{Y}_o = (\bar{Y}_{o,1} + \bar{Y}_{o,2} + \ldots + \bar{Y}_{o,No}) / N_o \tag{4.5.7}$$

$$SS_{PE} = \sum_{n=1}^{N_o} (\bar{Y}_{o,n} - \bar{Y}_o)^2 \tag{4.5.8}$$

and

$$MS_{PE} = [\sum_{n=1}^{N_o} (\bar{Y}_{o,n} - \bar{Y}_o)^2] / (N_o - 1) \tag{4.5.9}$$

of which N_o is the number of replications at the origin for checking the LOF.

This concept also applied to the two-level factorial design, the contrast between the resulting mean at the origin and the resulting mean of the factorial points provides a test for the "LOF" in a 2^K design. Consider the regression on the K main effects in full factorial design 2^K with no replication, but having N_o replicated runs at the origin. Let N_d represent the number of distinct responses that result from all the treatments. Then,

$$N_d = 2^K + 1 = (N - N_o) + 1 = N - (N_o - 1) \tag{4.5.10}$$

where

N_d is the number of distinct responses
K is the number of factors that render 2^K treatments
N is the total number of observations including replicated runs at the origin
N_o is the number of replicated runs at the origin

The $2^K + 1$ treatments are rendered by K factors at two levels each plus the treatment at the origin. The DOFs for individual components (i.e., error, pure error and LOF) are then

$$DOF_E = N - K - 1 \qquad \text{(for } 2^K \text{ design only)} \qquad (4.5.11)$$

$$DOF_{PE} = N_o - 1 = N - N_d \qquad (4.5.12)$$

and

$$DOF_{LOF} = DOF_E - DOF_{PE} = N_d - K - 1 \qquad (4.5.13)$$

For example, if three replicated runs are done at the origin (center point) for the 2^4 design, then $K = 4$, $N_o = 3$, $N = 16 + 3 = 19$, and $N_d = 16 + 1 = 17$ (2^4 runs + 1 at the origin). The mean squares for LOF are finally calculated as:

$$MS_{LOF} = SS_{LOF} / (N_d - K - 1) = (SS_E - SS_{PE}) / (N_d - K - 1) \qquad (4.5.14)$$

where
  N_d is the distinct design points
  K is the number of independent variables.

Thus, the null hypothesis for testing the LOF is tested out using the F-statistic that:

$$F_{Nd-K-1,\, No-1} = MS_{LOF} / MS_{PE} \qquad (4.5.15)$$

In summary, the ANOVA for the LOF of the 2^K design with N_o replicated runs at the origin is given in Table 4.5.2.

TABLE 4.5.2 Analysis of variance of LOF of the 2^K design with no replication, but having N_o replicated runs at the origin.

Variance	SS	DOF	MS	$F_{u,v}$
Lack of fit	SS_{LOF}	$N_d - K - 1$	$MS_{LOF} = SS_{LOF}/(N_d - K - 1)$	$F_{(Nd-K-1),\,(N-Nd)} = MS_{LOF}/MS_{PE}$
Pure Error	SS_{PE}	$N - N_d$	$MS_{PE} = SS_{PE}/N - N_d$	
Residual	SS_E	$N - K - 1$		

© SAE International.

4.6. Product Variability, Process Capability, and Gage R&R

Product variability refers to how much the "spread-out or dispersion" of a statistical distribution for a group of products is. There are two different cases of "not being good" as presented in Figure 4.6.1(a) and 4.6.1(b). The two frequently used measures of variability are bias (i.e., mean shift from the target (center of the circle) and dispersion (sample variance).

© SAE International.

FIGURE 4.6.1 Variability of measurements relative to the target value—biased and wider dispersion.

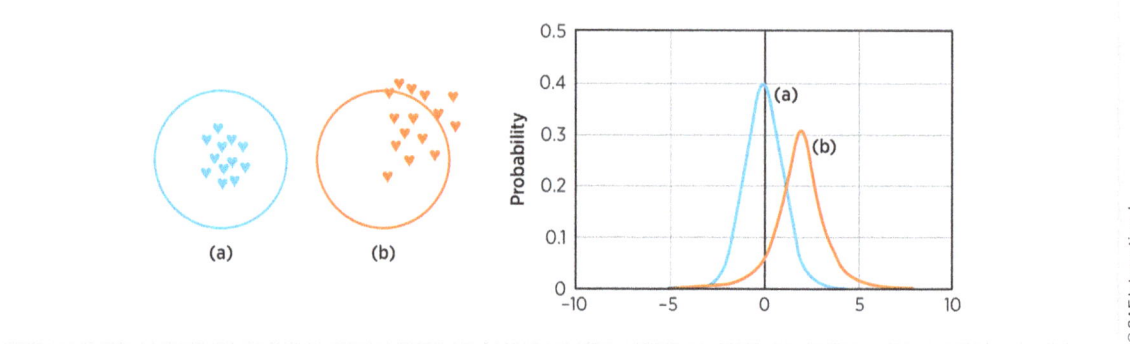

4.6.1. Measured Variability

The single biggest source of quality problems in plants, mills, and refineries is process variability. While typical causes of variability are often regarded as normal aspects of the operation, they often go undiagnosed. The process capability study [Gunter 1989a, 1989b] based on the DOE can be implemented to identify the root causes and resolve the issue.

Three major factors involved in a measurement are products (parts), operators, and measurement errors. The measurement error is usually assessed using the error of replication. Each observation (measured data) involved with these two factors (part and operator) can be formulated as:

$$Y_{ijn} = \bar{Y} + A_i + B_j + A_i B_j + \epsilon_{ijn} \qquad (4.6.1)$$

where

    $\bar{Y}$ is the grand mean of all observations
    A_i is the product (part) i
    B_j is the operator j
    ϵ_{ijn} is the random error of operator j on part i, at the nth measurement, $\epsilon_{ijn} - N(0, \sigma\epsilon)$

The variances corresponding to Equation (4.6.1) are then

$$\sigma_Y^2 = \sigma_{AA}^2 + \sigma_{BB}^2 + \sigma_{AB}^2 + \sigma_\epsilon^2 \qquad (4.6.2)$$

where

    σ_Y^2 is the variance of observations
    σ_{AA}^2 is the variance due to the main effect of product (parts), namely $\sigma^2_{Product}$
    σ_{BB}^2 is the variance due to the main effect of operator
    σ_{AB}^2 is the covariance due to the interaction between part and operator
    σ_ϵ^2 is the variance due to measurement error, and thus σ_ϵ is sample standard deviation

Assume that the operators are so well-trained that their operations are independent of parts and therefore $\sigma_{AB}^2 = 0$. Then, an observed variation in measurement consists of two major components only, of which one is the product variation in itself and the other is gage variation due to mainly the operator error, as:

$$\sigma_Y^2 = \sigma_{AA}^2 + \sigma_{BB}^2$$

i.e. $\sigma^2_{Observed} = \sigma^2_{Product} + \sigma^2_{Gage} \qquad (4.6.3)$

where

$\sigma^2_{Observed}$ is the observed variability

$\sigma^2_{Product}$ is the product variability, e.g., due to production tools

σ^2_{Gage} is the variability due to the operator's error

The gage variance can be further decomposed into two categories: repeatability and reproducibility of measurement.

$$\sigma^2_{Gage} = \sigma^2_{Repeatability} + \sigma^2_{Reproducibility} \qquad (4.6.4)$$

Repeatability is the uncertainty among replications of a certain number of measurements of a given part made by the same operator. Since part and operator are fixed, the variance component for the repeatability is the random error, i.e., σ^2_ε. The repeatability of measurement indicates the precision of the measurement facility including gages.

When different operators, instruments, or laboratories measure the same or replicate part, the variation induced is called reproducibility, distinguishing the operator's capability of using the gages. In other words, the variances embedded in Equation (4.6.1) can be written as [Searle et al. 1992]:

$$\sigma^2_{Observed} = \sigma^2_{Product} + \sigma^2_{Repeatability} + \sigma^2_{Reproducibility} \qquad (4.6.5)$$

Therefore, there are four sources of variation that an experimenter would primarily follow: repeatability (random error), reproducibility (operating error), product (part), and the observed (i.e., the total).

4.6.2. Process Capability Indices

Various process capability indices (PCIs) such as C_p, C_{pk}, C_{pm}, and P_{pk} have been in use to measure process capabilities quantitatively, including potential and performance. Process capability refers to the statistical position of the normal distribution compared to the product or process specification. Let UCL (upper control limit) and LCL (lower control limit) be the two limits specified for a variable in the drawing and S be the standard deviation measured. The difference between the UCL and the LCL is called tolerance band and the customary standard error range, 6S, is called manufacturing precision. The ratio of the specified tolerance band to the customary standard error range forms a process capability index [Kane 1986], namely C_p, i.e.,

$$C_P \equiv \frac{UCL - LCL}{6\,S} \qquad (4.6.6)$$

where

UCL is the upper control limit

LCL is the lower control limit

S is the sample standard deviation of the process

Nevertheless, the statistical distribution function for the variable may not be perfectly normally distributed. To account for a "abnormal" process, such as having a potential mean shift and/or asymmetric probability density function, an improved process capability index, called C_{pk}, is defined as [Kane 1986]:

$$C_{pk} \equiv \frac{\min (UCL - \mu, \mu - LCL)}{3\,S} \qquad (4.6.7)$$

where

> min is the minimum value of the two values in the parenthesis
> μ is the central tendency or process mean

PCIs C_p and C_{pk} have been in use in the US automotive industry such as Ford Motor Company and General Motors and in Japan. At least, a manufacturing process would have to demonstrate the potential to meet the specification only if $C_{pk} \geq 1$. When $C_{pk} = 1$, a process is rated to be barely capable [Hoffman 1993]. According to AIAG production part approval process (PPAP) (4th edition), the process capability for an automotive manufacturing process in control is divided into the following two categories:

(a) Requirement for preliminary process capability: A $C_{pk} \geq 1.67$ is mandatory for preliminary results; applied to a new production process (less than 30 production days), or it may be a chronically unstable process.

(b) Requirement for stable processes with normally distributed data: A $C_{pk} \geq 1.33$ should be achieved while $C_{pk} \geq 1.67$ is desired. 100% inspection is required for a process having $1.67 \geq C_{pk} \geq 1.33$, unless it is superseded by a control plan. If $C_{pk} < 1.33$, it is not a qualified supplier.

It is shown that a test for variability is equivalent to a test for process capability [Kane 1986]. Special causes have to be identified and corrected for a process having gone out of control regardless of its C_{pk} value. General test methodologies in practice in the automotive industry [AIAG 2010] are given in Table 4.6.1. For a normally distributed manufacturing process, defective parts per million (PPM) of parts produced is related to C_{pk} as shown in Table 4.6.2. The ±5σ (C_{pk} = 1.67) production precision is widely accepted by the automotive industry, although the ±6σ theory (C_{pk} = 2) indicates how to reach the goal of making the design and production process perfect. The ±6σ process capability may not be achieved in practice, but the production process will be improved as manufacturers keep on looking for innovative ways to re-engineer the product and process to get there.

TABLE 4.6.1 General test methodologies in practice in the automotive industry [AIAG 2010].

Phase	Category	Methodology	Acceptance criteria
DV (Design Verification)	Functional	Statistical tolerance limits	$C_{pk} \geq 1.33$
	Environmental	Statistical tolerance limits	$C_{pk} \geq 1.33$
	Durability	Accelerated testing	(R, C, N) trio
	Abuse test	Test to bogey	No failure prior to specified values
PV (Process Validation)	Functional	Statistical tolerance limits	$C_{pk} \geq 1.67$
	Environmental	Statistical tolerance limits	$C_{pk} \geq 1.67$
	Durability	Accelerated testing	(R, C, N) trio
	Abuse test	Test to bogey	No failure prior to specified values
CI & CC	Characteristics	Control plan	$C_{pk} \geq 1.67$

© SAE International.

Note: The C_{pk} acceptance criteria (TS 2.2.11.3) mentioned above are based on the normality and a two-sided specification (target in the center).

TABLE 4.6.2 Relationship of C_{pk} to (UCL/LCL) and PPM based on normal distribution.

C_{pk}	UCL/LCL	Range	PPM	Defective
0.10	± 0.3-sigma	0.6-sigma	76,420	76.4%
0.50	± 1.5-sigma	3-sigma	13,400	13.4%
0.8583	± 2.575-sigma	5.15-sigma	10,000	1%
1.00	*± 3-sigma*	*6-sigma*	*2700*	*0.27%*
1.33	*± 4-sigma*	*8-sigma*	*63.3*	*63.3 × 10^{-6}*
1.50	± 4.5-sigma	9-sigma	6.80	6.8 × 10^{-6}
1.67	*± 5-sigma*	*10-sigma*	*0.573*	*0.573 × 10^{-6}*
2.00	± 6-sigma	12-sigma	0.00197	1.97 × 10^{-9}
3.00	± 9-sigma	18-sigma	0	2.26 × 10^{-17}

Index P_{pk} is an alias index of C_{pk} but more sources of variation are considered in the process over a longer period of time. In other words, C_{pk} builds a robust process and P_{pk} leads to a reliable process. P_{pk} refers to a process capability study based on a historical analysis rather than a predictive analysis, used to evaluate process improvements [Cheng 1994]. When the process is stable, $P_{pk} = C_{pk}$, i.e., the actual performance will match the predicted potential performance. If the process is unstable, e.g., there is a shift or drift over time, $P_{pk} < C_{pk}$.

Similar to C_{pk}, C_{pm} is an index that takes into account the influence of the mean shift into the denominator and proves to be more statistically robust when dealing with a departure of the process mean μ from the process target [Chan et al. 1998, Phillips 1994]:

$$C_{pm} \equiv \frac{UCL - LCL}{6 [S^2 + (\mu - \mu_T)^2]^{1/2}} = \frac{C_p}{\{1 + [(\mu - \mu_T)/ S]^2\}^{1/2}} \tag{4.6.8}$$

of which μ_T is the process target, e.g., nominal value of a dimension. In light of indices C_{pk} and C_{pm}, another process index is proposed as [Pearn et al. 1992]:

$$C_{pkm} \equiv \min\left\{\frac{UCL - \mu}{6 [S^2 + (\mu - \mu_T)^2]^{1/2}}, \frac{\mu - LCL}{6 [S^2 + (\mu - \mu_T)^2]^{1/2}}\right\} \tag{4.6.9}$$

4.6.3. Sample Size Determination for the Estimation of PCIs

It is imperative to find the smallest sample size N such that the lower confidence bound for the ratio of the true capability index of interest to its estimated value. Let p-value α = 1 – C, where C is the desired confidence level. Then, the equations for the lower confidence bound of the capability indices are shown below [Wu and Kuo 2004, Statgraphics 2013]:

$$C_p: \left(\frac{\chi^2_{\alpha,N-1}}{N - 1}\right)^{1/2}$$

where

$\chi^2_{\alpha, N-1}$ is the value evaluated with the cumulative chi-square distribution (detailed in Section 8.5, Volume II)

$N - 1$ is the DOFs

α is the p-value as desired, e.g., 5%, 10%, etc.

$$C_{pk}: \quad 1 - \frac{z_\alpha}{\hat{C}_{pk}} \left[\frac{1}{9\,N} + \frac{\hat{C}_{pk}^{\,2}}{2\,(N-1)} \right]^{1/2}$$

where

z_α is the value at which the cumulative standard normal distribution equals $1 - \alpha$
$\hat{C}_{pk}$ is the estimated C_{pk}, e.g., 1.33, 1.67, etc.

$$C_{pm}: \quad \left\{ \frac{N\,\chi^2_{\alpha,N-1}}{N-1} \left[\frac{1 + 2\,(\mu - \mu_T)^2}{N\,[1 + (\mu - \mu_T)^2]^2} \right] \right\}^{1/2}$$

where

μ is the mean value as calculated from sample data
μ_T is the target mean value

4.6.4. ANOVA for Two Factors

Although factorial design with two factors (e.g., L^2 or $L_A^1 L_B^1$) appears to be simple, it is very useful. The ANOVA for two-factor designs described here is applicable to a factorial design of any level. Each experimental outcome can be expressed mathematically as:

$$Y_{ij} = \mu + A_i + B_j + A_i B_j + \epsilon_{ij} \qquad (i = 1, 2, ..., I;\ j = 1, 2, ..., J) \tag{4.6.10}$$

With replications, the equation for an experimental outcome (Y_{ijn}) can be broken into the following components:

$$Y_{ijn} = \bar{Y} + (\bar{Y}_{A,i} - \bar{Y}) + (\bar{Y}_{B,j} - \bar{Y}) + (\bar{Y}_{AB,ij} - \bar{Y}_{A,i} - \bar{Y}_{B,j} + \bar{Y}) + (Y_{ijn} - \bar{Y}_{AB,ij}) \tag{4.6.11}$$

i.e.,

$$Y_{ijn} - \bar{Y} = (\bar{Y}_{A,i} - \bar{Y}) + (\bar{Y}_{B,j} - \bar{Y}) + (\bar{Y}_{AB,ij} - \bar{Y}_{A,i} - \bar{Y}_{B,j} + \bar{Y}) + (Y_{ijn} - \bar{Y}_{AB,ij}) \tag{4.6.12}$$

where

$\bar{Y}$ is the grand average
$\bar{Y}_{A,i} - \bar{Y}$ is the main effect of A at level i, i = 1, 2, ..., I
I is the number of levels of factor A
$\bar{Y}_{B,j} - \bar{Y}$ is the main effect of B at level j, j = 1, 2, ..., J
J is the number of levels of factor B
n is the index of replication, n = 1, 2, ..., N
N is the number of replications
$\bar{Y}_{AB,ij} - \bar{Y}_{A,i} - \bar{Y}_{B,j} + \bar{Y}$ is the interactive effect of A at level i and B at level j
$Y_{ijn} - \bar{Y}_{AB,ij}$ is the residual, also called error or error effect generated by replication

Each average (sample mean) given above is herein represented by the averaged value for each repeated measurement corresponding to the individually identifiable main factor or two-factor interaction. The averages (sample means) of different combinations of factors A and B given above are calculated as follows:

$$\bar{Y} = \left[\sum_{i=1}^{I} \sum_{j=1}^{J} \sum_{n=1}^{N} Y_{ijn} \right] / (I\,J\,N) \qquad (4.6.13)$$

$$\bar{Y}_{A,i} = \left[\sum_{j=1}^{J} \sum_{n=1}^{N} Y_{ijn} \right] / (J\,N) \qquad (4.6.14)$$

$$\bar{Y}_{B,j} = \left[\sum_{i=1}^{I} \sum_{n=1}^{N} Y_{ijn} \right] / (I\,N) \qquad (4.6.15)$$

and

$$\bar{Y}_{AB,ij} = \left[\sum_{n=1}^{N} Y_{ijn} \right] / N \qquad (4.6.16)$$

If the null hypothesis is true, all the treatment combinations should yield the same result numerically. Thus, $\bar{Y}$ is a fair estimate of every experimental outcome and all the effects are attributed to the measured random variation. Assume that the variance (or standard deviation) of each measurement is the same. The remaining $\varepsilon = Y_{ijmn} - \bar{Y}_{ABC,ijm}$ is supposed to be the error as long as the "gage repeatability" persists. It means that the data from the process at hand behave as if they come from random drawings and have the same statistical distribution with a fixed location and fixed variation [Montgomery and Runger 1993a, 1993b]. An ANOVA for two-factor design is given in Table 4.6.3.

TABLE 4.6.3 ANOVA for two-factor design.

Variance	SS	DOF	MS	F$_{u,v}$
Regression	SS$_R$	I J – 1		
Factor A	SS$_A$	I – 1	MS$_A$ = SS$_A$/(I – 1)	F$_{(I-1),\,(N-1)\,IJ}$ = MS$_A$/MS$_E$
Factor B	SS$_B$	J – 1	MS$_B$ = SS$_B$/(J – 1)	F$_{(J-1),\,(N-1)\,IJ}$ = MS$_B$/MS$_E$
AB	SS$_{AB}$	(I – 1) (J – 1)	MS$_{AB}$ = SS$_{AB}$/[(I – 1) (J – 1)]	F$_{(I-1)\,(J-1),\,(N-1)\,IJ}$ = MS$_{AB}$/MS$_E$
Error	SS$_E$	I J (N – 1)	MS$_E$ = SS$_E$/[(N – 1) I J]	—
Subtotal	SS$_T$	I J N – 1	—	—
Grand Ave.	SS$_{ave}$	1	—	—
Total	SS	I J N	—	—

The sums of squares that have contributions to the sum of squares of the least-squares regression (SS$_R$) and the total sum of squares without accounting for the average (SS$_T$) based on Equation (4.6.12) can be summarized as, respectively:

$$SS_R = SS_A + SS_B + SS_{AB} \qquad (4.6.17)$$

and

$$SS_T = SS_A + SS_B + SS_{AB} + SS_E \tag{4.6.18}$$

where

$$SS_T = \sum_{i=1}^{I} \sum_{j=1}^{J} \sum_{n=1}^{N} (Y_{ijn} - \bar{Y})^2 \tag{4.6.19}$$

$$SS_A = J N \sum_{i=1}^{I} (\bar{Y}_{A,i} - \bar{Y})^2 \tag{4.6.20}$$

$$SS_B = I N \sum_{j=1}^{J} (\bar{Y}_{B,j} - \bar{Y})^2 \tag{4.6.21}$$

$$SS_{AB} = N \sum_{i=1}^{I} \sum_{j=1}^{J} (\bar{Y}_{AB,ij} - \bar{Y}_{A,i} - \bar{Y}_{B,j} + \bar{Y})^2 \tag{4.6.22}$$

and

$$SS_E = \sum_{i=1}^{I} \sum_{j=1}^{J} \sum_{n=1}^{N} (Y_{ijn} - \bar{Y}_{AB,ij})^2 \tag{4.6.23}$$

The sum of squares attributed to the grand average (SS_{ave}) and the grand total sum of squares (SS) can be calculated as, respectively:

$$SS_{Ave} = I J N \bar{Y}^2 \tag{4.6.24}$$

$$SS = SS_{Ave} + SS_T \tag{4.6.25}$$

4.6.5. Gage R&R Based on ANOVA

The problem of measurement error is so widely recognized that many different solutions in various fields of interest have been proposed [Wheeler 2013]. One solution that has been widely promoted in the automotive world and beyond is known as a Gauge R&R (Gage Repeatability and Reproducibility) study. It is an important tool for practicing the Six-Sigma methodology. The traditional indices to obtain the measurement error are its repeatability and reproducibility, namely gage R&R, that can be assessed using the ANOVA as a full factorial design of two factors. Examples based on factorial design $L_A{}^1 L_B{}^1$ using ANOVA for computing gage R&R will be given herein as a demonstration.

Repeatability measures the amount of variability induced in each measurement by the measurement system itself, and compares it to the total variability observed to determine the viability of the measurement system, including the gage or instrument itself and all mounting blocks, supports, fixtures, load cells, etc. [Burdick et al. 2005]. Reproducibility quantifies the ability of an operator to conduct the proper experimental test in each measurement, e.g., how the operator follows written or verbal instructions. The variance of gage R&R is the sum of the variance of repeatability and variance of reproducibility, as shown by Equation (4.6.4).

In practice, a ratio of the variance of gage R&R to the total variance in an individual measurement, denoted by $M_{R\&R}$, is defined as the measurement capability, denoted by $M_{R\&R}$. In combining the findings [Wheeler 2009, 2013] and doctrines [AIAG 2010], the sampling acceptance of an individual measurement is described as follows:

1. If $M_{R\&R} < 10\%$, the measurement system and operators are fully capable of determining whether a measured part meets the tolerance specification.
2. If $10\% < M_{R\&R} < 20\%$, the measurement system and operators are fairly acceptable.
3. If $20\% < M_{R\&R} < 30\%$, the measurement system and operators are marginally acceptable. Actions are needed for a further improvement.
4. If $30\% < M_{R\&R}$, the measurement system and operators are not acceptable. Actions must be taken immediately.

Before measuring equipment is issued for use, it must pass the gage R&R requirement. If the supplier's measurement is untruthful, how can the buyer trust the data provided by the supplier? When a robust measurement system is in need, e.g., implementing statistical process control (SPC) or conducting designed experiments, gage R&R has to be examined in advance. Sample parts used for inspection using gage R&R must be production parts and they shall be randomly measured and ideally "anonymous" to the appraiser to prevent any bias. Gage R&R study is a requirement for a PPAP documentation package [AIAG 2006].

EXAMPLE 4.6.1

There are 10 rubber O-rings that are selected for examining the process capability for sealing of engine oil. The O-ring diameter is independently measured by two lab engineers (Operator 1 and Operator 2) at random with two replications and the recorded data are listed as follows:

Run	Part	Operator	Y_1	Y_2	Y_{ave}	S_i^2
1	1	1	21	20	20.5	0.5
2	2	1	24	23	23.5	0.5
3	3	1	20	21	20.5	0.5
4	4	1	27	27	27	0
5	5	1	19	18	185	0.5
6	6	1	23	21	22	2
7	7	1	22	21	21.5	0.5
8	8	1	19	17	18	2
9	9	1	24	23	23.5	0.5
10	10	1	25	23	24	2
11	1	2	19	21	20	2
12	2	2	23	24	23.5	0.5
13	3	2	20	22	21	2
14	4	2	27	28	27.5	0.5
15	5	2	18	21	19.5	4.5
16	6	2	23	22	22.5	0.5
17	7	2	22	20	21	2
18	8	2	19	18	18.5	0.5
19	9	2	24	24	24	0
20	10	2	24	25	24.5	0.5

Solution:

The ANOVA table for replicated two-way factorial arrangement without considering the interaction between factor A and factor B is calculated using Minitab as follows:

Variation	S. of squares	Degree of freedom	Mean square	F-ratio	$F_{u,v,10\%}$
A: Part	$SS_A = 270.1$	$\nu_A = I - 1 = 9$	$MS_A = 30.01$	$F_{9,29} = 193.4$	$F_{9,29,10\%} = 2.89$
B: Operator	$SS_B = 2.5$	$\nu_B = J - 1 = 1$	$MS_B = 2.5$	$F_{1,29} = 16.11$	$F_{1,29,10\%} = 1.86$
Error	$SS_E = 4.5$	$\nu_E = 29$	$MS_E = 0.1552$		
Subtotal	$SS_T = 277.1$	$\nu_T = 39$			
Grand ave.	SS_{ave}	$\nu_{ave} = 1$			
Total	SS	$\nu = 40$			

Note that 40 ($I \times J \times N = 10 \times 2 \times 2 = 40$) treatments (rows) of data that cover the two replicated data have been applied for calculating all the sums of squares given above. Because the mean square is the estimate of the variance for that corresponding source of variability, the gage repeatability and producibility are calculated as, respectively:

$$\sigma^2_{Repeatability} = MS_E = 0.1552 \quad \rightarrow \quad \sigma_{Repeatability} = (MS_E)^{1/2} = 0.394$$

Both the parts and operator have a significant effect on the results according to the F-ratios listed in the table.

Note that the relative amounts of variance computed in ANOVA present how much each individual "*group mean*" differs from the "*overall mean*" of all observations in the experimental design. Note that "group" stands for "treatment" as a general practice in the DOE.

The reproducibility is the variance at each individual measurement that comes from the mean square of the operators (with N = number of trials and I = number of parts) is calculated as:

$$\sigma^2_{Reproducibility} = \frac{MS_B - \sigma^2_{Repeatability}}{N \times I} = \frac{2.5 - 0.1552}{2 \times 10} = 0.1172$$

The part-to-part variance is the variance that comes from the mean square of the parts (with N = number of trials and J = number of operators) for each measurement, as:

$$\sigma^2_{Part-to-Part} = \frac{MS_A - \sigma^2_{Repeatability}}{N \times J} = \frac{30.01 - 0.1552}{2 \times 2} = 7.4637$$

The contributions to the three variances as allocated by the source of variation are listed as follows:

Source of Variance	Contribution	Percentage
Gage R&R		
Repeatability	0.1552	2.01%
Reproducibility	0.1172	1.52%
Part-to-part	7.4637	1.52%
Total Variance	7.7361	96.48%

The variance of gage R&R is then the sum of variances resulting from repeatability and reproducibility,

$$\sigma^2_{gage} = \sigma^2_{Reproducibility} + \sigma^2_{Repeatability} = 0.1172 + 0.1552 = 0.2724$$

which accounts for 3.52% (0.2724/7.7361) of the total variance from these three sources involved in each individual measurement. Since the gage R&R variance occupies only 3.52% of the total variance that is less than 10%, the measurement data are accepted.

4.7. Fractional Factorial Design with Three Levels

Alias equations, also called design generators, for the frequently three-level fractional factorial design matrices, are given in Table 4.7.1. Several widely used orthogonal arrays involving a three-level fractional factorial design are given in Tables 4.7.2–4.7.6. They are easy to use and information about the main effects in a relatively few treatments (run conditions) can be obtained.

In two-level designs, the interactions each have one DOF and consist only of +/− components, so it is simple to see how to formulate the confounding. Things are more complicated in three-level designs. Generally speaking, a q-factor interaction has 2^q DOFs. If a main effect (2 DOFs) is to be confounded with a two-factor interaction (4 DOFs) in a three-level design, one needs to partition the interaction into two orthogonal subsets with 2 DOFs each. The main effect can be confounded with either one of the two subsets, i.e., there are two choices. Similarly, if someone wants to confound a main effect with a three-factor interaction, the interaction has to be broken into four subsets with 2 DOFs each. Each subset of the interaction is to be treated like a super-factor with three levels.

TABLE 4.7.1 Design generators for $3_R{}^{K-P}$ [Xu 2005].

Fraction	Runs	Design generators	Design matrix
3^3 (Full)	27	—	Table 4.2.1
$3_{III}{}^{3-1}$	9	$C = A^2 B^2$	Table 4.7.2 (on A and B)
$3_{IV}{}^{4-1}$	27	$D = A B C$	Table 4.7.3 (on A, B, and C)
$3_{III}{}^{4-2}$	9	$C = A^2 B$	Table 4.7.4 (on A and B)
		$D = A B$	
$3_{III}{}^{5-2}$	27	$D = A B C$	Table 4.7.5 (on A, B, and C)
		$E = A B^2$	
$3_{III}{}^{9-6}$	27	$C = A B$	Table 4.7.6 (on A, B, and E)
		$D = A^2 B$	
		$F = A E$	
		$G = A^2 E$	
		$H = B^2 E$	
		$J = A B^2 E$	
$3_{III}{}^{11-8}$	27	$C = A B$	Table 4.7.7 (on A, B, and M)
		$D = A^2 B$	
		$E = B M$	
		$F = A B M$	
		$G = A^2 B M$	
		$H = B^2 M$	
		$J = A B^2 M$	
		$K = A^2 B^2 M$	
$3_{III}{}^{13-10}$ (L$_{27}$)	27	—	Table 4.7.8

There are three operating rules for alias equations when working on 3^{K-P} designs:

(1) A factor multiplied by itself twice is I (identity), e.g.,

$$I = A = A^3 \tag{4.7.1}$$

(2)   A factor multiplied by I is itself, e.g.,

$$A\,I = A \tag{4.7.2}$$

(3) A^2 is an individual "quadratic effect" and

$$A^2 = A \tag{4.7.3}$$

For a three-level design, words W and W^2 (e.g., $I = ABCD^2$ and $I = A^2B^2C^2D$) present the same contrast as justified by applying Equation (4.7.1). The convention in DOE practice is to set the first nonzero coefficient to be 1 (i.e., using $I = ABCD^2$ instead of $I = A^2B^2C^2D$) to clear ambiguity. Exploration of the design matrix of a fractional factorial design 3^{K-P} based on the (0, 1, 2) coding system will be demonstrated further.

4.7.1. Interactions in 3^K and 3^{K-P} Factorial Designs

The operations of interactions among factors at three levels are different from those at two levels. For example, there are two interactive effects between factors A and B, when operating at three levels, and they are AB and AB^2, of which each has 2 DOFs. Term B^2, which is the quadratic term for factor B, is to be treated like a "pseudo-factor," while B represents the linear term.

Assume that the levels of A and B are denoted by x_A and x_B, respectively. When coded using the (0, 1, 2) system, AB represents the contrasts among the response values as:

$$A\,B \rightarrow \mathrm{mod}_3(x_A + x_B), \text{ when coded as } x_A = 0, 1, 2 \text{ and } x_B = 0, 1, 2 \tag{4.7.4}$$

where $\mathrm{mod}_3(x_A + x_B)$ is the modulo-3 operation of the sum of x_A and x_B, i.e., finding the remainder after division of what is in the parenthesis by 3. For example, $\mathrm{mod}_3(0) = 0$, $\mathrm{mod}_3(1) = 1$, $\mathrm{mod}_3(2) = 2$, $\mathrm{mod}_3(3) = 0$, $\mathrm{mod}_3(4) = 1$, $\mathrm{mod}_3(5) = 2$, $\mathrm{mod}_3(6) = 0$, $\mathrm{mod}_3(7) = 1$, ..., etc. In the meanwhile, AB^2 represents the contrasts among the response values, of which

$$A\,B^2 \rightarrow \mathrm{mod}_3(x_A + 2x_B) \tag{4.7.5}$$

three-factor interactions of factors A, B, and C, having 8 DOFs, may have only four different appearances such as ABC, ABC^2, AB^2C and AB^2C^2, of which each contributing part (i.e., A, B, C, B^2, or C^2) has 2 DOFs. The contrasts among the response values of these four components can be written as:

$$A\,B\,C \rightarrow \mathrm{mod}_3(x_A + x_B + x_C) \tag{4.7.6}$$

$$A\,B\,C^2 \rightarrow \mathrm{mod}_3(x_A + x_B + 2x_C) \tag{4.7.7}$$

$$A\,B^2\,C \rightarrow \mathrm{mod}_3(x_A + 2x_B + x_C) \tag{4.7.8}$$

and

$$A B^2 C^2 \rightarrow \mod_3(x_A + 2x_B + 2x_C) \tag{4.7.9}$$

Note that the contrasts of response values for C (linear term) and C² (quadratic term), due to factor C, are denoted by x_C and $2x_C$ in the above equations, respectively. As a conventional practice, the coefficient for the first nonzero factor is 1 (i.e., x_A for factor A) will be taken for granted to avoid ambiguity without losing the generality.

4.7.2. 3_{III}^{3-1}

Assume that the three variables are A, B, and C and the (0, 1, 2) coding system applies. The design levels for the first two factors (i.e., A and B) are filled up using the 3^2 design, as the first two columns shown in Table 4.7.2. The alias equation used for generating the third factor in the table is

$$C = A^2 B^2 \qquad (I = A B C) \tag{4.7.10}$$

The second equation can be obtained when both sides of $C = A^2 B^2$ are multiplied by AB. It means that factor C is to be constructed from interaction A^2B^2, i.e., "quadratic levels" of the factors A and B. Since the shortest one of confounding terms is 3 as enlightened by Equation (4.7.10), the resolution is III. Assume that x_A, x_B, and x_C, stand for the levels to be applied to the three columns, respectively. Then, for each run (treatment) one has

$$x_C = \mod_3(2x_A + 2x_B) \tag{4.7.11}$$

For example, the fourth run (treatment) given in the design matrix of Table 4.7.2, $x_A = 1$ and $x_B = 0$, then $x_C = 2 + 0 \rightarrow 2$. Note that $0 \rightarrow 0, 1 \rightarrow 1, 2 \rightarrow 2, 3 \rightarrow 0, 4 \rightarrow 1, 5 \rightarrow 2, 6 \rightarrow 0, 7 \rightarrow 1$, etc., when the (0, 1, 2) coding system is applied.

TABLE 4.7.2 The fractional factorial design matrix of 3_{III}^{3-1} (C = A²B²).

Run	a	b	c
1	0	0	0
2	0	1	2
3	0	2	1
4	1	0	2
5	1	1	1
6	1	2	0
7	2	0	1
8	2	1	0
9	2	2	2

4.7.3. 3_{IV}^{4-1}

One fractional factorial design having four factors with three levels is 3_{IV}^{4-1}, which is a one-third fraction of 3^4 and its corresponding design matrix is listed in Table 4.7.3. The fourth column is coded using the following alias equation:

$$D = A\,B\,C \qquad (I = A\,B\,C\,D^2) \tag{4.7.12}$$

Following Equations (4.7.1)–(4.7.3), one has all aliased effects/interactions for the 3_{IV}^{4-1} design based on D = ABC as listed below:

$$\begin{aligned}
A &= B\,C\,D^2 = A\,B^2\,C^2\,D\\
B &= A\,C\,D^2 = A\,B^2\,C\,D^2\\
C &= A\,B\,D^2 = A\,B\,C^2\,D^2\\
D &= A\,B\,C = A\,B\,C\,D\\
A\,B &= C\,D^2 = A\,B\,C^2\,D\\
A\,B^2 &= A\,C^2\,D = B\,C^2\,D\\
A\,C &= B\,D^2 = A\,B^2\,C\,D\\
A\,C^2 &= A\,B^2\,D = B\,C^2\,D^2\\
A\,D &= A\,B^2\,C^2 = B\,C\,D\\
A\,D^2 &= B\,C = A\,B^2\,C^2\,D^2\\
B\,C^2 &= A\,B^2\,D^2 = A\,C^2\,D^2\\
B\,D &= A\,B^2\,C = A\,C\,D\\
\text{and} \quad C\,D &= A\,B\,C^2 = A\,B\,D
\end{aligned}$$

Since the shortest one of confounding terms is 4, the resolution is IV. Assume that x_A, x_B, x_C, and x_D stand for the levels to be applied to the four columns, respectively. Then, for each run (treatment combination) one has

$$x_D = \mathrm{mod}_3(x_A + x_B + x_C) \tag{4.7.13}$$

A case study based on design matrix 3_{IV}^{4-1} herein is presented. The experiment was set up to study the effect of the following four factors on the pull strength of truck seatbelts: (A) Pressure, (B) Die Flat, (C) Crimp Length, (D) Anchor Lot #, as extended from Table 4.2.2. The three levels taken for each individual factor are given as follows [Wu and Hamada 2009]:

Level	(A) Pressure	(B) Die flat	(C) Crimp length	(D) Anchor location
0: Low	7.58 MPa	10.0 mm	18 mm	#74
1: Middle	9.65 MPa	10.2 mm	23 mm	#75
2: High	11.72 MPa	10.4 mm	27 mm	#76

Factor D, i.e., anchor lot location, is purposely neglected in Table 4.2.3. For a reality check, the anchor location is here included to complete the study as given in Table 4.7.3.

TABLE 4.7.3 The 3_{IV}^{4-1} (D = ABC) design matrix for evaluating seat-belt pull strength [Wu and Hamada 2009].

Run	a	b	c	d	Strength (kN)		
1	0	0	0	0	23.0	29.0	26.5
2	0	0	1	1	23.8	27.2	23.2
3	0	0	2	2	13.7	16.8	18.9
4	0	1	0	1	24.7	29.2	28.1
5	0	1	1	2	21.1	19.6	24.2
6	0	1	2	0	24.6	18.0	20.1
7	0	2	0	2	25.3	27.8	27.6
8	0	2	1	0	25.5	27.9	26.0
9	0	2	2	1	25.6	21.3	24.1
10	1	0	0	1	30.4	30.7	30.9
11	1	0	1	2	29.1	28.1	21.3
12	1	0	2	0	27.4	25.9	26.5
13	1	1	0	2	30.5	30.3	30.7
14	1	1	1	0	30.2	29.8	30.2
15	1	1	2	1	29.0	28.9	29.2
16	1	2	0	0	28.8	31.0	29.9
17	1	2	1	1	30.4	31.3	22.5
18	1	2	2	2	22.1	25.3	25.6
19	2	0	0	2	31.8	30.8	27.7
20	2	0	1	0	30.7	31.4	31.8
21	2	0	2	1	30.8	32.0	29.7
22	2	1	0	0	32.1	31.9	31.2
23	2	1	1	1	31.2	31.3	32.0
24	2	1	2	2	27.6	27.8	28.9
25	2	2	0	1	31.8	30.6	31.0
26	2	2	1	2	31.9	32.3	30.9
27	2	2	2	0	31.4	31.4	30.9

The ANOVA table for the seat-belt strength based on 3^{4-1} design matrix with three replicated runs for each experimental treatment, resulting from the calculations using Minitab (Stat → DOE → Response Surface), is given as follows:

Variance	SS$_{adj}$	DOF	MS$_{adj}$	F$_{u,v}$	P-value
Model	1,116.64	17	65.685	17.15	0.0%
Linear	955.35	5	191.069	49.89	0.0%
A	669.22	1	669.222	174.75	0.0%
B	17.114	1	17.114	4.47	3.8%
C	181.87	1	181.867	47.49	0.0%
D	87.14	2	43.571	11.38	0.0%
Square	23.69	3	7.897	2.06	11.4%
A^2	18.61	1	18.605	4.86	3.1%
B^2	1.62	1	1.620	0.42	51.8%
C^2	3.47	1	3.467	0.91	34.5%
Interaction	137.60	9	15.289	3.99	0.0%

Variance	SS$_{adj}$	DOF	MS$_{adj}$	F$_{u,v}$	P-value
AB	0.59	1	0.593	0.15	69.5%
AC	31.90	1	31.905	8.33	0.5%
AD	8.06	2	4.031	1.05	35.5%
BC	0.32	1	0.322	0.08	77.3%
BD	9.31	2	4.655	1.22	30.3%
CD	31.06	2	15.530	4.06	2.2%
Error	241.26	63	3.830		
Lack-of-fit	27.90	9	3.100	0.78	**63.1%**
Pure error	213.36	54	3.951		
Subtotal	1357.90	80			
Grand ave.		1			
Total		81			

There are two major differences in the problem solution given above and the one in Table 4.2.4. First, only factors A, B, and C (three continuous factors) are included in Table 4.2.4, while factor D (a categorial factor) is also considered additionally here. The other is the adjusted sum of squares and adjusted mean squares are taken into consideration instead of the sum of squares and mean squares. In general, adjusted sum of squares and adjusted mean squares should be used in reality checks.

Conclusion

The main effects of all these four factors on the seat-belt strength are significant, given that the statistical significance level is chosen to be 5%. Factor A (pressure) has the greatest impact on the seat-belt strength and their interactive effect is also significant. Factor C (crimp length) is the next influential factor, followed by factor D and factor B. Influences by two-factor interactions AC and CD are also significant. The quadratic term of factor A, i.e., A^2, is effective, too. The p-value of lack-of-fit is 63.1%, which means that the model is statistically well fit because of the insignificant lack-of-fit.

4.7.4. 3_{III}^{4-2}

Another simple but useful fractional factorial design with four factors at three levels is 3_{III}^{4-2}, which is a one-ninth fraction of 3^4 and its design matrix is listed in Table 4.7.4. It is also called L_9 design in the automotive industry. The two components of interaction chosen to do the design are AB^2 and AB. The design generators are then:

$$C = A^2 B \qquad (I = A B^2 C = B C D) \tag{4.7.14}$$

and

$$D = A B \qquad (I = A C^2 D = A B D^2) \tag{4.7.15}$$

Note that AC^2D and ABD^2 are two generalized interactions, which can be derived from $(AB^2C)(BCD) = AC^2D$ and $(AB^2C)(BCD)^2 = ABD^2$, given that $A^3 = B^3 = C^3 = I$ for operations of three-level design generators.

Assume that x_A, x_B, x_C, and x_D stand for the levels to be applied to the four columns, respectively. Then, for each run (treatment combination) one has

$$x_C = \text{mod}_3(2x_A + x_B) \tag{4.7.16}$$

and

$$x_D = \text{mod}_3(x_A + x_B) \tag{4.7.17}$$

TABLE 4.7.4 The fractional factorial design matrix of 3_{III}^{4-2} ($C = A^2B$ and $D = AB$).

Run	a	b	c	d
1	0	0	0	0
2	0	1	1	1
3	0	2	2	2
4	1	0	2	1
5	1	1	0	2
6	1	2	1	0
7	2	0	1	2
8	2	1	2	0
9	2	2	0	1

4.7.5. 3_{III}^{5-2}

The test matrix for a 3_{III}^{5-2} design given in Table 4.7.5 can be constructed in two steps. First, write down all possible $3^3 = 27$ level combinations for the first three columns (i.e., columns A, B, and C), of which each column takes on three deferent values such as 0, 1, 2. The next step is to define the last two columns (columns D and E) by letting

$$D = A B C \quad (\text{i.e. } I = A B C D^2) \tag{4.7.18}$$

and

$$E = A B^2 \quad (\text{i.e. } I = A B^2 E^2) \tag{4.7.19}$$

Assume that x_A, x_B, x_C, x_D, and x_E stand for the levels to be applied to the five columns, respectively. Then, for each run (treatment combination) one has

$$x_D = \text{mod}_3(x_A + x_B + x_C) \tag{4.7.20}$$

and

$$x_E = \text{mod}_3(x_A + 2x_B) \tag{4.7.21}$$

TABLE 4.7.5 The fractional factorial design matrix of 3_{III}^{5-2} (D = ABC and E = AB^2).

Run	a	b	c	d	e
1	0	0	0	0	0
2	0	0	1	1	0
3	0	0	2	2	0
4	0	1	0	1	2
5	0	1	1	2	2
6	0	1	2	0	2
7	0	2	0	2	1
8	0	2	1	0	1
9	0	2	2	1	1
10	1	0	0	1	1
11	1	0	1	2	1
12	1	0	2	0	1
13	1	1	0	2	0
14	1	1	1	0	0
15	1	1	2	1	0
16	1	2	0	0	2
17	1	2	1	1	2
18	1	2	2	2	2
19	2	0	0	2	2
20	2	0	1	0	2
21	2	0	2	1	2
22	2	1	0	0	1
23	2	1	1	1	1
24	2	1	2	2	1
25	2	2	0	1	2
26	2	2	1	2	2
27	2	2	2	0	2

© SAE International.

4.7.6. 3_{III}^{9-6}

One three-level design to deal with a high number of design factors is 3_{III}^{9-6}, of which the design matrix is given in Table 4.7.6. Factors A, B, and E are used to generate the design matrix as a full 3^3 design in the table. The defining equations for the other six factors are derived from the following six alias equations:

$$C = A\,B \qquad (I = A^2\,B^2\,C) \tag{4.7.22}$$

$$D = A^2\,B \qquad (I = A\,B^2\,D) \tag{4.7.23}$$

$$F = A\,E(I = A^2\,E^2\,F) \tag{4.7.24}$$

$$G = A^2\,E \qquad (I = A\,E^2\,G) \tag{4.7.25}$$

$$H = B^2\, E \qquad (I = B\, E^2\, H) \tag{4.7.26}$$

and

$$J = A\, B^2\, E \qquad (I = A^2\, B\, E^2\, J) \tag{4.7.27}$$

It takes only 27 runs to analyze the main effects of nine variables, including linear and nonlinear terms. Assume that x_A, x_B, x_C, x_D, x_E, x_F, x_G, x_H, and x_J stand for the levels to be applied to the five columns, respectively. Then, for each run (treatment combination) one has

$$x_C = mod_3(x_A + x_B) \tag{4.7.28}$$

$$x_D = mod_3(2x_A + x_B) \tag{4.7.29}$$

$$x_F = mod_3(x_A + x_E) \tag{4.7.30}$$

$$x_G = mod_3(2x_A + x_E) \tag{4.7.31}$$

$$x_H = mod_3(2x_B + x_E) \tag{4.7.32}$$

and

$$x_J = mod_3(x_A + 2x_B + x_E) \tag{4.7.33}$$

TABLE 4.7.6 The fractional factorial design matrix of 3_{III}^{9-6}.

Run	a	b	c	d	e	f	g	h	j
1	0	0	0	0	0	0	0	0	0
2	0	0	0	0	1	1	1	1	1
3	0	0	0	0	2	2	2	2	2
4	0	1	1	1	0	0	0	2	2
5	0	1	1	1	1	1	1	0	0
6	0	1	1	1	2	2	2	1	1
7	0	2	2	2	0	0	0	1	1
8	0	2	2	2	1	1	1	2	2
9	0	2	2	2	2	2	2	0	0

4.7.7. 3_{III}^{11-8}

Another three-level design to deal with a high number of design factors is 3_{III}^{11-8}, of which the design matrix is given in Table 4.7.7. Factors A, B, and L are used to generate the design matrix as a full 3^3 design in the table. The defining equations for the other eight factors are derived from the following eight alias equations:

$$C = A\,B \tag{4.7.34}$$

$$D = A^2\,B \tag{4.7.35}$$

$$E = B\,M \tag{4.7.36}$$

$$F = A\,B\,M \tag{4.7.37}$$

$$G = A^2\,B\,M \tag{4.7.38}$$

$$H = B^2\,M \tag{4.7.39}$$

$$J = A\,B^2\,M \tag{4.7.40}$$

and

$$K = A^2\,B^2\,M \tag{4.7.41}$$

TABLE 4.7.7 The fractional factorial design matrix of 3_{III}^{11-8}.

Run	a	b	c	d	e	f	g	h	j	k	m
1	0	0	0	0	0	0	0	0	0	0	0
2	0	1	1	1	1	1	1	2	2	2	0
3	0	2	2	2	2	2	2	1	1	1	0
4	1	0	1	2	0	1	2	0	1	2	0
5	1	1	2	0	1	2	0	2	0	1	0
6	1	2	0	1	2	0	1	1	2	0	0
7	2	0	2	1	0	2	1	0	2	1	0
8	2	1	0	2	1	0	2	2	1	0	0
9	2	2	1	0	2	1	0	1	0	2	0
10	0	0	0	0	1	1	1	1	1	1	1
11	0	1	1	1	2	2	2	0	0	0	1
12	0	2	2	2	0	0	0	2	2	2	1
13	1	0	1	2	1	2	0	1	2	0	1
14	1	1	2	0	2	0	1	0	1	2	1
15	1	2	0	1	0	1	2	2	0	1	1
16	2	0	2	1	1	0	2	1	0	2	1
17	2	1	0	2	2	1	0	0	2	1	1
18	2	2	1	0	0	2	1	2	1	0	1
19	0	0	1	0	2	2	2	2	2	2	2
20	0	1	1	1	0	0	0	1	1	1	2
21	0	2	3	2	1	1	1	0	0	0	2
22	1	0	1	2	2	0	1	2	0	1	2
23	1	1	2	0	0	1	2	1	2	0	2
24	1	2	0	1	1	2	0	0	1	2	2
25	2	0	2	1	2	1	0	2	1	0	2
26	2	1	0	2	0	2	1	1	0	2	2
27	2	2	1	0	1	0	2	0	2	1	2

4.7.8. 3_{III}^{13-10}

Another attractive three-level design is 3_{III}^{13-10}, of which the design matrix is given in Table 4.7.8 without derivation. It is an L_{27} design as known in the automotive industry.

TABLE 4.7.8 Fractional factorial design matrix of 3_{III}^{13-10} (L_{27}).

Run	a	b	c	d	e	f	g	h	j	k	l	m	n
1	0	0	0	0	0	0	0	0	0	0	0	0	0
2	0	0	0	0	1	1	1	1	1	1	1	1	1
3	0	0	0	0	2	2	2	2	2	2	2	2	2
4	0	1	1	1	0	0	0	1	1	1	2	2	2
5	0	1	1	1	1	1	1	2	2	2	0	0	0
6	0	1	1	1	2	2	2	0	0	0	1	1	1
7	0	2	2	2	0	0	0	2	2	2	1	1	1
8	0	2	2	2	1	1	1	0	0	0	2	2	2
9	0	2	2	2	2	2	2	1	1	1	0	0	0
10	1	0	1	2	0	1	2	0	1	2	0	1	2
11	1	0	1	2	1	2	0	1	2	0	1	2	0
12	1	0	1	2	2	0	1	2	0	1	2	0	1
13	1	1	2	0	0	1	2	1	2	0	2	0	1
14	1	1	2	0	1	2	0	2	0	1	0	1	2
15	1	1	2	0	2	0	1	0	1	2	1	2	0
16	1	2	0	1	0	1	2	2	0	1	1	2	0
17	1	2	0	1	1	2	0	0	1	2	2	0	1
18	1	2	0	1	2	0	1	1	2	0	0	1	2
19	2	0	2	1	0	2	1	0	2	1	0	2	1
20	2	0	2	1	1	0	2	1	0	2	1	0	2
21	2	0	2	1	2	1	0	2	1	0	2	1	0
22	2	1	0	2	0	2	1	1	0	2	2	1	0
23	2	1	0	2	1	0	2	2	1	0	0	2	1
24	2	1	0	2	2	1	0	0	2	1	1	0	2
25	2	2	1	0	0	2	1	2	1	0	1	0	2
26	2	2	1	0	1	0	2	0	2	1	2	1	0
27	2	2	1	0	2	1	0	1	0	2	0	2	1

4.8. DOE with Definitely Known Inferential Mechanisms

In the process of pursuing a predictive equation (model), multiple linear regression provides the following convenient filtering functionalities:

1. Detection of multicollinearity among coefficients/constants of interest in the metamodel and having them removed one by one until an effective and adequate model (i.e., predictive equation) appears. It means to have the most nonsignificant coefficient removed according to its p-value and variance inflation factor (VIF).

2. When the inferential mechanism for an experimental design is known, it can be utilized to refine its DOE modeling.

Both functionalities are illustrated using an experiment that is performed to assess the deflection of a simply supported circular beam as a function of beam radius A, beam span B, concentrated load C, and Young's modulus of elasticity E. The relevant dimensions are shown in Figure 4.8.1.

© SAE International.

FIGURE 4.8.1 Simply supported slender beam with a concentrated load at the middle point.

It is simply supported at both ends and a point load is applied at the center of the span where beam deflection is measured. Based on the closed-form solution, the deflection (Y) at the loading point is:

$$Y = C\,B^3 / (48\,E\,I) \qquad\qquad (4.8.1)$$

where

$$I = \pi\,A^4 / 4 \qquad\qquad (4.8.2)$$

or

$$Y = C\,B^3 / (12\,\pi\,E\,A^4) \qquad\qquad (4.8.3)$$

The above equation is valid for slender beam with small deformation. It means that span B >> radius A > 0.

There are four design factors (i.e., A, B, C and E), of which their design levels of interest are given in Table 4.8.1. The range of Young's modulus given in the table is the material elasticity of generically mentioned fiber-reinforced polyethylene terephthalate (PET), polystyrene (PS), and nylon (PA), which are widely used for structural parts of automotive interiors. The closed-form solution is also known as the inferential mechanism in the DOE.

The work task is then to conduct the response surface analysis with the DOE matrix given in Table 4.8.2, where a, b, c, and e are coded variables of factors A, B, C, and E, respectively. Two questions are to be answered:

1. What is the correlation between the measured data given in Table 4.8.2 and predicted values obtained from the DOE model?

2. What should the deflection be if no load is applied (C = 0)?

TABLE 4.8.1 Design factors and related design levels of a simply supported circular beam.

Factor	(1)	(2)	(3)	Coded variable
A: Radius (mm)	4	5	6	a = (A − 5)/1 + 2
B: Span (mm)	400	500	600	b = (B − 500)/100 + 2
C: Concentrated load (N)	1	2	3	c = (C − 2)/1 + 2
E: Young's modulus (MPa)	6000	8000	10000	e = (D − 8000)/2000 + 2

TABLE 4.8.2 Design matrix 3^{4-1} and measured deflections of a simply supported circular beam.

Run	A	B	C	E	a	b	c	e	Y (mm)
1	4	400	1	6000	1	1	1	1	1.11
2	4	400	2	8000	1	1	2	2	1.66
3	4	400	3	10000	1	1	3	3	1.99
4	4	500	1	8000	1	2	1	2	1.62
5	4	500	2	10000	1	2	2	3	2.59
6	4	500	3	6000	1	2	3	1	6.48
7	4	600	1	10000	1	3	1	3	2.24
8	4	600	2	6000	1	3	2	1	7.46
9	4	600	3	8000	1	3	3	2	8.39
10	5	400	1	8000	2	1	1	2	0.34
11	5	400	2	10000	2	1	2	3	0.54
12	5	400	3	6000	2	1	3	1	1.36
13	5	500	1	10000	2	2	1	3	0.53
14	5	500	2	6000	2	2	2	1	1.77
15	5	500	3	8000	2	2	3	2	1.99
16	5	600	1	6000	2	3	1	1	1.53
17	5	600	2	8000	2	3	2	2	2.29
18	5	600	3	10000	2	3	3	3	2.75
19	6	400	1	10000	3	1	1	3	0.13
20	6	400	2	6000	3	1	2	1	0.44
21	6	400	3	8000	3	1	3	2	0.49
22	6	500	1	6000	3	2	1	1	0.43
23	6	500	2	8000	3	2	2	2	0.64
24	6	500	3	10000	3	2	3	3	0.77
25	6	600	1	8000	3	3	1	2	0.55
26	6	600	2	10000	3	3	2	3	0.88
27	6	600	3	6000	3	3	3	1	2.21

4.8.1. Regression Based on Second-Ordered Response Surface Model

The second-ordered response surface model is usually the first metamodel to be tried out for response surface analysis, i.e.,

$$Y \approx \gamma_0 + \gamma_1\,A + \gamma_2\,B + \gamma_3\,C + \gamma_4\,E + \gamma_{11}\,A^2 + \gamma_{22}\,B^2 + \gamma_{33}\,C^2 + \gamma_{44}\,E^2$$

$$+ \gamma_{12}\,AB + \gamma_{13}\,AC + \gamma_{14}\,AE + \gamma_{23}\,BC + \gamma_{24}\,BE + \gamma_{34}\,CE + \varepsilon$$

(4.8.4)

The data analysis is done with Minitab (Stat ➔ Regression ➔ Regression ➔ Fit Regression Model). The "most insignificant" factor or interaction that comes with the largest p-value is weeded out one by one until all p-values are less than 10% that is here defined as the parting significance level based on F-distribution. By the ANOVA presented in Table 4.8.3, the predictive equation based on the factorial design is:

$$Y_p = 2.14 - 5.12\ A + 0.05671\ B + 2.431\ C - 0.000851\ E + 0.771\ A^2 - 0.00896\ A\ B$$

$$- 0.735\ A\ C + 0.000233\ A\ E + 0.00448\ B\ C - 0.000001\ B\ E$$

(4.8.5)

$$(R = 98.65\%,\ R_{adj} = 97.79\%,\ \text{and}\ R_{pred} = 95.07\%)$$

According to the model correlation, adjusted model correlation, and predictive correlation (i.e., R, R_{adj}, and R_{pred}) shown above, it is concluded that the regression is effective and adequate.

Nevertheless, substituting C = 0 (no force applied) in Equations (4.8.3) and (4.8.5) yields, respectively:

Equation (4.8.3): $Y = 0$

and   Equation (4.8.5): $Y = 2.14 - 5.12\ A + 0.05671\ B - 0.000851\ D + 0.771\ A^2$

$$- 0.00896\ A\ B + 0.000233\ A\ D - 0.000001\ B\ D$$

The discrepancy between the above two results occurs because of the high p-value (75.4%) for the constant as exhibited in Table 4.8.4. It means that the intercept (constant) of Equation (4.8.5) is not statistically significant and invokes room for improvement in modeling. Note that the constant (intercept) of a metamodel has to be always checked to make sure that its p-value (significance level) is small, e.g., less than 10% (or 5%).

It is also possible that the adjusted correlation R_{adj} for a regression model is great and even the overall F-test statistic is also significant, but some of the individual coefficients or constants (intercept in a polynomial) are statistically insignificant. This scenario can be attributed to multicollinearity that affects the coefficients or constants and their corresponding p-values, but it does not affect the goodness-of-fit statistics or the overall model significance. This can be explained by Equation (4.4.10). The inverse of matrix $[u]^T_{V\times Q}\ [u]_{Q\times V}$ in the equation may suffer from different levels of multicollinearity even it is not exactly singular. Multicollinearity refers to the situation in which several independent variables in a model are significantly correlated in the regression analysis. Multicollinearity inflates the variances of coefficients or constants and causes type II errors. It means that mutual influences among coefficients and constants make the regression analysis doubtful.

As shown in Tables 4.8.4 and 4.8.6, each VIF is used to measure the effect of multicollinearity among the applied design factors (predictors). Here, the VIF of each design factor is calculated as [Heckman 2015]:

$$\text{VIF} = 1\ /\ (1 - R_i^2),\ \text{where}\ i = A,\ B,\ C,\ \text{and}\ D$$

(4.8.6)

Note that R_i is the unadjusted correlation of one individual design factor i the model of one individual coefficient (or constant) against all the others, while R, R_{adj}, and R_{pred} in Equation (4.8.5) are measures for the overall coefficients of the regression model. More specifically, the VIF measures how much the variance of an estimated regression coefficient increases if design factors are correlated.

The square root of the variance inflation factor, $(\text{VIF})^{1/2}$, reveals how much larger the standard error is as compared to if the design factor of concern is not correlated with any other design factor. Some comments:

1. VIF = 1 (R_i = 0): The estimated variance for the coefficient of interest is not correlated with others at all.

2. VIF > 5.26 (R_i > 90%): Conventional wisdom has it that the coefficient estimates may not be trusted and the statistical significance is questionable with high R_i (>90%).

3. VIF > 10.26 (R_i > 95%): The estimated coefficient of interest is definitely not statistically significant without any doubt.

Multicollinearity can be resolved sometimes by removing a redundant term from the model. A proper data transformation is also an effective way of reducing the VIFs.

TABLE 4.8.3 ANOVA for assessing the deflection at the middle point of a simply supported beam.

Source	SS$_{adj}$	DOF	MS$_{adj}$	F$_{u,v}$	p-value
Regression	114.80	10	11.4807	57.85	0.0%
A	1.274	1	1.2735	6.42	2.2%
B	8.211	1	8.2110	41.37	0.0%
C	1.313	1	1.3134	6.62	2.0%
D	0.643	1	0.6430	3.24	9.1%
A^2	3.563	1	3.5630	17.95	0.1%
AB	9.631	1	9.6310	48.53	0.0%
AC	6.071	1	6.0711	30.59	0.0%
AD	2.441	1	2.4406	12.30	0.3%
BC	2.257	1	2.2568	11.37	0.4%
BD	0.651	1	0.6507	3.28	8.9%
Error	3.175	16	0.1985		
Subtotal	117.982	26			
Grand ave.	—	1			
Total	—	27			

© SAE International.

TABLE 4.8.4 Significance levels and VIFs of terms (constant and coefficients).

Term	Coefficient	SE coeff	t-Value	P-value	VIF
Constant	2.14	6.69	0.32	75.4%	—
A	−5.12	2.02	−2.53	2.2%	370.50
B	0.05671	0.00882	6.43	0.0%	70.50
C	2.431	0.945	2.57	2.0%	81.00
D	−0.000851	0.000473	−1.80	9.1%	81.00
A^2	0.771	0.182	4.24	0.1%	301.00
AB	−0.00896	0.00129	−6.97	0.0%	76.00
AC	−0.735	0.133	−5.53	0.0%	47.47
AD	0.000233	0.000066	3.51	0.3%	66.67
BC	0.00448	0.00133	3.37	0.4%	47.47
BD	−0.000001	0.000001	−1.81	8.9%	66.67

© SAE International.

where
SE coeff: Standard error of coefficient
t-value = Coefficient/SE coeff
VIF: variance inflation factor

4.8.2. Regression Based on Observance of Inferential Mechanism

As Equation (4.8.5) is LOF to the measured data because of high VIFs and the p-value of constant (Table 4.8.4), the experimenter may seek opportunities to re-assert the metamodel. One possibility is to resort to the inferential mechanism, i.e., Equation (4.8.3). It seems that taking logarithmic transformations of Equation (4.8.3) will decouple the higher-order terms and yield a simple linear model. Repeating the procedure for deriving Equation (4.8.5), with Minitab (Stat → Regression → Regression → Fit Regression Model), one can do the regression analysis again using the following metamodel:

$$\ln(Y) = \text{Constant} + \gamma_a \ln(A) + \gamma_b \ln(B) + \gamma_c \ln(C) + \gamma_e \ln(E) \tag{4.8.7}$$

Pugging data of $\ln(Y)$, $\ln(A)$, $\ln(B)$, $\ln(C)$, and $\ln(E)$ into Minitab leads to the following predictive equation:

$$\ln(Y_p) = -3.5345 - 4.00211\ \ln(A) + 2.99787\ \ln(B) + 0.99983\ \ln(C) - 1.00873\ \ln(D)$$

$$(R = 99.87\%,\ R_{adj} = 99.85\%,\ \text{and}\ R_{pred} = 99.79\%) \tag{4.8.8}$$

$$Y_p = C^{0.99983}\ B^{2.99787} / (10.911\ \pi\ E^{1.00873}\ A^{4.00211}) \tag{4.8.9}$$

Thus, $Y_p = 0$, when $C = 0$ (no load applied). The related significance levels and VIFs of terms (constant and coefficients) obtained are exhibited in Tables 4.8.5 and 4.8.6, respectively. The constant in Equation (4.8.8) is statistically significant with p-value = 0.00% and VIF = 1.0.

TABLE 4.8.5 ANOVA for assessing the logarithmic deflection at the middle point of a simply supported beam.

Source	SS_{adj}	DOF	MS_{adj}	$F_{u,v}$	p-value
Regression	25.3155	4	6.3289	640,032.36	0.0%
ln(A)	11.8894	1	11.8894	1,202,368.65	0.0%
ln(B)	6.6713	1	6.6713	674,661.43	0.0%
ln(C)	5.5535	1	5.5535	561,624.80	0.0%
ln(D)	1.2012	1	1.2012	121,474.57	0.0%
Error	0.0002	22	0.00001		
Subtotal	25.3157	26			
Grand ave.	—	1			
Total	—	27			

© SAE International.

TABLE 4.8.6 Significance levels of terms (constant and coefficients) on logarithmic scale.

Term	Coefficient	SE coeff.	t-Value	P-value	VIF
Constant	−3.5345	0.0349	−101.17	0.0%	—
ln(A)	−4.00211	0.00365	−1096.53	0.0%	1.00
ln(B)	2.99787	0.00365	821.38	0.0%	1.00
ln(C)	0.99983	0.00133	749.42	0.0%	1.00
ln(D)	−1.00873	0.00289	−348.53	0.0%	1.00

© SAE International.

References

AIAG, *Product Part Approval Process*, 4th ed. (Southfield, MI: Automotive Industry Action Group, 2006).

AIAG, *Measurement Systems Analysis*, 4th ed. (Southfield, MI: Automotive Industry Action Group, 2010).

Annigeri, M.A. and Raju, G.U., "Effect of Process Parameters on Tensile Properties of Tamarind Particles/MWCNT Hybrid Polymer Nanocomposites," *AIP Conference Proceedings* 2247 (2020): 050018.

Box, G.E.P. and Behnken, D.W., "Some New Three Level Designs for the Study of Quantitative Variables," *Technometrics* 2, no. 4 (1960): 455-495.

Box, G.E.P. and Hunter, W.G., "Multi-Factor Experimental Designs for Exploring Response Surfaces," *Annals of Mathematical Statistics* 28 (1957): 195-241.

Burdick, R.K., Borror, C.M., and Montgomery, D.C., *Design and Analysis of Gauge R and R Studies: Making Decisions with Confidence Intervals in Random and Mixed ANOVA Models* (Philadelphia, PA: American Statistical Association and the Society for Industrial and Applied Mathematics, 2005), ISBN:0898715881.

Chan, L.K., Cheng, S., and Spring, F., "A New Measurement of Process Capability, C_{pm}," *Journal of Quality Technology* 20 (1998): 162-175.

Cheng, S.W., "Practical Implementation of the Process Capability Indices," *Quality Engineering* 7, no. 2 (1994): 239-253.

Federal Motor Vehicle Safety Standard (1994), "Seat Belt Assembly Anchorages," FMVSS No. 210, Federal Motor Vehicle Safety Standard, Washington, DC.

Fwa, T.F. et al., "Effectiveness of Tire-Tread Patterns in Reducing the Risk of Hydroplaning," *Journal of the Transportation Research Board* 2094 (2009): 91-102.

Gunter, B. (1989a), "The Use and Abuse of C_{pk}, Part 2," *Quality Progress*, March 1989, pp. 108-109.

Gunter, B. (1989b), "The Use and Abuse of C_{pk}, Part 3," *Quality Progress*, May 1989, pp. 79-80.

Heckman, E. (2015), "What in the World Is a VIF?" Minitab Blog, July 20, 2015, https://blog.minitab.com/en/starting-out-with-statistical-software/what-in-the-world-is-a-vif.

Hinkelmann, K. and Kempthorne, O. (2008), *Design and Analysis of Experiments Set*, Wiley, Hoboken, NJ, 1411 pages; ISBN: 978-0-470-38551-7.

Hoffman, L.L., "A General Approach for Testing Process Capability Index," *Quality and Reliability Engineering International* 9 (1993): 445-449.

Jou, Y. et al., "Integrating the Taguchi Method and Response Surface Methodology for Process Parameter Optimization of the Injection Molding," *Applied Mathematics and Information Sciences* 8, no. 3 (2014): 1277-1285.

Kane, V.E., "Process Capability Indices," *Journal of Quality Technology* 18, no. 1 (1986): 41-52.

Khuri, A.I. and Cornell, J.A. (2019), *Response Surfaces Design and Analyses*, Taylor & Francis Group, Boca Raton, FL, 536 pages; ISBN: 978-0367401252.

Li, X., Sudarsanam, N., and Frey, D.D., "Regularities in Data from Factorial Experiments," *Complexity* 11, no. 5 (2006): 32-45.

Montgomery, D. and Runger, G.C., "Gage Capability and Designed Experiments, Part I: Basic Methods," *Quality Engineering* 6, no. 1 (1993a): 115-135.

Montgomery, D.C. and Runger, G.C., "Gage Capability and Designed Experiments, Part II: Experimental Design Models and Variance Component Estimation," *Quality Engineering* 6, no. 2 (1993b): 289-305.

Myers, R.H., Montgomery, D.C., and Anderson-Cook, C.M., *Response Surface Methodology*, 3rd ed. (New York: John Wiley & Sons, 2009).

Pearn, W., Kotz, S., and Johnson, N., "Distributional and Inferential Properties of Process Capability Indices," *Journal of Quality Technology* 24 (1992): 216-232.

Phillips, G.P., "Target Ratio Simplifies Capability Index System Makes It Easy to Use C_{pm}," *Quality Engineering* 7, no. 2 (1994): 299-313.

Rhinehart, R.R. (2016), *Nonlinear Regression Modeling for Engineering Applications: Modeling, Model Validation, and Enabling Design of Experiments*, John Wiley & Sons, Chichester, UK, 400 pages; ISBN: 978-1-118-59796-5.

Ross, S.M., *Introduction to Probability and Statistics for Engineers and Scientists* (New York: John Wiley & Sons, 1987).

Sahai, H. and Ageel, M.I., *The Analysis of Variance* (Boston, MA: Birkhauser, 2000), ISBN:978-1-4612-1344-4.

Searle, S., Casella, G., and McCulloch, C., *Variance Components* (New York: John Wiley & Sons, 1992).

Statgraphics (2013), "Sample Size Determination (Capability Indices)," StatPoint Technologies, Inc., Revised September 16, 2013.

Wheeler, D.J. (Revised 2009), "An Honest Gauge R&R Study," in *ASQ/ASA 2006 Fall Technical Conference*, Manuscript No. 189.

Wheeler, D.J. (2013), "Gauge R&R Methods Compared. How Do the ANOVA, AIAG, and EMP Approaches Differ?" *Quality Digest Daily*, Manuscript 262, December 2013.

Wu, C.F.J. and Hamada, M.S., *Experiments: Planning, Analysis and Parameter Design Optimization*, 2nd ed. (New York: Wiley, 2009), ISBN:978-0471699460.

Wu, C.C. and Kuo, H.L., "Sample Size Determination for the Estimate of Process Capability Indices," *Information and Management Sciences* 15, no. 1 (2004): 1-12.

Xu, H., "A Catalogue of Three-Level Regular Fractional Factorial Designs," *Metrika* 62 (2005): 259281.

Problems

P4.1: An experiment concerning the tire life of a passenger vehicle because of wear has been conducted. Each tire life (kilometer) because of wear is measured as a function of tire pressure and vehicle speed. How is the tire life related to the two operating factors? Test data are listed as follows:

Run	Pressure (kPa)	Speed (km/hr)	Life (km)
1	180	100	59,000
2	180	120	46,000
3	180	140	19,000
4	200	100	61,000
5	200	120	50,000
6	200	140	30,000
7	220	100	64,000
8	220	120	54,000
9	220	140	51,000

P4.2: Three operators (A, B, and C) are asked to measure the thickness (mm) of five gaskets, of which each is measured twice (i.e., R = 1 and R = 2). The data are given as follows:

	A	B	C
R = 1	4.24, 5.33, 4.75, 4.80, 3.96	3.94, 5.23, 4.62, 4.67, 3.63	3.86, 5.23, 4.57, 4.57, 3.71
R = 2	4.11, 5.41, 4.65, 4.98, 3.73	3.99, 5.05, 4.55, 4.52, 3.61	3.94, 5.16, 4.60, 4.62, 3.91

(a) How much are the variances of repeatability, reproducibility, and gage R&R?

(b) Is this gaging process acceptable?

P4.3: Write down the fractional factorial design matrix of 3_{III}^{3-1} using $C = AB^2$ as the alias equation.

P4.4: Identify the effects (main and interactive) that have influences on the seat-belt strength in pull based on the data given in Table 4.7.3.

P4.5: Hydroplaning speeds V (km/h) of tires with six different tire tread groove patterns (factor C) and five tread groove depths (factor B) rolling on water with different film thickness (factor A) are given as follows [Fwa et al. 2009]:

Run	A (mm)	B (mm)	V_{C1}	V_{C2}	V_{C3}	V_{C4}	V_{C5}	V_{C6}
1	10	9.8	100.1	126.7	104.8	112.3	110.0	128.9
2	10	7.0	95.41	123.7	101.2	108.5	106.7	123.6
3	10	5.0	92.39	120.3	97.20	104.7	103.3	118.8
4	10	3.0	88.47	117.2	93.24	100.3	99.08	113.4
5	10	2.0	86.91	113.6	89.28	96.19	95.66	109.7
6	10	1.0	85.61	111.2	87.48	92.34	92.14	106.0
7	10	0.5	84.82	109.4	86.40	90.54	90.58	104.5
8	7.0	9.8	103.5	130.0	107.6	115.0	113.6	132.6
9	7.0	7.0	99.05	127.0	104.4	111.6	109.0	127.3
10	7.0	5.0	95.33	123.4	99.36	107.1	105.0	122.7
11	7.0	3.0	91.19	120.2	94.68	103.3	101.2	118.1
12	7.0	2.0	89.48	116.6	91.44	98.82	98.22	115.1
13	7.0	1.0	88.08	114.1	89.64	95.18	96.25	112.0
14	7.0	0.5	87.26	111.7	88.38	92.88	94.75	109.7
15	5.0	9.8	105.41	133.2	111.2	118.3	117.0	135.3
16	5.0	7.0	101.81	130.9	107.3	114.7	112.9	131.3
17	5.0	5.0	98.10	126.6	101.2	110.9	108.6	126.2
18	5.0	3.0	94.15	123.3	96.48	106.6	104.5	122.0
19	5.0	2.0	92.46	119.8	93.60	101.7	100.4	118.5
20	5.0	1.0	90.86	117.0	91.80	97.74	98.38	114.8
21	5.0	0.5	90.00	114.4	90.54	95.04	96.85	112.4
22	3.0	9.8	107.4	136.8	115.4	123.1	120.2	138.4
23	3.0	7.0	104.2	134.5	111.9	119.3	116.4	134.7
24	3.0	5.0	100.4	130.4	106.3	115.7	112.0	129.5
25	3.0	3.0	97.16	126.8	101.2	112.0	108.2	124.7
26	3.0	2.0	95.38	123.3	97.56	106.6	104.0	120.9
27	3.0	1.0	94.18	120.5	95.76	101.4	101.0	116.6
28	3.0	0.5	93.74	117.7	94.50	98.82	99.00	114.4
29	1.0	9.8	110.3	140.7	120.5	127.8	123.9	141.2
30	1.0	7.0	107.0	138.4	116.5	124.6	120.0	137.5
31	1.0	5.0	103.7	134.1	112.3	121.3	115.6	132.2
32	1.0	3.0	100.7	130.3	106.5	118.1	111.9	127.1
33	1.0	2.0	99.07	126.7	102.4	112.7	108.6	123.1
34	1.0	1.0	97.67	124.1	100.3	107.3	106.1	120.0
35	1.0	0.5	96.73	120.9	98.93	103.0	104.7	116.5

where
C1: Longitudinal grooves

C2: Transverse grooves

C3: V-20° grooves

C4: V-40° grooves

C5: Longitudinal + transverse grooves

C6: Longitudinal + V-20° grooves

Please find out the influences of water film thickness, tire groove depth, and tread groove pattern on the hydroplaning speed.

P4.6: Given that A, B, and C are the three basic design factors in Table 4.7.8, please give the 10 possible defining equations (alias equations) for factors D, E, F, G, H, J, K, L, M, and N.

P4.7: Composite Laminae (230 mm × 160 mm × 7 mm) made of epoxy reinforced with natural tamarind seed particulates and multiwalled carbon nano tubes (MWCNT), are prepared with the following three-level variations based on (0, 1, 2) system for an experimental design:

Factor	(0)	(1)	(2)	Coded variable
A: Particle size, nominal (μm)	112.5	187.5	262.5	$a = (C - 187.5)/75 + 1$
B: MWCNT (%)	0.1	0.2	0.3	$b = (A - 0.2)/0.1 + 1$
C: Filler weight (%)	10	20	30	$c = (B - 20)/10 + 1$
D: Epoxy (%) = 1 − A	—	—	—	—

The tensile test of each lamina is carried out according to ASTM D638. Please conduct multiple linear regression to derive the predictive equation. Data of the obtained ultimate tensile strengths are given as follows [Annigeri and Raju 2020]:

Run	a	b	c	A	B	c	σ_{uts} (MPa)
1	0	0	0	112.5	0.1	10	22.83
2	0	0	1	112.5	0.1	20	28.28
3	0	0	2	112.5	0.1	30	17.91
4	0	1	0	112.5	0.2	10	34.32
5	0	1	1	112.5	0.2	20	18.35
6	0	1	2	112.5	0.2	30	20.59
7	0	2	0	112.5	0.3	10	28.75
8	0	2	1	112.5	0.3	20	16.23
9	0	2	2	112.5	0.3	30	19.3
10	1	0	0	187.5	0.1	10	19.3
11	1	0	1	187.5	0.1	20	23.33
12	1	0	2	187.5	0.1	30	19.27
13	1	1	0	187.5	0.2	10	27.95
14	1	1	1	187.5	0.2	20	23.3
15	1	1	2	187.5	0.2	30	22.6
16	1	2	0	187.5	0.3	10	32.71
17	1	2	1	187.5	0.3	20	26.52
18	1	2	2	187.5	0.3	30	17.0-3
19	2	0	0	262.5	0.1	10	31.83
20	2	0	1	262.5	0.1	20	27.25
21	2	0	2	262.5	0.1	30	22.31
22	2	1	0	262.5	0.2	10	33.48
23	2	1	1	262.5	0.2	20	24.65
24	2	1	2	262.5	0.2	30	21.06
25	2	2	0	262.5	0.3	10	26.45
26	2	2	1	262.5	0.3	20	24.32
27	2	2	2	262.5	0.3	30	20.33

5

Composite Designs

Composite designs have many desirable features and are effective for both screening design in a single experiment and response surface modeling in a sequential follow-up experiment. They are often the choice in successive sampling of parametric space in such a way as to construct a second-order response surface model with both quadratic and interactive terms for progressive optimization. Each composite design is in gear with nonlinear models on how to get proper information with less experimentation.

5.1. Composite Factorial Designs

If lack of fit is significant, one will need to use a more complex model, such as a composite design (CD) or an orthogonal 3^{K-P} model, which are usually solved using ANOVA (analysis of variance) or by regression. However, a three-level orthogonal 3^{K-P} design is often prohibitive in terms of the number of runs and thus in terms of cost and effort. CDs are then the concise approach for rendering the successive sampling of parameter space in such a way as to construct a second-order polynomial function. For k quantitative factors, the second-order model to be sought after can be written as:

$$Y = \beta_0 + \sum_{i=1}^{K} \beta_i x_i + \sum_{i=1}^{K} \beta_{ii} x_i^2 + \sum_{i=1}^{K-1} \sum_{j=i+1}^{K} \beta_{ij} x_i x_j + \varepsilon \qquad (5.1.1)$$

where β_0, β_i, β_{ii}, and β_{ij} are the intercept and coefficients of linear, quadratic, and interactive (bilinear) terms, respectively, while $\varepsilon \sim N(0, \sigma^2)$ is the error term. A single DOE with two levels cannot provide adequate information to fit Equation (5.1.1). For example, if there are three factors of concern (K = 3), it will take at least 10 observations for exploring the 10 parameters in Equation (5.1.1), including β_0, β_1, β_2, β_3, β_{11}, β_{22}, β_{33}, β_{12}, β_{13}, and β_{23}. This is not payable on demand with eight runs inherent in the 2^3 design. On the other hand, it takes more than the need (10 runs) when employing 27 runs inherent in the 3^3 design. Practically, a CD is more realistic in search of a second-order model.

It is called a CD when the DOE is made of two or more distinct design categories, such as central composite design (e.g., 2_{III}^{K-P} augmented by additional runs mainly at center and axial points), design with mixed levels, orthogonal array composite design (e.g., 2_{IV}^{K-P} augmented by an orthogonal array), nested design (one factor nested in another factor), crossed design (special case of nested design), and split-plot design [Jones and Nachtsheim 2009, Anderson-Cook 2007].

5.2. Central Composite Designs

The original central composite design was first introduced by Box and Wilson to seek an economical way to do the DOE analysis when a first-order model shows any evidence of lack of fit. Some more accurate central composite designs have been discovered ever since. For example, a two-level design with center points, namely central composite designs (CCDs), is much less expensive while it still is a very good and simple way to establish the presence or absence of curvature.

CCDs are basically two-level full-factorial 2^K or fractional factorial 2^{K-P} designs augmented by a number of central and axial points, as well as other chosen runs, which is the most widely used experimentation strategy for the second-order model fitting. These designs allow the best estimation of all the regression parameters required to fit a complete rotatable second-order model to a desired response. "Rotatable" refers to the variance of the response function. CCDs are the most widely used experimental design as a RSM as these additional points, also called star points, are able to clear main effects of aliasing with two-factor interactions. These designs could be particularly useful for a design that contains four or five factors and only the experimental units from one lot are chosen (no blocking). Very large fractional factorials and CCDs are addressed in Sanchez and Sanchez [2005].

The run of "center points" at the design center such as the origin (0, 0, 0) for k = 3 and at a distance d along the axis from the design center give two add-on parameters in the CCD design relative to its corresponding 2^{K-P} design. The center runs provide information for exhibiting the response surface curvature as long as it is significant, while the additional axial points allow for the experimenter to obtain an estimation of the quadratic terms at the least effort. A CCD whose factorial portion has resolution III is a second-order design once augmented by the center and axial points [Box and Draper 2007]. These additional points are called star points in practice. The implementation of CCD for estimating a second-order polynomial model comes with the following five principal needs:

- Orthogonal: The property that allows individual effects of the k factors to be estimated independently without confounding. Also, orthogonality provides minimum variance estimates of the model coefficients that are not correlated. The sums of squares for the main effects (although not for the interaction or residual) will vary depending on the order in which the factors are entered in to the model, if the design is not orthogonal; it is common to use adjusted sums of squares instead.

- Rotatable: The property of rotating points of the design about the center of the factor space. The moments of the distribution of the design points are constant. A CD is chosen to produce rotatability. It means that the predicted response is capable of being estimated with uniform (equal) variance regardless of the direction from the center of the design space.

- Uniform: The third property of CCD designs used to control the number of center points is uniform precision.

- Replication: The CCD template calls for replication of the center point a few times, ideally six or more for the proper predictive characteristics in the middle region of experimentation.

- Random: The actual run order, including center points, should always be done at random. Otherwise, the effects may turn out to be biased by order and/or time-related lurking variables, consequently confounding true cause-and-effect relationships.

If a minimum or maximum value is not reached, the experimenter needs redo the experiments with runs (treatment combinations) that are closer to the optimum. The augmentation of a design matrix to make a CCD is illustrated here using 2^3. There are several different ways to make a CCD, including face-centered design and spherical CD.

5.2.1. Face-Centered Design (FCD)

FCD is a factorial or fractional factorial design with center points, augmented with a group of face points (Figure 5.2.1) that let you estimate curvature. It is particular CCD results from its basic full or fractional factorial design. For example, FCD based on full-factorial design 2^3 may be constructed with additional runs at the following points:

$$(-1, 0, 0), (1, 0, 0), (0, -1, 0), (0, 1, 0), (0, 0, -1), (0, 0, 1), \text{ and } (0, 0, 0)$$

The treatments listed above are attached to the original 2^3 design, as shown in Figure 5.2.1. These additional run points are referred to as the axial points (or star points), representing runs where all but one of the factors are set at their mid-levels at a time. Thus, the number of axial points is 2k (one positive and one negative) in a CCD for arguing 2^K design, e.g., six axial points for the 2^3 design as identified in the block 2 of Table 5.2.1.

FIGURE 5.2.1 Additional design contrasts with FCD arguing 2^3.

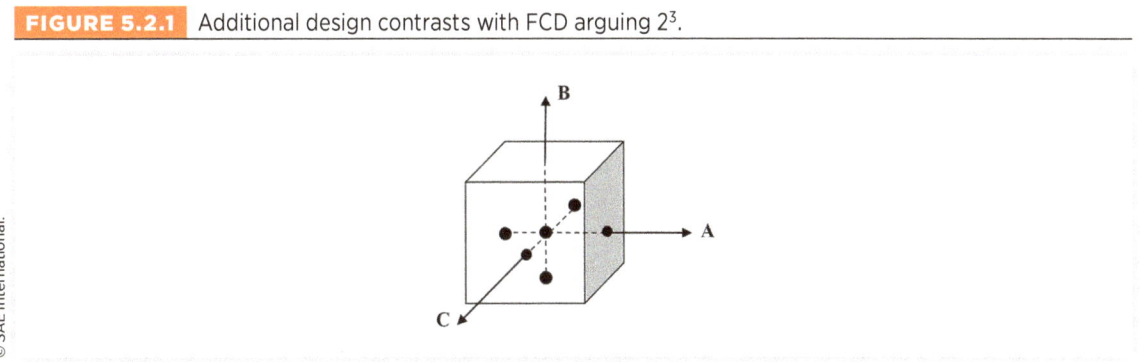

TABLE 5.2.1 16 runs (treatments) for the face CD based on 2^3.

Block	Run	A	B	C	
1	1	−1	−1	−1	
1	2	1	−1	−1	
1	3	−1	1	−1	
1	4	1	1	−1	
1	5	−1	−1	1	
1	6	1	−1	1	
1	7	−1	1	1	
1	8	1	−1	1	
2	9	−1	0	0	
2	10	1	0	0	
2	11	0	−1	0	
2	12	0	1	0	
2	13	0	0	−1	
2	14	0	0	1	
2	15	0	0	0	(for formulating predictive equation)
2	16	0	0	0	(replicated run for checking lack of fit)

This design will not have rotatability and will have a degree less orthogonality for the quadratic terms but the issue is more of a theoretical concern than a practical one. FCD designs provide relatively good predictions over the entire design space and do not require using points outside the original factor range. However, the precision for estimating pure quadratic coefficients is low.

5.2.2. Spherical Composite Design-Circumscribed (SCD-C)

Each factor is run at five levels, i.e., $-\lambda$, -1, 0, 1, and λ for the SCD-C (spherical composite design-circumscribed) algorithm, instead of the three levels of -1, 0, and 1 for the original 2^3 design based on the FCD. The additional design contrasts are depicted in Figure 5.2.2. The reason for running the spherical CCD is to have a rotatable design around the origin $(0, 0, 0)$ and the SCD-C has a balanced equivalent directional variation with each factor. An example spherical CD is given in Table 5.2.2.

As each response surface design applies, it is necessary to consider the potential "block effect" as to eliminating nuisance variables. Such a problem may occur when the higher order is assembled sequentially from the lower order. The necessity arises because of various reasons. For example, test environment may have changed considerably as time elapses between the running of the first eight runs (first-order terms) and the running of the supplemental experiments (runs 9–15). Note that runs 16–20 are replicated at the center point for checking the lack of fit.

The variance of outcome Y will be consistent in both the size and pattern [Oehlert 2000], if the SCD-C is to be rotatable around the origin (center) with the following λ value:

$$\lambda = 2^{K/4} \qquad \text{(for full factorial designs)} \tag{5.2.1}$$

or

$$\lambda = 2^{(K-P)/4} \qquad \text{(for fractional factorial designs)} \tag{5.2.2}$$

The λ-values are to be measured in reference to the dimensionless scale, (−1, 1). For example, $\lambda = 1.414$ for design 2^2, $\lambda = 1.682$ for design 2^3, $\lambda = 2.000$ for design 2^4, $\lambda = 2.378$ for design 2^5, $\lambda = 2.828$ for design 2^6, $\lambda = 3.364$ for design 2^7, $\lambda = 1.682$ for design $2^{7\text{-}4}$, and $\lambda = 2.000$ for design $2^{8\text{-}4}$. Rotatable SCD-C designs provide high quality predictions over the entire design space. A spherical CD means a SCD-C design by default, unless otherwise addressed.

An example design matrix based on the 2^3 design is demonstrated in Table 5.5.2(I). Consider the case that A = 20 at a = (A − 15)/5 = 1. Thus, at a = 1.682, A = 23.4. This means the additional point moves away from the center point further by 3.4 (=23.4 − 20).

FIGURE 5.2.2 Additional design contrasts with the spherical design arguing 2^3 ($\lambda = 2^{3/4}$).

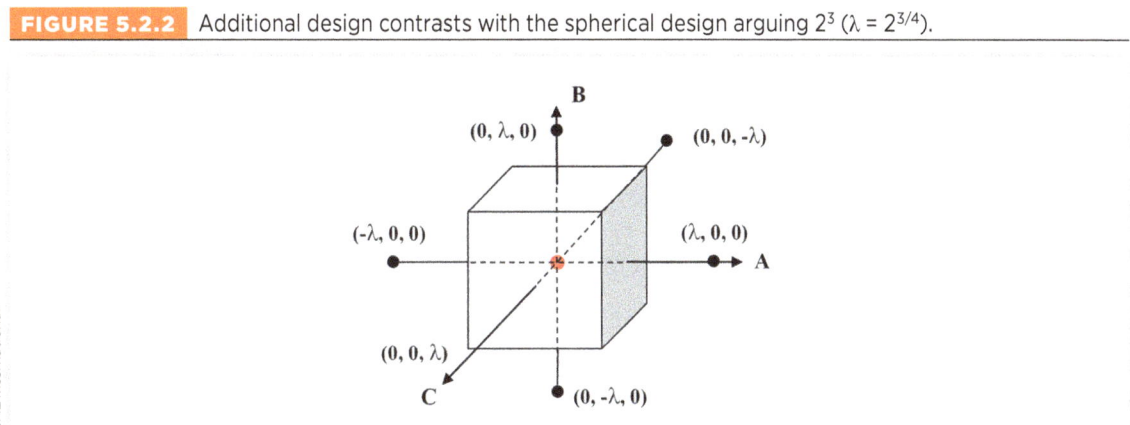

Spherical composite designs, in contrast to the FCD, generates rotatable or near-rotatable design matrices that improve estimation of quadratic terms and reduce the prediction error.

Nevertheless, it is sometimes cumbersome for an experimenter to go through the complexity of operating each variable at five levels.

TABLE 5.2.2 Constructions of treatments for spherical CDs based on 2^3.

(I) SCD—Circumscribed with λ = 1.682:									
Level			Run	Coded variable			Experimentation		
Factor	(−1)	(+1)		a	b	c	A	B	C
A	10	20	1	−1	−1	−1	10	40	25
B	40	80	2	1	−1	−1	20	40	25
C	25	35	3	−1	1	−1	10	80	25
			4	1	1	−1	20	80	25
			5	−1	−1	1	10	40	45
			6	1	−1	1	20	80	45
			7	−1	1	1	10	40	45
			8	1	1	1	20	80	45
			9	−λ	0	0	6.6	60	35
			10	λ	0	0	23.4	60	35
			11	0	−λ	0	15	26.36	35
			12	0	λ	0	15	93.64	35
			13	0	0	−λ	15	60	18.18

(Continued)

TABLE 5.2.2 (Continued) Constructions of treatments for spherical CDs based on 2^3.

Level			Run	Coded variable			Experimentation		
(I) SCD—Circumscribed with $\lambda = 1.682$:									
Factor	(−1)	(+1)		a	b	c	A	B	C
			14	0	0	λ	15	60	51.82
			15	0	0	0	15	60	35
			16	0	0	0	15	60	35
			17	0	0	0	15	60	35
			18	0	0	0	15	60	35
			19	0	0	0	15	60	35
			20	0	0	0	15	60	35

Level			Run	Coded variable			Experimentation		
(II) SCD—Inscribed with $\lambda = 1.682$:									
Factor	(−λ)	(+λ)		a	b	c	A	B	C
A	10	20	1	−1	−1	−1	13.4	46.8	28.4
B	40	80	2	1	−1	−1	16.6	46.8	28.4
C	25	35	3	−1	1	−1	13.4	53.2	28.4
			4	1	1	−1	16.6	53.2	28.4
			5	−1	−1	1	13.4	46.8	41.6
			6	1	−1	1	16.6	53.2	41.6
			7	−1	1	1	13.4	46.8	41.6
			8	1	1	1	16.6	53.2	41.6
			9	−λ	0	0	10	60	35
			10	λ	0	0	20	60	35
			11	0	−λ	0	15	40	35
			12	0	λ	0	15	80	35
			13	0	0	−λ	15	60	25
			14	0	0	λ	15	60	45
			15	0	0	0	15	60	35
			16	0	0	0	15	60	35
			17	0	0	0	15	60	35
			18	0	0	0	15	60	35
			19	0	0	0	15	60	35
			20	0	0	0	15	60	35

Notes:

(1) Run 15 is the run at the center point included in the SCD.

(2) Runs 16–20 are replicated runs for checking the potential lack of fit.

5.2.3. Spherical Composite Design-Inscribed (SCD-I)

In the spherical composite design-inscribed (SCD-I), the specified low and high values become the star points as illustrated in Table 5.2.2(II). The coding for the SCD-I is the same as SCD-C, but the physical values of star points are "inscribed" inside the boundary of the original (−1, +1) for each factor. Analysis of variance for various options of SCD-I is demonstrated in [Nwanya and Dozie 2020].

SCD-I designs use only points within the factor ranges originally specified on the condition that interpolation is less risky that extrapolation in the real-world experimentation. However, the quality of response surface model based on SCD-I is usually not as good as SCD-C. Predictive capabilities with different replicated runs for SCD-I is presented by Nwanya and Dozie [2020].

5.2.4. Study on Warpage of Injection-Molded Plastics Using Spherical Composite Design

The trend of injection molding of plastics is set to produce complex free-form parts with thin walls such as housings or casings that meet both the engineering needs and aesthetic criteria [Chiang and Chang 2007, Tang et al. 2007]. This is often accomplished using the combination of DOE and finite element methods. For example, on the injection molding defects of PPS (Polyphenylsulfone—amorphous thermoplastics), a screening design with seven process factors, i.e., melt temperature, mold temperature, injection time, injection pressure, packing time, packing pressure and filling-to-packing switchover point, was conducted via finite element analysis using Moldflow and only four of them appear to be statistically significant [Andrisano et al. 2011]. An optimization study on the reduction of injection molding defects with respect to these four process parameters, as shown in Table 5.2.3, is then applied utilizing the CCD. The basis of variation in the design is the distance of the axial points measured from the center of the design following the following five levels: $(-\lambda, -1, 0, 1, \lambda)$, of which $\lambda = 2^{K/4} = 2^{4/4} = 2$ according to Equation (5.2.1) as denoted by the (−1, 1) coding system. The design levels of the four factors are identified in Table 5.2.4 accordingly.

TABLE 5.2.3 Design levels of process parameters for injection molding of plastics [Andrisano et al. 2011].

Factor	Level					Coding
	−2	−1	0	1	2	
A: Melt temperature (°C)	350	360	370	380	390	a = (A − 370)/10
B: Mold temperature (°C)	130	140	150	160	170	b = (B − 150)/10
C: Packing time (s)	0.4	0.5	0.6	0.7	0.8	c = (C − 0.6)/0.1
D: Packing pressure (MPa)	90	105	120	135	150	d = (D − 120)/15

Courtsey of Francesco Gherardini.

TABLE 5.2.4 Spherical CD based on 2^4 design for injection molding of plastics [Andrisano et al. 2011].

Run	a	b	c	d	A	B	C	D	Y (µm)	Y_p (µm)
1	−1	−1	−1	−1	360	140	0.5	105	57	59.4
2	1	−1	−1	−1	380	140	0.5	105	64	65.0
3	−1	1	−1	−1	360	160	0.5	105	59	59.4
4	1	1	−1	−1	380	160	0.5	105	66	65.0
5	−1	−1	1	−1	360	140	0.7	105	58	59.4
6	1	−1	1	−1	380	140	0.7	105	65	65.0
7	−1	1	1	−1	360	160	0.7	105	60	59.4
8	1	1	1	−1	380	160	0.7	105	67	65.0
9	−1	−1	−1	1	360	140	0.5	135	33	34.1
10	1	−1	−1	1	380	140	0.5	135	39	41.3
11	−1	1	−1	1	360	160	0.5	135	35	34.1
12	1	1	−1	1	380	160	0.5	135	41	41.3
13	−1	−1	1	1	360	140	0.7	135	34	34.1
14	1	−1	1	1	380	140	0.7	135	40	41.3
15	−1	1	1	1	360	160	0.7	135	36	34.1
16	1	1	1	1	380	160	0.7	135	42	41.3
17	−2	0	0	0	350	150	0.6	120	43	43.1
18	2	0	0	0	390	150	0.6	120	56	55.9
19	0	−2	0	0	370	130	0.6	120	47	49.5
20	0	2	0	0	370	170	0.6	120	51	49.5
21	0	0	−2	0	370	150	−0.8	120	52	49.5
22	0	0	2	0	370	150	0.8	120	50	49.5
23	0	0	0	−2	370	150	0.6	90	76	75.8
24	0	0	0	2	370	150	0.6	150	27	26.8
25	0	0	0	0	370	150	0.6	120	49	49.5

where Y (µm): Global warpage, defined as the averaged displacements of finite element mesh nodes.

Courtesy of Francesco Gherardini.

The global warpage responses, i.e., Y values listed in Table 5.2.4, resulting from finite element analyses are utilized to find the relationship between the global warpage and these four factors using DOE. Effects of the four selected parameters and their two-factor interactions on global warpage are examined using the CCD based on ANOVA regression. After several iterations for removing insignificant effects one by one, the F-ratios and their significant levels (denoted as p-values) derived using Minitab are listed in Table 5.2.5. The final regression model is:

$$Y = 175.4 - 2.267\,D + 0.00194\,D^2 + 0.002663\,A\,D$$

with which the associated correlative functions are correlation R = 99.4%, adjusted correlation R_{adj} = 99.5%, predictive correlation R_{pred} = 99.3%. The predicted values of all treatments using the regression model, i.e., Y_p, are listed in Table 5.2.4. Model effectiveness and adequacy are supported by the prediction plot (Column Y versus Column Y_p) and residual plots (Y_p–Y versus Column Y_p), as demonstrated in Figure 5.2.3. The most important process factor in reducing warpage of the thin-shelled plastic part is factor D, i.e., packing pressure. The higher the packing pressure is, the less warpage the part can hold up to.

TABLE 5.2.5 Analysis of variance on global warpage of injection molding of thin-shelled plastics.

Source of variation	DOF	SS$_{Adj}$	MS$_{Adj}$	F-ratio	P-value
Regression	3	3854	1285	632	0.000
D (Packing pressure)	1	103.4	103.4	50.9	0.000
D^2	1	4.740	4.74	2.33	0.142
AD	1	247.7	247.7	122	0.000
Error	21	42.72	2.03	—	—
Lack of fit	5	7.920	1.58	0.73	0.613
Pure error	16	34.80	2.17	—	—
Subtotal	24	3897			
Grand ave.	1				
Total	25				

© SAE International.

FIGURE 5.2.3 Model checking of warpage of injection-molded thin-shelled plastics.

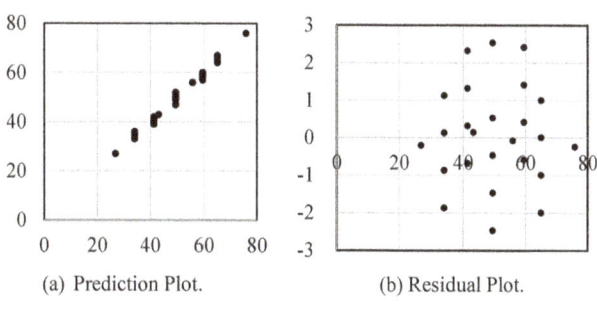

(a) Prediction Plot.   (b) Residual Plot.

© SAE International.

5.2.5. Between-Face-and-Spherical CD

How to select the right CCD is a noteworthy art [Oyejola and Nwanya 2015]. An extrapolation of factors from face CD (taking three levels) to spherical CD (taking five levels) based on Equations (5.2.1) and (5.2.2) are sometimes not feasible. Some automotive researchers are more concerned with the desired upper and lower factorial limits in the physical sense [Westlake 1965], whether the extrapolation is statistically rotatable or not. In this case, the axial points will correspond to the design space minimum and maximum and the min/max points of the factorial will be something greater or less than the overall minimum and maximum. Because these CCDs are slightly non-orthogonal, some exact properties may be deteriorated. A measure of orthogonality for CCDs is addressed in [Ohaegbulem and Chigbu 2022]. Nevertheless, as long as these points are nicely spread out in space, the fitted regression model can still approximate the reality closely, but slightly with a difference in variance. A framework for generating, with a limited number of design points, a design which is nearly orthogonal and also nearly balanced for any mix of factor types (categorical, numerical discrete, and numerical continuous) and/or mix of factor levels is discussed in [Vieira et al. 2011].

5.2.6. Box-Behnken Design

This is the first CCD that has ever been proposed [Box and Behnken 1960], but it is a unique factorial design that has nothing to do with the orthogonal 2^{K-P} designs. Consider a design example that has three factors for the purpose of illustration. The origin (0, 0, 0) and all the treatments (runs) at the middle points of edges are used to create quadratic terms. Thirteen runs are suggested for a three-factor design to utilize such a CCD as demonstrated graphically in Figure 5.2.4. Box-Behnken design offers some advantage in requiring a fewer number of runs for three factors, but this advantage disappears for four or more factors.

FIGURE 5.2.4   Thirteen runs for exploring quadratic terms of a three-factor design [Box and Behnken 1960].

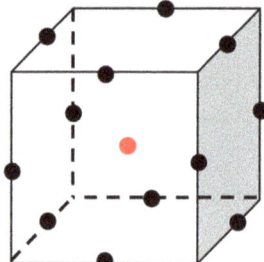

© SAE International.

Box-Behnken design (BBD) has limited orthogonal blocking when compared to other CCDs, but it is convenient to use. It is nearly an orthogonal Resolution V design, and all linear effects, quadratic effects, and bilinear effects (due to two-factor interactions) can be estimated. In general, slight nonorthogonality is not an issue. Nevertheless, because of this slight-orthogonality, the response surface modeling accuracy based on a BBD is usually not as good as its corresponding face-center design, which in turn is not as good as spherical CD because of the rotatability.

An application study of BBD is conducted by Perincek and Colak [2013] in order to investigate harmonic currents induced by various electric appliances such as heaters, cell phone chargers, lamps, and television sets in an electric distribution system as installed in an electric vehicle or office. Harmonic currents are considered as harmonic pollution that may result in (a) overheating or derating of transformer, (b) overheating of wiring, (c) damaging of capacitor banks, (d) resonance, (e) communication interference, and (f) distorted supply voltage.

In the context of DOE, the amounts of induced third and fifth harmonic currents, i.e., responses Y_3 (Ampere) and Y_5 (Ampere) in Table 5.2.4, respectively, are examined against the following three loading elements in the single-phased mode: incandescent lamp (40 W, 50/60 Hz, 20–230 V), electric heaters (530 W, 50/60 Hz, 20–230 V) and compact fluorescent lamps (20 W, 50/60 Hz, 20–240 V), which are thus chosen as the three independent design factors. The design matrix and test results are listed in Table 5.2.6. Following the regression analysis, one can formulate the predictive equations for the third and fifth harmonic currents as follows [Perincek and Colak 2013]:

$$Y_3 = 0.0161 + 0.0101\ a + 0.0288\ b - 0.0206\ c + 0.0207\ a^2 + 0.0139\ b^2 + 0.0122\ c^2$$

$$- 0.0344\ a\ b + 0.00155\ a\ c - 0.00378\ b\ c$$

and

$$Y_5 = -0.00066 + 0.0845\ a + 0.0329\ b - 0.0045\ c + 0.0098\ a^2 + 0.00122\ b^2 + 0.0065\ c^2$$

$$- 0.008628\ a\ b - 0.0029\ a\ c - 0.001275\ b\ c$$

TABLE 5.2.6 Thirteen runs for BBD with three factors + two replicates at the center.

Run	a	b	c	Y_3	Y_5	
1	−1	−1	0	0.105	0.201	
2	1	−1	0	0.0025	0.0048	
3	−1	1	0	0.122	0.0723	
4	1	1	0	0.0869	0.234	
5	−1	0	−1	0.0569	0.0318	
6	1	0	−1	0.0949	0.229	
7	−1	0	1	0.0568	0.0434	
8	1	0	1	0.1010	0.229	
9	0	−1	−1	0.0540	0.0919	
10	0	1	−1	0.0842	0.146	
11	0	−1	1	0.0644	0.106	
12	0	1	1	0.0795	0.155	
13	0	0	0	0.0498	0.116	(for formulating predictive equation)
14	0	0	0	0.0428	0.118	(for checking lack of fit)
15	0	0	0	0.041	0.117	(for checking lack of fit)

5.2.7. **Small CD**

A typical small CD (SCD) for a quadratic response model resulting from a DOE with four factors is demonstrated in Table 5.2.7 that has 20 runs based on 2^{4-1} (D = ABC). In addition to Equation (5.2.3), the λ-value is also constrained by another equation as [Hartley 1959]:

$$\lambda < (K - P)^{1/2} \tag{5.2.3}$$

TABLE 5.2.7 Twenty runs for the small CD based on 2^{4-1} (D = ABC).

Run	a	b	c	d	
1	−1	−1	−1	−1	
2	1	−1	−1	1	
3	−1	1	−1	1	
4	1	1	−1	−1	
5	−1	−1	1	1	
6	1	−1	1	−1	
7	−1	1	1	−1	
8	1	1	1	1	
9	−λ	0	0	0	
10	λ	0	0	0	
11	0	−λ	0	0	
12	0	λ	0	0	
13	0	0	−λ	0	
14	0	0	λ	0	
15	0	0	0	−λ	
16	0	0	0	λ	
17	0	0	0	0	(center point for predictive equation)
18	0	0	0	0	(replicated for checking the lack of fit)
19	0	0	0	0	(replicated for checking the lack of fit)
20	0	0	0	0	(replicated for checking the lack of fit)

© SAE International.

5.2.8. Incomplete Small CD

For a large-scaled second-order response surface model, a small CD still requires many runs. A new subsaturated sampling method, called incomplete small CD-I (ISCD-I), was proposed by Kim and Heo [2003].

One algorithm of the incomplete small CD is to reduce the number of runs that result in the cubic terms and three-factor interactions of small CD. The number of runs can be reduced using incomplete small CD as long as the number of factors for building a response surface model is equal to five or above.

The other algorithm, called incomplete small design-II (ISCD-II), is to further scratch some points out of ISCD-I design, as demonstrated in Figure 5.2.5. Note that both design orthogonality and design rotatability suffer from such a reduction after scratching.

FIGURE 5.2.5 Going from small CD to ISCD-II.

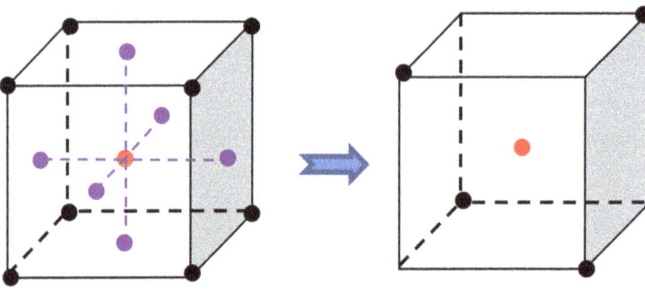

© SAE International.

5.3. Design with Mixed Two-Level and Three-Level Factors

A DOE with mixed two-level and three-level is needed occasionally. These designs have many desirable features and are effective for factor screening and response surface modeling. Three frequently used designs are identified in Table 5.3.1 and their design matrices are listed in Tables 5.3.1–5.3.3.

TABLE 5.3.1 Design matrices for designs with mixed levels.

Design	Runs	Design generators	Matrix	Resolution
$2^1 \times 3^1$ (L_8)	8	—	Table 5.3.2	—
$2^1 \times 3^{7-5}$ (L_{18})	18	—	Table 5.3.3	III
$2^{11-9} \times 3^{12-10}$ (L_{36})	36	—	Table 5.3.4	III

© SAE International.

5.3.1. $2^1 \times 3^1$

Consider a design that consists of one two-level factor (factor A) and one three-level factor (factor Z). The three-level factor situated in the design matrix can be formed using two additional emulation two-level factors (i.e., factors B and C in Table 5.3.2) by setting the following mapping [Montgomery 2019]:

$$(B = -1, C = -1) \rightarrow Z = Z_1 \text{ (Low Level)}$$
$$(B = 1, C = -1) \rightarrow Z = Z_2 \text{ (Medium Level)}$$
$$(B = -1, C = 1) \rightarrow Z = Z_2 \text{ (Medium Level)}$$
$$(B = 1, C = 1) \rightarrow Z = Z_3 \text{ (High Level)}$$

Now that factor Z has 2 DOFs (degrees of freedom) as formed above, it has the following two terms: Z_L (linear term) and Z_Q (quadratic term). According to the ANOVA, one has the following final SS (sums of squares) for factors A and Z, their interaction, and the error (residual):

$$SS_A = SS_A \tag{5.3.1}$$

$$SS_Z = SS_{ZL} + SS_{ZQ} \tag{5.3.2}$$

$$SS_{AZ} = SS_{A\,ZL} + SS_{A\,ZQ} \tag{5.3.3}$$

and

$$SS_{error} = \text{Obtained from replicated runs}$$

Note that run 5 is a replicate of run 3, while run 6 is a replicate of run 4 in Table 5.3.2. Thus, there are two DOFs for the error as obtained from these two sets of two replications each.

TABLE 5.3.2 Forming a $2^1 \times 3^1$ design (factor A—2 levels and factor Z—3 levels) like 2^3.

Run	A	Z	B(Z_L)	C(Z_Q)	AB(AZ_L)	AC(AZ_Q)	BC($Z_L Z_Q$)	ABC($AZ_L Z_Q$)
1	−1	Z_1	−1	−1	1	1	1	−1
2	1	Z_1	−1	−1	−1	−1	1	1
3	−1	Z_2	1	−1	−1	1	−1	1
4	1	Z_2	1	−1	1	−1	−1	−1
5	−1	Z_2	−1	1	1	−1	−1	1
6	1	Z_2	−1	1	−1	1	−1	−1
7	−1	Z_3	1	1	−1	−1	1	−1
8	1	Z_3	1	1	1	1	1	1

5.3.2. $2^1 \times 3^{7-5}$ (L$_{18}$) with Resolution III

It is a good idea for an experimenter to do fractional factorial design $2^1 \times 3^{7-5}$ (L$_{18}$) as shown in Table 5.3.3 for gathering information on the main effects in relatively only a few runs as a screening design. It releases no design content about the interactions.

TABLE 5.3.3 The fractional factorial design matrix of $2^1 \times 3^{7-5}$ (L_{18}) with resolution III.

Run	a	b	c	d	e	f	g	h
1	1	1	1	1	1	1	1	1
2	1	1	2	2	2	2	2	2
3	1	1	3	3	3	3	3	3
4	1	2	1	1	2	2	3	3
5	1	2	2	2	3	3	1	1
6	1	2	3	3	1	1	2	2
7	1	3	1	2	1	3	2	3
8	1	3	2	3	2	1	3	1
9	1	3	3	1	3	2	1	2
10	2	1	1	3	3	2	2	1
11	2	1	2	1	1	3	3	2
12	2	1	3	2	2	1	1	3
13	2	2	1	2	3	1	3	2
14	2	2	2	3	1	2	1	3
15	2	2	3	1	2	3	2	1
16	2	3	1	3	2	3	1	2
17	2	3	2	1	3	1	2	3
18	2	3	3	2	1	2	3	1

© SAE International.

5.3.3. $2^{11-9} \times 3^{12-10}$ (L_{36}) with Resolution III

An experimenter will gather a lot of information on the main effects in a relatively few runs using fractional factorial design $2^{11-9} \times 3^{12-10}$, as shown in Table 5.3.4. It is capable of testing out whether nonlinear terms are needed in the model, as long as the three-level factors are mandated. Nevertheless, the information about interactions based on such a design is limited.

TABLE 5.3.4 The fractional factorial design matrix of $2^{11-9} \times 3^{12-10}$ (L_{36}).

Run	a	b	c	d	e	f	g	h	i	j	k	l	m	n	o	p	q	r	s	t	u	v	w
1	1	1	1	1	1	1	1	1	1	1	1	1	1	1	1	1	1	1	1	1	1	1	1
2	1	1	1	1	1	1	1	1	1	1	1	2	2	2	2	2	2	2	2	2	2	2	2
3	1	1	1	1	1	1	1	1	1	1	1	3	3	3	3	3	3	3	3	3	3	3	3
4	1	1	1	1	1	2	2	2	2	2	2	1	1	1	1	2	2	2	2	3	3	3	3
5	1	1	1	1	1	2	2	2	2	2	2	2	2	2	2	3	3	3	3	1	1	1	1
6	1	1	1	1	1	2	2	2	2	2	2	3	3	3	3	1	1	1	1	2	2	2	2
7	1	1	2	2	2	1	1	1	2	2	2	1	1	2	3	1	2	3	3	1	2	2	3
8	1	1	2	2	2	1	1	1	2	2	2	2	2	3	1	2	3	1	1	2	3	3	1
9	1	1	2	2	2	1	1	1	2	2	2	3	3	1	2	3	1	2	2	3	1	1	2
10	1	2	1	2	2	1	2	2	1	1	2	1	1	3	2	1	3	2	3	2	1	3	2
11	1	2	1	2	2	1	2	2	1	1	2	2	1	3	2	1	3	1	3	2	1	3
12	1	2	1	2	2	1	2	2	1	1	2	3	3	2	1	3	2	1	2	1	3	2	1
13	1	2	2	1	2	2	1	2	1	2	1	1	2	3	1	3	2	1	3	3	2	1	2
14	1	2	2	1	2	2	1	2	1	2	1	2	3	1	2	1	3	2	1	1	3	2	3
15	1	2	2	1	2	2	1	2	1	2	1	3	1	2	3	2	1	3	2	2	1	3	1

(Continued)

© SAE International.

TABLE 5.3.4 (Continued) The fractional factorial design matrix of $2^{11-9} \times 3^{12-10}$ (L_{36}).

Run	a	b	c	d	e	f	g	h	i	j	k	l	m	n	o	p	q	r	s	t	u	v	w
16	1	2	2	2	1	2	2	1	2	1	1	1	2	3	2	1	1	3	2	3	3	2	1
17	1	2	2	2	1	2	2	1	2	1	1	2	3	1	3	2	2	1	3	1	1	3	2
18	1	2	2	2	1	2	2	1	2	1	1	3	1	2	1	3	3	2	1	2	2	1	3
19	2	1	2	2	1	1	2	2	1	2	1	1	2	1	3	3	3	1	2	2	1	2	3
20	2	1	2	2	1	1	2	2	1	2	1	2	3	2	1	1	1	2	3	3	2	3	1
21	2	1	2	2	1	1	2	2	1	2	1	3	1	3	2	2	2	3	1	1	3	1	2
22	2	1	2	1	2	2	2	1	1	1	2	1	2	2	3	3	1	2	1	1	3	3	2
23	2	1	2	1	2	2	2	1	1	1	2	2	3	3	1	1	2	3	2	2	1	1	3
24	2	1	2	1	2	2	2	1	1	1	2	3	1	1	2	2	3	1	3	3	2	2	1
25	2	1	1	2	2	2	1	2	2	1	1	1	3	2	1	2	3	3	1	3	1	2	2
26	2	1	1	2	2	2	1	2	2	1	1	2	1	3	2	3	1	1	2	1	2	3	3
27	2	1	1	2	2	2	1	2	2	1	1	3	2	1	3	1	2	2	3	2	3	1	1
28	2	2	2	1	1	1	1	2	2	1	2	1	3	2	2	2	1	1	3	2	3	1	3
29	2	2	2	1	1	1	1	2	2	1	2	2	1	3	3	3	2	2	1	3	1	2	1
30	2	2	2	1	1	1	1	2	2	1	2	3	2	1	1	1	3	3	2	1	2	3	2
31	2	2	1	2	1	2	1	1	1	2	2	1	3	3	3	2	3	2	2	1	2	1	1
32	2	2	1	2	1	2	1	1	1	2	2	1	1	1	3	1	3	3	2	3	2	2	2
33	2	2	1	2	1	2	1	1	1	2	2	3	2	2	2	1	2	1	1	3	1	3	3
34	2	2	1	1	2	1	2	1	2	2	1	1	3	1	2	3	2	3	1	2	2	3	1
35	2	2	1	1	2	1	2	1	2	2	1	2	1	2	3	1	3	1	2	3	3	1	2
36	2	2	1	1	2	1	2	1	2	2	1	3	2	3	1	2	1	2	3	1	1	2	3

5.3.4. Orthogonal Array CD (OACD)

The CCD employs a one-factor-at-a-time approach for the additional points because each axial point has only one nonzero component. These additional axial points in CCD provide no information on interactive terms and thus resolution IV designs cannot be used as the two-level portion. A new class of CDs that combine a two-level full or fractional factorial design and a three-level orthogonal array, referred as OACD, was proposed by Xu et al. [2014]. One advantage of OACD is that one can utilize a resolution IV design in the two-level portion and potentially continue the design process with sequential experiments [Xu et al. 2014]. Each OACD may provide an inferential procedure on quadratic terms in addition to linear and interactive (bilinear terms) terms right after fractional factorial design 2_{IV}^{K-P} is applied.

OACDs can have different run sizes as obtained by combining a two-level (fractional) factorial designs and three-level orthogonal arrays. For example, the design matrix given in Table 5.3.5 consists of a fractional factorial design matrix with two levels (i.e., runs 1–16), i.e., 2_{IV}^{5-1} (I = ABCDE), and 18 additional runs that form a three-level orthogonal array, which has one run (treatment) at the center point (i.e., run 18), namely OA_{18} (K = 5). Note that K is the number of factors, which will be also the number of columns in the orthogonal arrays. There are 68 data available, as two replicates are taken for each run (treatment). After the regression model is undertaken, a frequency plot of effects that account for intersect, linear, bilinear and quadratic terms in Equation (5.1.1) is reserved in Figure 5.3.1 [Xu et al. 2014]. The first 16 runs based on 2_{IV}^{5-1} cannot accommodate the quadratic coefficients and the last 18 runs based on OA_{18} (K = 5) cannot do the interactive terms. Nevertheless, the complete 34 runs are able to generate all the inferential information on all the effects for Equation (5.1.1). This is crucial when there coexists a significant quadratic term and a significant interactive term as unveiled in Figure 5.3.1, where the significance of each term is assessed according to t-statistics.

TABLE 5.3.5 Fractional factorial design 2_{IV}^{5-1} (runs 1–6; A = BCDE) + three-level orthogonal array (OA: runs 17–34) [Xu et al. 2014].

Design	Run	a	b	c	d	e	Y_1	Y_2
2_{IV}^{5-1}	1	1	−1	−1	−1	−1	69.8	72.0
	2	−1	1	−1	−1	−1	66.4	67.4
	3	−1	−1	1	−1	−1	83.0	68.6
	4	−1	−1	−1	1	−1	16.2	23.4
	5	−1	−1	−1	−1	1	46.1	33.6
	6	1	1	1	−1	−1	68.6	65.5
	7	1	1	−1	1	−1	6.8	7.2
	8	1	1	−1	−1	1	15.6	19.1
	9	1	−1	1	1	−1	11.1	7.0
	10	1	−1	1	−1	1	19.8	20.3
	11	1	−1	−1	1	1	3.7	4.7
	12	−1	1	1	1	−1	5.8	3.9
	13	−1	1	−1	1	1	2.6	4.0
	14	−1	1	1	−1	1	42.2	23.2
	15	−1	−1	1	1	1	1.8	5.2
	16	1	1	1	1	1	3.1	3.4
OA_{18} (K = 5)	17	−1	−1	−1	−1	−1	78.6	81.9
	18	0	0	0	0	0	13.3	16.7
	19	1	1	1	1	1	3.4	3.8
	20	−1	−1	0	0	1	21.4	25.2
	21	0	0	1	1	−1	8.6	4.4
	22	1	1	−1	−1	0	18.0	27.3
	23	−1	0	−1	1	0	7.3	2.4
	24	0	1	0	−1	1	17.9	23.7
	25	1	−1	1	0	−1	52.9	54.3
	26	−1	1	1	0	0	13.2	8.8
	27	0	−1	−1	1	1	2.1	4.5
	28	1	0	0	−1	−1	73.4	73.9
	29	−1	0	1	−1	1	19.6	14.6
	30	0	1	−1	0	−1	59.1	41.7
	31	1	−1	0	1	0	1.4	2.6
	32	−1	1	0	1	−1	7.3	4.8
	33	0	−1	1	−1	0	22.3	24.0
	34	1	0	−1	0	1	14.1	18.3

where Y_1 and Y_2: Replicated runs for each treatment.

FIGURE 5.3.1 Frequency plot of all effects with the data given in Table 5.3.3 [Xu et al. 2014].

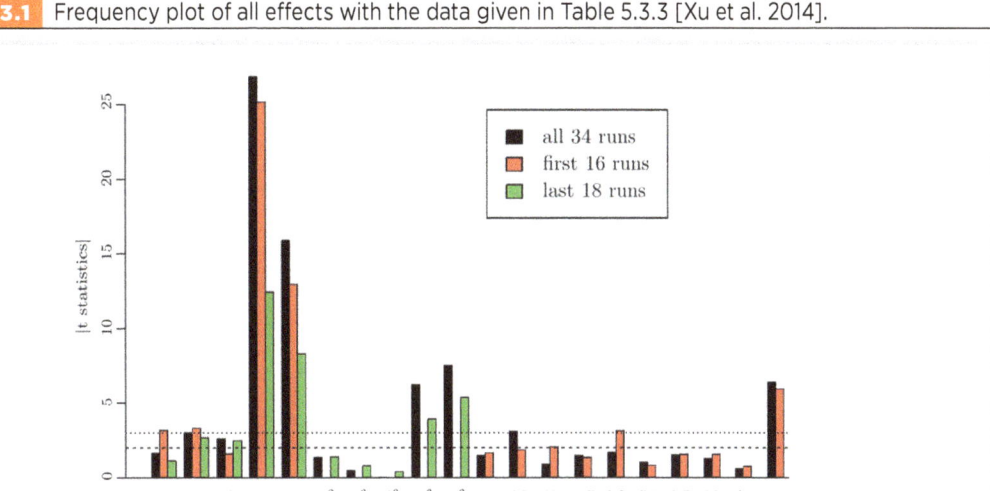

Orthogonal arrays OA_9 (K = 4) and OA_{18} (K = 7) are given in Table 5.3.6 [Xu et al. 2014], while more three-level orthogonal arrays can be found in Dean et al. [2017].

TABLE 5.3.6 Three-level orthogonal arrays [Xu et al. 2014]: OA_9 (K = 4) and OA_{18} (K = 7).

OA₉ (K = 4):	Run	a	b	c	d		
	1	−1	−1	−1	−1		
	2	−1	0	0	1		
	3	−1	1	1	0		
	4	0	−1	0	0		
	5	0	0	1	−1		
	6	0	1	−1	1		
	7	1	−1	1	1		
	8	1	0	−1	0		
	9	1	1	0	−1		
OA₁₈ (K = 7):	**Run**	**a**	**b**	**c**	**d**	**e**	**f**	**g**
	1	−1	−1	−1	−1	−1	−1	−1
	2	−1	0	0	0	0	0	0
	3	−1	1	1	1	1	1	1
	4	0	−1	−1	0	0	1	1
	5	0	0	0	1	1	−1	−1
	6	0	1	1	−1	−1	0	0
	7	1	−1	0	−1	1	0	1
	8	1	0	1	0	−1	1	−1
	9	1	1	−1	1	0	−1	0
	10	−1	−1	1	1	0	0	−1
	11	−1	0	−1	−1	1	1	0
	12	−1	1	0	0	−1	−1	1
	13	0	−1	0	1	−1	1	0
	14	0	0	1	−1	0	−1	1
	15	0	1	−1	0	1	0	−1
	16	1	−1	1	0	1	−1	0
	17	1	0	−1	1	−1	0	1
	18	1	1	0	−1	0	1	−1

5.4. Design with Mixed Two-Level and Four-Level Factors

A four-level factorial design matrix is employed when there is a cubic relation between the response and a factor. When m four-level factors and K two-level factors are combined in an experiment, the experiment is called a 4^m2^{K-P} design, where P is the degree of fractionation and there will be 4^m2^{K-P} treatments (runs). Some of such mixed factorial designs involving four levels and two levels are given in Table 5.4.1. A four-level factor can be formulated in nature by combining two two-level factors into it, such as $4^1 \times 2^2$ with three factors shown in Table 5.4.2: Z, C and D, of which:

$$(A, B) \rightarrow Z$$

or

$$(A = -1, B = -1) \rightarrow Z = 1$$
$$(A = 1, B = -1) \rightarrow Z = 2$$
$$(A = -1, B = 1) \rightarrow Z = 3$$
$$(A = 1, B = 1) \rightarrow Z = 4$$

TABLE 5.4.1 Matrices for designs with four levels.

Design	Factors	Runs	Design generators	Matrix	Resolution
$4^1 \times 2^2$	Z, C, and D	16	$(A, B) \rightarrow Z$	Table 5.4.1	Full
$(4^1 \times 2^3)_{III}$	Z, C, D, and E	8	$(A, B) \rightarrow Z$; D = ABC, E = AC	Table 5.4.2 (first 8 runs)	III
$(4^1 \times 2^3)_{IV}$	Z, C, D, and E	16	$(A, B) \rightarrow Z$; I = ABCD	Table 5.4.2 (Complete)	IV
$(4^2 \times 2^3)_{III}$	X, Y, E, F, and G	16	$(A, B) \rightarrow X, (C, D) \rightarrow Y$; E = AD, F = BC, G = ABCD, and G = ABCD	Table 5.4.3 (first 16 runs)	III
$(4^2 \times 2^3)_{IV}$	X, Y, E, F, and G	32	$(A, B) \rightarrow X, (C, D) \rightarrow Y$; I = ABCDEF, I = BCEG, and I = ADFG	Table 5.4.4 (Complete)	IV
$(4^1 \times 2^4)_{III}$	Z, C, D, E, F	8	$(A, B) \rightarrow Z$; D = Z_2C, E = Z_3C, and F = Z_4C	Table 5.4.5	III

© SAE International.

5.4.1. 4^1x2^2 (Full-Factorial Design)

Z gets the combined effect of A and B, including main effects A and B, and their interaction, as shown in Table 5.4.2. The 15 sums of square for F-test are calculated according to the 2^4 (A, B, C, and D) full-factorial design, namely SS_A, SS_B, SS_C, SS_D, SS_{AB}, SS_{AC}, SS_{AD}, SS_{BC}, SS_{BD}, SS_{CD}, SS_{ABC}, SS_{ABD}, SS_{ACD}, SS_{BCD}, and SS_{ABCD}. The final sums of square for factors Z, C, and D are:

$$SS_Z = SS_A + SS_B + SS_{AB} \tag{5.4.1}$$

$$SS_C = SS_C \tag{5.4.2}$$

$$SS_D = SS_D \tag{5.4.3}$$

$$SS_{CD} = SS_{CD} \tag{5.4.4}$$

$$SS_{ZC} = SS_{AC} + SS_{BC} + SS_{ABC} \tag{5.4.5}$$

$$SS_{ZD} = SS_{AD} + SS_{BD} + SS_{ABD} \tag{5.4.6}$$

and

$$SS_{ZCD} = SS_{ACD} + SS_{BCD} + SS_{ABCD} \tag{5.4.7}$$

In general, each four-level factor requires two columns and each two-level factor requires one column, and thus the base design for a 4^m2^{n-p} design will be a 2^{K-P} design where K = 2m + n. For example, the base design for $4^1\times2^2$ must have a total of four columns (A, B, C, and D in Table 5.4.2).

TABLE 5.4.2 The factorial design matrix of $4^1\times2^2$ for three factors: Z (A×B; 4 levels), C (2 levels), and D (2 levels).

Run	Z	(a, b)	c	d	ab	ac	ad	bc	bd	cd	...	abcd
1	1	(−1, −1)	−1	−1	1	1	1	1	1	1	...	1
2	2	(1, −1)	−1	−1	−1	−1	−1	1	1	1	...	−1
3	3	(−1, 1)	−1	−1	−1	−1	1	−1	−1	1	...	−1
4	4	(1, 1)	−1	−1	1	−1	−1	−1	−1	1	...	1
5	1	(−1, −1)	1	−1	1	−1	1	−1	1	−1	...	−1
6	2	(1, −1)	1	−1	−1	1	−1	−1	1	−1	...	1
7	3	(−1, 1)	1	−1	−1	−1	1	1	−1	−1	...	1
8	4	(1, 1)	1	−1	1	1	−1	1	−1	−1	...	−1
9	1	(1, 1)	1	1	1	1	1	1	1	1	...	−1
10	2	(−1, 1)	1	1	−1	−1	−1	1	1	1	...	1
11	3	(1, −1)	1	1	−1	1	1	−1	−1	1	...	1
12	4	(−1, −1)	1	1	1	−1	−1	−1	−1	1	...	−1
13	1	(1, 1)	−1	1	1	−1	1	−1	1	−1	...	1
14	2	(−1, 1)	−1	1	−1	1	−1	−1	1	−1	...	−1
15	3	(1, −1)	−1	1	−1	−1	1	1	−1	−1	...	−1
16	4	(−1, −1)	−1	1	1	1	−1	1	−1	−1	...	1

5.4.2. $(4^1\times2^3)_{IV}$

The initial base design, first eight runs in Table 5.4.3, shows that an experiment with one four-level factor, Z, and three two-level factors, C, D, and E, where the generators for the base design are D = ABC and E = AC. When the initial base design is folded over (i.e., C, D, and E columns) on A and B, the defining relation of the combined base design becomes I = ABCD and consequently the combined design matrix with 16 runs is of resolution IV.

TABLE 5.4.3 Factorial design matrix of $4^1 \times 2^3$ for four factors: Z (A×B; 4 levels), C (2 levels), D (2 levels), and E (2 levels) [Ankenman 1999]—$(4^1 \times 2^3)_{III}$ and $(4^1 \times 2^3)_{IV}$.

Run	a	b	Z	c	d	e
1	−1	−1	1	−1	−1	1
2	1	−1	2	−1	1	−1
3	−1	1	3	−1	1	1
4	1	1	4	−1	−1	−1
5	−1	−1	1	1	1	−1
6	1	−1	2	1	−1	1
7	−1	1	3	1	−1	−1
8	1	1	4	1	1	1
			------- Foldover on factors A and B -----------			
9	1	1	4	1	1	−1
10	−1	1	3	1	−1	1
11	1	−1	2	1	−1	−1
12	−1	−1	1	1	1	1
13	1	1	4	−1	−1	1
14	−1	1	3	−1	1	−1
15	1	−1	2	−1	1	1
16	−1	−1	1	−1	−1	−1

Note: Resolution = III for the first 8 runs and Resolution = IV for the combined runs.

Reprinted by permission of Taylor & Francis Ltd, http://www.tandfonline.com. From "Design of Experiments with Two-Level and Four-Level Factors," Ankenman, Bruce 1999. Journal of Quality Technology.

5.4.3. $(4^2 \times 2^3)_{IV}$

The first 16 runs for the $4^2 \times 2^3$ design with resolution III, as shown in Table 5.4.4, is based on the following design generators: E = AD, F = BC, and G = ABCD. Factors X and Y are obtained from A, B, C, and D, respectively as follows:

$$(A = -1, B = -1) \rightarrow X = 1 \text{ and } (C = -1, D = -1) \rightarrow Y = 1;$$
$$(A = 1, B = -1) \rightarrow X = 2 \text{ and } (C = 1, D = -1) \rightarrow Y = 2;$$
$$(A = -1, B = 1) \rightarrow X = 3 \text{ and } (C = -1, D = 1) \rightarrow Y = 3;$$
$$(A = 1, B = 1) \rightarrow X = 4 \text{ and } (C = 1, D = 1) \rightarrow Y = 4;$$

After the first 16 runs are folded over on factors A and F, the combined 32 runs make a resolution IV design based on the following generators: I = ABCDEF = BCEG = ADFG.

TABLE 5.4.4 Factorial design matrix of $(4^2 \times 2^3)_{IV}$ for five factors: X (A×B; 4 levels), Y (C×D; four levels), E (2 levels), F (2 levels), and G (2 levels) [Ankenman 1999].

Run	a	b	X	c	d	Y	e	f	g
1	−1	−1	1	−1	−1	1	1	1	1
2	1	−1	2	−1	−1	1	−1	1	−1
3	−1	1	3	−1	−1	1	1	−1	−1
4	1	1	4	−1	−1	1	−1	−1	1
5	−1	−1	1	1	−1	2	1	−1	−1
6	1	−1	2	1	−1	2	−1	−1	1
7	−1	1	3	1	−1	2	1	1	1
8	1	1	4	1	−1	2	−1	1	−1
9	−1	−1	1	−1	1	3	−1	1	−1
10	1	−1	2	−1	1	3	1	1	1
11	−1	1	3	−1	1	3	−1	−1	1
12	1	1	4	−1	1	3	1	−1	−1
13	−1	−1	1	1	1	4	−1	−1	1
14	1	−1	2	1	1	4	1	−1	−1
15	−1	1	3	1	1	4	−1	1	−1
16	1	1	4	1	1	4	1	1	−1
---------------------- Foldover on factors A and F ------------------------										
17	1	−1	2	−1	−1	1	1	−1	1
18	−1	−1	1	−1	−1	1	−1	−1	−1
19	1	1	4	−1	−1	1	1	1	−1
20	−1	1	3	−1	−1	1	−1	1	1
21	1	−1	2	1	−1	2	1	1	−1
22	−1	−1	1	1	−1	2	−1	1	1
23	1	1	4	1	−1	2	1	−1	1
24	−1	1	3	1	−1	2	−1	−1	−1
25	1	−1	2	−1	1	3	−1	−1	−1
26	−1	−1	1	−1	1	3	1	−1	1
27	1	1	4	−1	1	3	−1	1	1
28	−1	1	3	−1	1	3	1	1	−1
29	1	−1	2	1	1	4	−1	1	1
30	−1	−1	1	1	1	4	1	1	−1
31	1	1	4	1	1	4	−1	−1	−1
32	−1	1	3	1	1	4	1	−1	−1

Note: Resolution = III for the first 16 runs and Resolution = IV for the combined runs.

5.4.4. $(4^1 \times 2^4)_{III}$

The eight runs for the $4^1 \times 2^4$ design with resolution III, as shown in Table 5.4.5, is based on the following design generators: D = AB, E = AC, and F = BC. Factors Z is obtained from A and B as follows:

$$(A = -1, B = -1) \rightarrow Z = 1$$
$$(A = 1, B = -1) \rightarrow Z = 2$$
$$(A = -1, B = 1) \rightarrow Z = 3$$
$$(A = 1, B = 1) \rightarrow Z = 4$$

TABLE 5.4.5 The factorial design matrix of $(4^1{\times}2^4)_{III}$ for three factors: Z (A×B; 4 levels), C (2 levels), D (2 levels), E (2 levels), and F (2 levels).

Run	Z	(a, b)	c	d = ab	e = ac	f = bc
1	1	(−1, −1)	−1	1	1	1
2	2	(1, −1)	−1	−1	−1	1
3	3	(−1, 1)	−1	−1	1	−1
4	4	(1, 1)	−1	1	−1	−1
5	1	(−1, −1)	1	1	−1	−1
6	2	(1, −1)	1	−1	1	−1
7	3	(−1, 1)	1	−1	−1	1
8	4	(1, 1)	1	1	1	1

© SAE International.

5.4.5. 4^{4-2}_{IV} with Example on Voiding on Ball Grid Arrays

The design matrix of full-factorial design 4^2, having 16 runs with variables A (4 levels) and B (4 levels), are given in the first two columns as shown in Table 5.4.6. This can be obtained utilizing the following relationships:

$$
\begin{aligned}
(\alpha = -1, \beta = -1) &\rightarrow A = 1 \\
(\alpha = 1, \beta = -1) &\rightarrow A = 2 \\
(\alpha = -1, \beta = 1) &\rightarrow A = 3 \\
(\alpha = 1, \beta = 1) &\rightarrow A = 4
\end{aligned}
\quad \text{and} \quad
\begin{aligned}
(\gamma = -1, \delta = -1) &\rightarrow B = 1 \\
(\gamma = 1, \delta = -1) &\rightarrow B = 2 \\
(\gamma = -1, \delta = 1) &\rightarrow B = 3 \\
(\gamma = 1, \delta = 1) &\rightarrow B = 4
\end{aligned}
$$

Next, consider fractional factorial design 2_{IV}^{8-4}, one may choose the following confounding patterns:

$$
\begin{aligned}
e &= \pm\beta\gamma\delta & (\pm B\,C\,D) \\
f &= \pm\alpha\gamma\delta & (\pm A\,C\,D) \\
g &= \pm\alpha\beta\gamma & (\pm A\,B\,C) \\
h &= \pm\alpha\beta\delta & (\pm A\,B\,D)
\end{aligned}
$$

After variates e, f, g, and h are obtained from any one set of the above equations (8 sets in total), one may obtain the design levels for factors C and D following the rules:

$$
\begin{aligned}
(e = -1, f = -1) &\rightarrow c = 1 \\
(e = 1, f = -1) &\rightarrow c = 2 \\
(e = -1, f = 1) &\rightarrow c = 3 \\
(e = 1, f = 1) &\rightarrow c = 4
\end{aligned}
$$

$$
\text{and} \quad
\begin{aligned}
(g = -1, h = -1) &\rightarrow d = 1 \\
(g = 1, h = -1) &\rightarrow d = 2 \\
(g = -1, h = 1) &\rightarrow d = 3 \\
(g = 1, h = 1) &\rightarrow d = 4
\end{aligned}
$$

The design matrix given in Table 5.4.6. was provided by Taguchi using one of the eight set of equations. For example, given that a = 1 and b = 1 for run 1, one has $(\alpha = -1, \beta = -1)$ and $(\gamma = -1, \delta = -1)$. Thus, $e = \beta\gamma\delta = -1$, $f = \alpha\gamma\delta = -1$, $g = \alpha\beta\gamma = -1$, and $h = \alpha\beta\delta = -1$, which resulting in c = 1 and d = 1.

TABLE 5.4.6 Factorial design matrix of 4^{4-2}_{IV} for four factors, taking 4 levels each factor [Lasky et al. 2008].

Run	a	b	c	d	Y_1 (%)	Y_2 (%)
1	1	1	1	1	19	21
2	1	2	2	2	16	16
3	1	3	3	3	11	13
4	1	4	4	4	15	13
5	2	1	2	3	21	24
6	2	2	1	4	22	23
7	2	3	4	1	19	18
8	2	4	3	2	9	9
9	3	1	3	4	23	25
10	3	2	4	3	24	25
11	3	3	1	2	14	13
12	3	4	2	1	12	12
13	4	1	4	2	29	25
14	4	2	3	1	17	21
15	4	3	2	4	19	14
16	4	4	1	3	13	9

Note: Y_1 and Y_2 are replicated runs for the percentage of voids in the ball grid array.

Voiding on ball grid arrays (BGAs) causes a reliability problem on an assembly line— shortening the lifespan of IC chips. The 4^{4-2} factorial design, of which the four processing variables and their design levels are listed in Table 5.4.7, was performed to minimize BGA voiding derived from observed industry trends [Lasky et al. 2008]. Two replicated tests are conducted for each treatment and their results (voids in percentage) are listed in Table 5.4.7. The analysis to identify the effects of individual factors and their interactions based on the data given in this table is to be left to the students as a homework exercise.

TABLE 5.4.7 List of factors and design levels for experimental tests on voiding of BGA [Lasky et al. 2008].

Factors	Levels (4 levels per factor)	Coded variable
A: Solder Paste Brand	A_1, A_2, A_3, A_4	a
B: Temperature Profile	B_1, B_2, B_3, B_4	b
C: Storage Humidity (Relative)	0, 20%, 40%, 60%	c
D: Solder Paste Type	D_1, D_2, D_3, D_4	d

5.5. Reliability and Confidence Level

Product reliability comes with a certain confidence level, because the runs of experimental tests is always a finite number. A confidence level may be presented by a confidence interval, which is defined by an upper and lower boundary for the value of a variable of interest and it aims to aid in assessing the uncertainty associated with a measurement, usually in experimental context with random errors. In other words, a confidence interval is by way of calculating the range of values that contains the true value of an estimate.

5.5.1. Confidence Interval

A confidence interval refers to a range of values that is likely to contain the value of an unknown population parameter, such as the mean, based on data sampled from that population. A sample prediction is incomplete without some statement of the uncertainty in the predicted value. A prediction interval is a range that is likely to contain the response value of an individual new observation under specified settings of your predictors. To predict a value using a regression model, based on either DOE regression or maximum likelihood estimate, is a simple matter, but such a prediction should be presented with its corresponding confidence interval, denoted by CI.

The two-sided CI is defined by two limits: an upper confidence limit (UCL) and a lower confidence limit (LCL). These limits are constructed so that the designated proportion (confidence level) of such intervals will include the true population value. Any value inside the CI cannot be rejected, thus when the null hypothesis of interest is covered by the interval it cannot be rejected. The above essentially means that the values outside the interval are the ones we can draw an inference from. The wider a CI is, the more uncertainty there is in the estimate [Fletcher and Dillingham 2010].

A CI is constructed based on a particular required confidence level, which is also the coverage probability of the interval. The confidence level, $1 - \alpha$, can be also interpreted as follows: If a great number (e.g., thousands) of samples of N items are drawn from a population using simple random sampling and a CI is calculated for each sample, the proportion of those intervals that will include the true population parameter is $1 - \alpha$. For a variety of regression models, the associated p-value computes predicted outcomes along with CIs for these predictions. The predicted CI for the unbiased Y at some confidence level C can be calculated as follows:

$$Y_p - z_{\alpha/2}\, S_Y \leq Y \leq Y_p + z_{1-\alpha/2}\, S_Y \tag{5.5.1}$$

and

$$C = 1 - \text{p-value} = 1 - \alpha$$
$$= \text{Probability}(Y_p + z_{1-\alpha/2}\, S_Y \leq Y \leq Y_p + z_{\alpha/2}\, S_Y) \tag{5.5.2}$$

where
   Y_p is the predicted value of response variable Y
   S_Y is the sample standard deviation to the predicted value, i.e., sample error
   z is the standard statistic variate, e.g., $N(0, 1)$ for the standard normal distribution
   $z_{1-\alpha/2}$ and $z_{\alpha/2}$ are z values at two CI bounds, i.e., $1 - \alpha/2$ and $\alpha/2$, respectively

The z-variate presented above is a general representation of the statistic of concern, as it can be a t-statistic, lognormal statistic, Weibull statistic, or exponential statistic. A CI provides the plausible bounding values with physical units (e.g., °C for temperature) for the underlying population average. For models with categorical outcomes, the probability of each outcome is computed for clarifying modeling precision.

Often, the value $1 - \alpha = 90\%$, 95%, or 99% is applied. When CI = 95%, for example, it will contain the true value of interest 95% of the time with the desired reliability.

5.5.2. Sampling CI in t-Distribution

One may use either p-values or CIs to determine whether the DOE regressions are statistically significant. If a hypothesis test produces both, these results will agree. For the purpose of a comprehensive understanding of probability associated with CI, parameter C is employed to stand for confidence level, or confidence (for simplicity), which is the probability defined as $C = 1 - \alpha = 1 - $ p-value. Since $t_{v,\alpha/2} = -t_{v,1-\alpha/2}$ for a symmetric t-distribution, the following range for the unbiased estimate of Y applies:

$$Y_p - t_{v,1-\alpha/2}\, S_Y \leq Y \leq Y_p + t_{v,1-\alpha/2}\, S_Y \tag{5.5.3}$$

In case of a two-sided comparison, when $\alpha = 10\%$, i.e., saying that the CI is 90% ($C = 1 - \alpha = 90\%$), the estimated range will provide a "bootstrap plot" of the predicted values; named after the fact that the plotted curves shape like a bootstrap [Franklin and Wasserman 1992].

On a one-sided comparison, one can apply the following two algorithms:

1. $Y_p - t_{v,1-\alpha/2}\, S_Y$ is applied to the case in need of minimum requirement with confidence level $C = 1 - \alpha$. For example, one would expect the minimum strength of one kind of steel to be greater than 400 MPa. Then, $Y_p - t_{v,1-\alpha/2}\, S_Y > 400$ MPa is an anticipated warrant on the criterion. Note that $Y_p > 400$ MPa is only to warrant that the mean strength of the steel is greater than 400 MPa.

2. $Y_p + t_{v,1-\alpha/2}\, S_Y$ is applied to the case in need of maximum requirement with confidence level $C = 1 - \alpha$. For example, one would expect the maximum smog concentration of PM10 particles in the air to be less $50 \times 10^{-6}\ \mu g/m^3$. Then, $Y_p + t_{v,1-\alpha/2}\, S_Y < 50 \times 10^{-6}\ \mu g/m^3$ is an anticipated warrant on the criterion. Note that $Y_p < 50 \times 10^{-6}\ \mu g/m^3$ is only to warrant that the mean smog pollution is less than $50 \times 10^{-6}\ \mu g/m^3$.

There tends to be a great focus on p-values and simply detecting a significant effect or difference in statistical analyses. However, a statistically significant effect is not necessarily meaningful in the real world. For instance, the effect might be too small to be of any practical value. This can be detected using the CIs. A CI would allow an experimenter to assess these important characteristics along with the statistical significance, since it yields solid decision-making information rather than the simple comparison resulting from hypothesis tests, as a range of plausible values for the unknown parameter is estimated. Example 5.5.1 is presented for a single treatment only in order to further explain the meaning of CIs.

EXAMPLE 5.5.1

Assume that the seat-belt strength is known to be normally distributed. Three test data (18.9 kN, 19.9 kN, and 20.9 kN) are submitted by the supplier for assuring the production part approval. The approval requirement of an automotive seat-belt strength is 17.9 kN mandated by the automotive manufacturer. Is it a qualified product?

$$Y_{ave} = 19.9\ kN$$

$$S_Y = \{[(18.9 - 19.9)^2 + (19.9 - 19.9)^2 + (20.9 - 19.9)^2]\,/\,2\}^{1/2} = 1$$

Then, the statistical t-test for the seat-belt strength to exceed 17.9 kN leads to:

$$t_2 = \frac{Mean - Target}{S_Y} = \frac{19.9 - 17.9}{1} = 2.0$$

Thus, $\alpha = $ p-value = 9% (**Table 2.5.1**)

In other words, 9% of the future product will be defective. This is definitely not accepted according to the general practice in the automotive industry.

5.5.3. CI of Sample Population by DOE-F

The CI of the sample population for DOE-F in the DOE based on an F-distribution is [Ross 1988]:

$$CI = \{F_{1,\,DOFerror;\alpha}\ S_{error}^2\ [\frac{1 + DOF_{effects}}{N_0}]\}^{1/2} \qquad (5.5.4)$$

where
    α is the p-value, i.e., statistical significance level in DOE
    DOF_{error} is the DOFs used to estimate the random error
    $DOF_{effects}$ is the DOFs associated with all effects, without sample mean, in the predictive equation
    $F_{1,\,DOFerror;\,\alpha}$ is the F-ratio having the DOFs that u = 1, v = DOF_{error}
    S_{error}^2 is the sample variance of random error ε
    N_0 is the total number of observations (data) resulting from experimental tests

Assume that α = 5% is taken for granted, the calculated values using the above equation form the upper and lower bounds of Y at confidence level C = 1 − α = 95%. In advance, there is a 95% chance of obtaining an interval that contains the population average. This is to be further illustrated using an example in Section 5.5.3.

5.5.4. CI of Sample Population by DOE-F with Confirmation Runs

The CI for the Y value on a confirmation treatment, that may be replicated, in an experimental design based on F-distribution can be evaluated as [Ross 1988]:

$$CI = \{F_{1,\,DOFerror;\,\alpha}\ S_{error}^2\ [\frac{1 + DOF_{effects}}{N_0} + \frac{1}{N_C}]\}^{1/2} \qquad (5.5.5)$$

of which N_C is the number of replications for the confirmation treatment. It can be understood that the more the confirmation runs are, the more precise the prediction is.

5.5.5. Factorial Design $2^1 \times 3^{7-5}$ (L_{18}) for Designing Magneto-Rheological Dampers

A typical automotive magneto-rheological damper consists mainly of magneto-rheological fluids, cylinders, pistons, valves, energizing coils, and other supporting components, as shown in Figure 5.5.1. Analytical solutions to the idealized electric energization process have been examined by researchers, while how the damper performance (e.g., damping force) in practice varies with respect to the following geometric parameters is hard to determine without DOE: pole length, distance between poles, piston radius, clearance between piston and cylinder, cylinder thickness, piston rod radius, clearance between piston rod and coil, etc. Mangal and Kumar [2015] did the system analysis using the finite element models, which were validated by experimental tests.

FIGURE 5.5.1   Components of a basic magneto-rheological damper [Mangal and Kumar 2015]. (a) Components of the magneto-rheological damper. (b) Grooved portion of piston with insulator material. (c) Assembled piston and its rod with copper winding. (d) Copper winding covered with a cloth. (e) Assembly of piston, piston rod, and lid. (f) Complete assembly of the optimized MR damper.

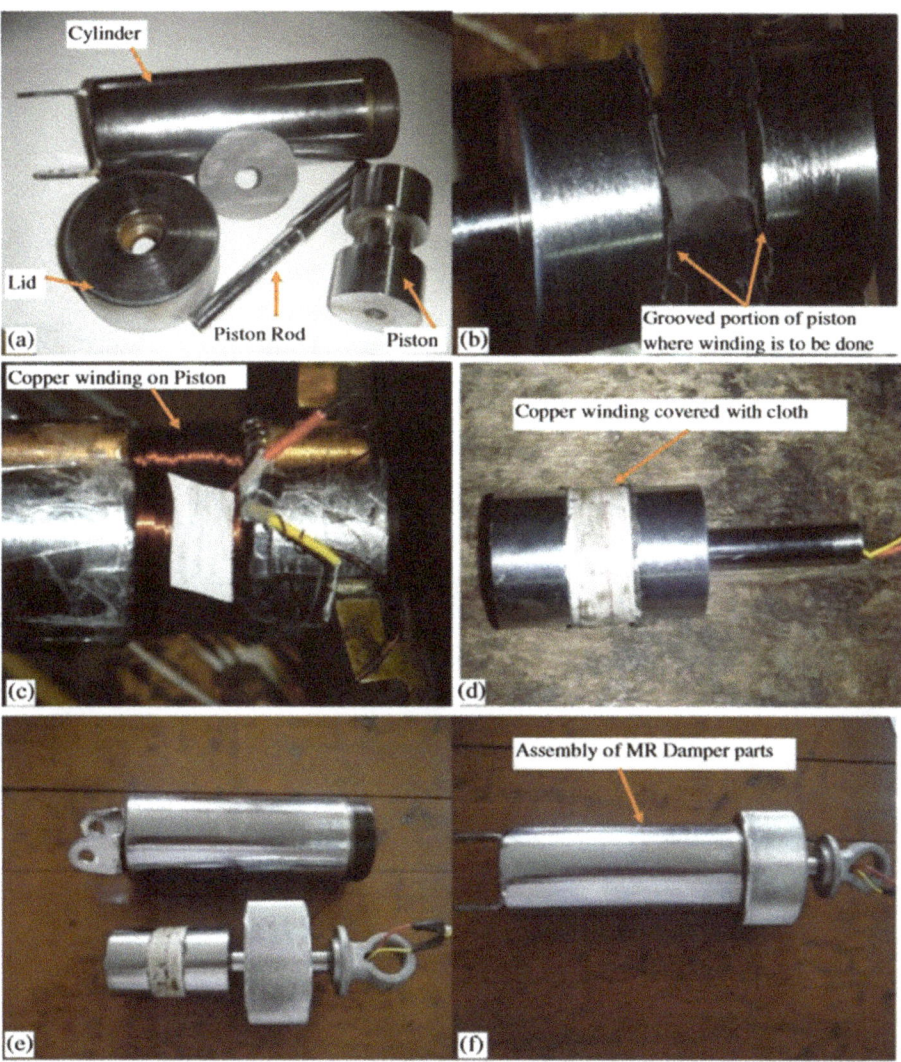

Four of the geometric parameters of an experimental magneto-rheological damper, as exhibited in Figure 5.5.2, which is Figure 1 in Mangal and Kumar [2015], are selected to do the DOE using ANOVA in order to explore their impact on the damping force. The design levels of these four selected parameters are listed in Table 5.5.1. Design matrix $2^1 \times 3^{7-5}$ (L_{18}) in Section 5.4.2 was employed to identify the main effects that are potentially influential (Table 5.5.2). Extra dummy factors, i.e., factors E, F, G, and H, are then regarded random errors since their design levels are fixed in the design matrix.

FIGURE 5.5.2 Cross-sectional view of MR (magneto-rheological) damper [Mangal and Kumar 2015].

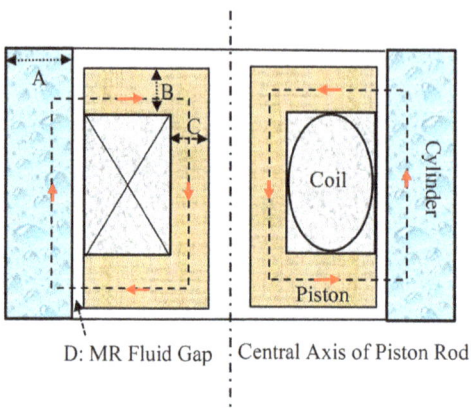

TABLE 5.5.1 Factors and their design levels for the operation of magneto-rheological dampers.

Factor	Level (1)	Level (2)	Level (3)
A: Cylinder thickness	6 mm	—	8 mm
B: Pole length	18 mm	23 mm	28 mm
C: Radial clearance between piston rod and coils	5 mm	7 mm	9 mm
D: Radial clearance between piston and cylinder	0.8 mm	1.0 mm	1.2 mm

TABLE 5.5.2 Factorial design $2^1 \times 3^{7-5}$ (L_{18}) for four active factors on magneto-rheological damping force Y [Mangal and Kumar 2015].

Run	a	b	c	d	Y (N)
1	1	1	1	1	216.81
2	1	1	2	2	164.11
3	1	1	3	3	127.73
4	1	2	1	1	262.26
5	1	2	2	2	195.52
6	1	2	3	3	150.01
7	1	3	1	2	232.26
8	1	3	2	3	183.49
9	1	3	3	1	318.92
10	2	1	1	3(1)	138.38
11	2	1	2	1(2)	213.13
12	2	1	3	2(3)	199.35
13	2	2	1	2	204.34
14	2	2	2	3	161.97
15	2	2	3	1	298.53
16	2	3	1	3	191.52
17	2	3	2	1	295.99
18	2	3	3	2	253.58

Note that the design matrix of $2^1 \times 3^{7-5}$ (L_{18}), as presented in Table 5.5.3, is used here for modeling four active factors (i.e., A, B, C, and D) only, while factors E, F, G, H are dummy and their "effects" can be regarded as random errors.

The sum of squares of sample error can be estimated by subtracting the sums of squares for the grand average and individual main effects from the total sum of squares. The contribution ratio is calculated using MS (mean square) values as the percentage of the contribution merely from individual factors (A–H) and error, while the sum of squares of grand average is excluded. Factor A has only one degree of freedom, but each of factors B–H has two DOFs because of taking three levels in the design. Factor D (clearance between piston and cylinder) has the greatest influence, followed by factor B (pole length). The combined individual main effects of factor D and factor B have a dominant influence on the damping force with a combined contribution ratio of 93.5%.

TABLE 5.5.3 ANOVA of multivariable influences on magneto-rheological damping force.

Source of variance	SS	DOF	MS	F-ratio	p-value	Contribution
Factor A	$SS_A = 620.5$	1	$MS_A = 620.5$	$F_{1,10} = 4.92$	5.08%	1.16%
Factor B	$SS_B = 14{,}442$	2	$MS_B = 7{,}221$	$F_{2,10} = 57.3$	<0.01%	27.0%
Factor C	$SS_C = 1{,}634.8$	2	$MS_C = 817.4$	$F_{2,10} = 6.49$	1.56%	3.05%
Factor D	$SS_D = 35{,}586$	2	$MS_D = 17{,}793$	$F_{2,10} = 141.2$	<0.01%	66.5%
Errors, Sample	$SS_{error} = 1{,}260$	10	$MS_{error} = 126$	—	—	0.86%
Factor E	$SS_E = 459.3$	2	$MS_E = 229.6$	—	—	0.86%
Factor F	$SS_F = 230.2$	2	$MS_F = 115.1$	—	—	0.43%
Factor G	$SS_G = 155.5$	2	$MS_G = 77.74$	—	—	0.29%
Factor H	$SS_H = 189.7$	2	$MS_H = 94.85$	—	—	0.35%
Residual	$SS_e = 226$	2	$MS_e = 112.9$	—	—	0.42%
Subtotal	$SS_T = 53{,}544$	17				
Grand average	SS_{ave}	1				
Total	SS	18				

The CI for the sample population without any replication can be calculated using Equation (5.5.4) as:

$$CI = \left\{ F_{1,DOFerror;\alpha} \; S_{error}^2 \left[\frac{1 + DOF_{effects}}{N_o} \right] \right\}^{1/2}$$

$$= \{ F_{1,10;5\%} \; 126 \, [\, (1 + 7) \, / \, 18 \,] \}^{1/2}$$

$$= \{ (4.96) \; 126 \, (8 \, / \, 18) \}^{1/2}$$

$$= \pm 16.67$$

of which $F_{1,10;5\%} = 4.96$ by looking up the F-distribution table. Assume that one predicted value is Y_p, then one has 95% (=1 − 5%) confidence that the Y value will fall between the upper and lower limits as:

$$Y_p - 16.67 < Y < Y_p + 16.67 \tag{5.5.6}$$

Similarly, one has 95% confidence that the bounds for each individual confirm run Y should fall in a range that depends on the replication size (N_o) of DOE treatments, as:

$$Y_{C,ave} - 16.67 < Y_C < Y_{C,ave} + 16.67 \qquad \text{when } N_0 = 1$$

$$Y_{C,ave} - 11.78 < Y_C < Y_{C,ave} + 11.78 \qquad \text{when } N_o = 2$$

$$Y_{C,ave} - 9.62 < Y_C < Y_{C,ave} + 9.62 \qquad \text{when } N_o = 3$$

$$Y_{C,ave} - 8.33 < Y_C < Y_{C,ave} + 8.33 \qquad \text{when } N_o = 4$$

$$Y_{C,ave} - 7.45 < Y_C < Y_{C,ave} + 7.45 \qquad \text{when } N_o = 5$$

$$Y_{C,ave} - 6.80 < Y_C < Y_{C,ave} + 6.80 \qquad \text{when } N_o = 6$$

$$Y_{C,ave} - 5.27 < Y_C < Y_{C,ave} + 5.27 \qquad \text{when } N_o = 10$$

$$\text{and} \quad Y_{C,ave} - 3.73 < Y_C < Y_{C,ave} + 3.73 \qquad \text{when } N_o = 20$$

Next, consider the CI for confirmation runs. Assume N_C confirmation runs are replicated for one confirmation treatment, then the CI for the confirmation runs can be calculated using Equation (5.5.5). When $N_C = 1$,

$$CI = \left\{ F_{1,\,DOFerror;\alpha}\ S_{error}^2 \left[\frac{1 + DOF_{effects}}{N_o} + \frac{1}{N_C} \right] \right\}^{1/2}$$

$$= \left\{ F_{1,10;5\%}\ 126\ [\ (1 + 7)\ /\ 18 + 1\ /\ 1] \right\}^{1/2}$$

$$= \left\{ (4.96)\ 126\ (8\ /\ 18 + 1) \right\}^{1/2}$$

$$= \pm 30.05$$

Thus, one has 95% confidence that the bounds for each individual confirm run Y_C should fall in a range that depends on the replication size (N_C) within the confirmation treatment, as:

$$Y_{C,ave} - 30.05 < Y_C < Y_{C,ave} + 30.05 \qquad \text{when } N_C = 1$$

$$Y_{C,ave} - 24.29 < Y_C < Y_{C,ave} + 24.29 \qquad \text{when } N_C = 2$$

$$Y_{C,ave} - 22.05 < Y_C < Y_{C,ave} + 22.05 \qquad \text{when } N_C = 3$$

$$Y_{C,ave} - 20.83 < Y_C < Y_{C,ave} + 20.83 \qquad \text{when } N_C = 4$$

$$Y_{C,ave} - 20.07 < Y_C < Y_{C,ave} + 20.07 \qquad \text{when } N_C = 5$$

$$Y_{C,ave} - 19.54 < Y_C < Y_{C,ave} + 19.54 \qquad \text{when } N_C = 6$$

$$Y_{C,ave} - 18.45 < Y_C < Y_{C,ave} + 18.45 \qquad \text{when } N_C = 10$$

$$\text{and} \quad Y_{C,ave} - 16.85 < Y_C < Y_{C,ave} + 16.85 \qquad \text{when } N_C = 100$$

Note that $Y_{C,ave}$ is the averaged value, denoting the mean value theoretically, of all the N_C conformation runs. It is shown above that the more conformation runs conducted within the given (desired) confirmation treatment, the confirmation runs will be closer to the population prediction, i.e., Equation (5.5.6). In other words, replicated runs will reduce the effect of uncontrolled variation and consequently enhance precision.

5.6. Identification of Statistic Distribution

The statistical distribution of a set of data of interest has to be identified before the CI for a desired reliability can be found for this specific set of data. There are two widely applied methods for statistic distribution identification:

1. Nonparametric Method: The bootstrap is a nonparametric method for estimating the sampling distribution of a statistic. It is often presented using a bootstrap plot, of which the horizontal axis is the subsample number and the vertical axis is the computed value of the desired statistic for the subsample.

2. Parametric Method: It is to fit the data directly with different potentially feasible probability density functions (or distribution functions) using the least square method, maximum likelihood method, etc. Then, one may select the one that fits best using statistical criteria (e.g., p-value, likelihood ratio test, and Anderson-Darling statistic).

5.6.1. Bootstrap Plot for a Statistic

A bootstrap plot is designed for an estimation of the required statistical information from the sample population. Among all statistics, the most common uncertainty calculation is generating a CI for the mean, such as the range for the unbiased estimate of Y given by Equation (5.5.3). In general, a bootstrap plot would provide the following statistical information [NIST 2022]:

(a) What does the sampling distribution for the statistic look like? The reason to choose the bootstrap is to figure out the sample distribution. The bootstrap technique is appropriate for characterizing most distribution functions of statistics such as Weibull, lognormal, and normal distributions. Because the bootstrap plot is demonstrated in the uniform random numbers case study and thus not suitable for a statistic that is heavily dependent on the tails, such as the range [Efron and Gong 1983].

(b) Where is the reliability (R) given a CI (C) for the statistic? For example, it would tell the subsample number that corresponds to R = 90% and C = 90%. It is feasible to use the sample standard deviation from the bootstrap distribution to compute the appropriate percentiles based on a specified statistical distribution. The percentile method determines the α and $1 - \alpha$ percentile from the bootstrap distribution without knowing the shape of that distribution in advance.

(c) Which statistic has a sampling distribution with the smallest variance? Or, which statistic generates the narrowest CI? The bootstrap technique is widely employed to generate a CI for the uncertainty of a statistic of concern with a specified empirical distribution in general-purpose statistical software programs.

When the uncertainty formula for assessing Equation (5.5.3) is mathematically intractable, the bootstrap provides a method for calculating the uncertainty in these cases. Learning by example, one is hereupon to calculate the sample mean Y_{ave} as a point estimate of Y_{mean}. But how can one find a CI for the uncertainty of Y_{mean} around Y_{ave}?

When attempting to calculate a two-sided CI with C = 90%, e.g., mean here, from a batch of 1,000 samples $(Y_1, Y_2, ..., Y_{1000})$, one may make a bootstrap plot by taking the following steps [Guan 2003, Long 1997]:

1. Subsampling: For generating a bootstrap estimate of a given response variable (Y) from a set of data of the same statistical distribution function, a subsample of a size less than or equal to the size of data is generated. For example, the thickness of newly produced steel plates is in quest and there are 2,000 samples available. Ten subsamples $(Y_1, Y_2, ..., Y_{10})$ are then randomly taken from the given 2,000 samples.

2. Identification: Identify the statistic of concern (e.g., mean) of these 10 samples. For the mean, it is numerically implemented as $Y_{ave} = (Y_1 + Y_2 + \cdots + Y_{10})/10$.

3. Sampling with Replacement: Put the drawn 10 samples back into the batch and there are still 2,000 samples.

4. Repeat steps (1) and (3) for 500 times. Then, there are 500 data, i.e., Y_{ave1}, Y_{ave2}, ..., and Y_{ave500}. The sample variance of these bootstrap samples is an estimate of variance of the sample distribution of the statistic of interest. Often this approach is simply referred to as the bootstrap estimate, and for a sufficient amount of data (e.g., 500 or above), there is little difference between the bootstrap estimator and the Monte Carlo approximation. Given that the number of bootstrap subsampling data is somewhat chosen arbitrarily, 500 subsampling is considered to be sufficient in practice although 1,000 subsampling is preferred.

5. Sorting: The collected averages, i.e., Y_{ave1}, Y_{ave2}, ..., and Y_{ave500}, are then sorted in the ascending order.

6. Inscribing Confidence Level: For example, the value of the 25th mean (Y_{ave25}) is taken as the LCL, while the value of the 475th (Y_{ave475}) mean is the UCL. Then, these two limits correspond to C = 90%.

7. Making Bootstrap Plot: The bootstrap plot is formed using the computed values (Y_{ave1}, Y_{ave2}, ..., Y_{ave500}) versus the subsample number.

Then, the statistical histogram, probability density function, cumulative distribution function of failure, and cumulative distribution function of reliability can be obtained, as demonstrated by Examples 5.6.1 and 9.2.1 (Volume II).

Slightly different values of the CI for the parameter estimation will be obtained each time because of the random nature of the picking of bootstrap samples, namely statistical noise. Nevertheless, just like Monte Carlo simulations, bootstrap method is supported by the central limit theorem that the distribution of the mean of a random sample from a population with finite variance is approximately normally distributed when the sample size is large. In other words, the empirical distribution of Y_{ave} is a valid approximation of the distribution of Y_{mean}. Thus, the variation of Y_{mean} can be well-approximated by the variation of Y_{ave}. Although the bootstrap method is appropriate for a lot of distributions and statistics, there are some exceptions, especially for long-tailed statistical distributions [Efron 1981].

EXAMPLE 5.6.1

Following the steps of the bootstrap technique, one can obtain the final 500 data for evaluating the ultimate tensile strength of 2,000 bolts in the same batch using the microindentation equipment. They are classified into 16 groups when rounded to a whole number with an interval of 5 MPa apart as listed in Table 5.6.1. Construct the statistical histogram, probability density function, cumulative distribution function of failure, and cumulative distribution function of reliability.

Solution:

1. 500 test data, namely Y_{ave1}, Y_{ave2}, ..., and Y_{ave500} as obtained from Step (5), are available. They are classified into 16 groups when rounded to a whole number with an interval of 5 MPa apart, for the determination of the related statistical distribution. Column n of Table 5.6.1 contains the number of data available in each class.

2. A plot of column n versus column Y in the order of class yields the histogram, as shown in Figure 5.6.1(a).

3. Based on the data given in columns Y and n, one is able to approximate the probability density function using f = n/N/w. The plot of column f versus column Y yields the probability density function, as shown in Figure 5.6.1(b).

4. Plots of statistical distribution functions F and R versus column Y are given in Figure 5.6.1(c), where the data in columns F and R are calculated using the following cumulative equations:

$$F_1 = f_1 \, w_1 \, / \, 2$$

$$F_q = \sum_{i=1}^{q-1} f_i \, w_i + \frac{f_q \, w_q}{2}, \text{ where } q = 2, 3, ..., \text{ and } 16$$

and $R_q = 1 - F_q$, where q = 1, 2, 3, ..., and 16

TABLE 5.6.1 Bootstrap technique for the identification of statistical distribution.

q	Y	n	f = n/(Nw)	f w	F	R
1	590	1	0.0004	0.002	0.001	0.999
2	595	6	0.0024	0.012	0.008	0.992
3	600	9	0.0036	0.018	0.023	0.977
4	605	21	0.0084	0.042	0.053	0.947
5	610	41	0.0164	0.082	0.115	0.885
6	615	54	0.0216	0.108	0.210	0.790
7	620	69	0.0276	0.138	0.333	0.667
8	625	76	0.0304	0.152	0.478	0.522
9	630	69	0.0276	0.138	0.623	0.377
10	635	65	0.026	0.13	0.757	0.243
11	640	41	0.0164	0.082	0.863	0.137
12	645	25	0.01	0.05	0.929	0.071
13	650	14	0.0056	0.028	0.968	0.032
14	655	6	0.0024	0.012	0.988	0.012
15	660	2	0.0008	0.004	0.996	0.004
16	665	1	0.0004	0.002	0.999	0.001

where
q: Class
Y (MPa): Ultimate tensile strength
n: Number of specimens in the class
N: Total number of specimens; N = 500
w (MPa): 5 MPa
F: Statistical distribution function of failure
R: Reliability

FIGURE 5.6.1 Histogram, probability density function, and statistical distribution function of ultimate tensile strength of steel bolts as sampled with replacement from 2,000 bolts.

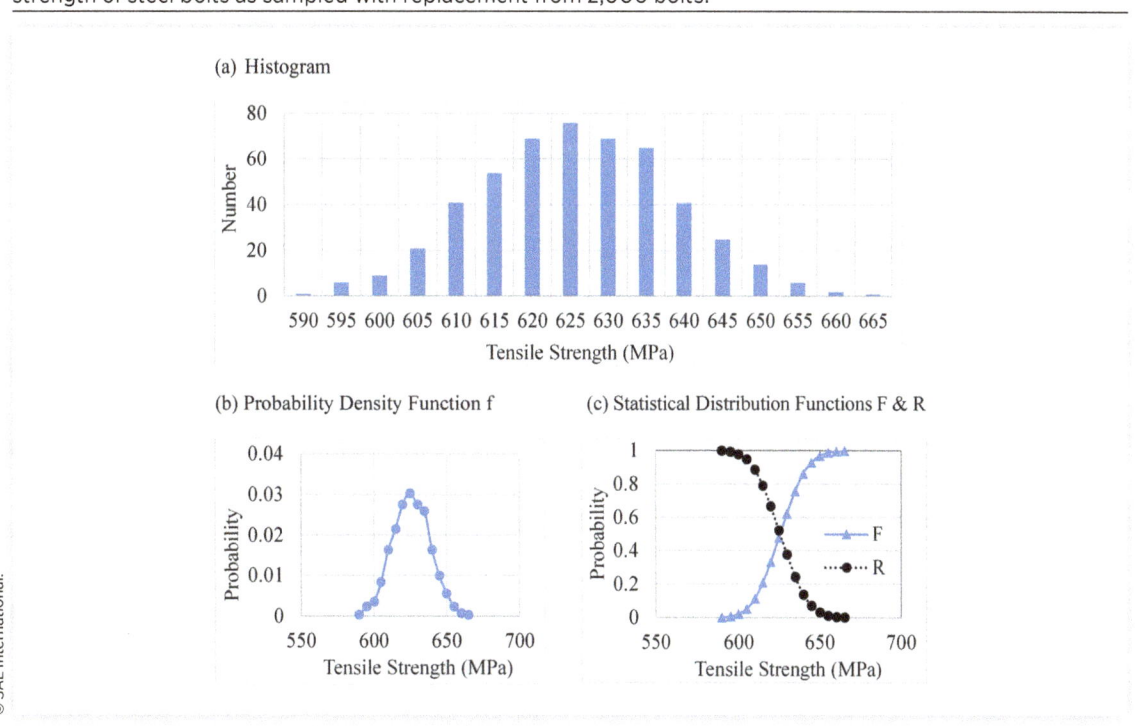

5.6.2. Bootstrap Plot of Model-Averaged CIs for a DOE-t

Some calculations of CIs, such as those for DOE-t in practice, achieve their nominal coverage only approximately; their coverage is not guaranteed while being approximate. Model-averaging schemes that summarize the statistical meaning of a factorial DOE in light of individual treatment means has been proven to be a practical algorithm to approach the DOE-t, as exhibited in Chapters 2 and 3. One of the model-averaging schemes is the bootstrap method, which has been widely accepted ever since its discovery [Buckland et al. 1997].

The bootstrap method often points out problems of linear models that might easily be overlooked. Bootstrapping can be used for the estimation of parameter variances, and it is straightforward to be implemented but computationally demanding compared with other methods for parameter error estimation. It is not bound to any restrictions such as the distribution of measurement errors such as the assumption of normality. Because of the possible asymmetry of the probability densities of the parameters, the parameter estimation errors acquired by bootstrapping are likely to be more accurate. It relies on picking random subsamples with replacement of measurement data and fitting the same model to those subsamples over and over again using, for example, the least squares regression.

It is proposed to assign subjects so as to minimize the discrepancies in centered first and second moments, where the assignment is gleaned via integer optimization [Bertsimas et al. 2015]. After the discrepancy is minimized, statistics such as the mean difference in subject responses are far more precise, and concentrated tightly around their nominal value, while still being unbiased estimates. However, the statistics after optimization will no longer follow their original distributions, which are wider, and traditional tests that rely on knowledge of this distribution, like the student's t test, no longer apply. A hypothesis test for applying bootstrap techniques to draw inferences on the mean differences between treatments is proposed and experimental evidence shows that these inferences can be more powerful than is usually possible [Bertsimas et al. 2015].

5.6.3. Identification of Statistical Distribution by Parametric Method Using Minitab

It is recommended to use computer software (self-developed or commercial) for identifying a statistical distribution for a set of data of interest by parametric methods because of lengthy calculations. Here is the procedure for using Minitab (Version 18) to identify the suitable statistical distribution: Stat → Quality Tools → Individual Distribution Identification.

A lot of fitted statistical distributions pop up as the output in both the Minitab Session Window and Graphics. There are three typical measures for each fitted statistical distribution: Anderson-Darling statistic (adjusted), P-value, and likelihood ratio test (LRT), for each fitted distribution function. These three measures are to be explained further in Section 8.10.1, Volume II.

As the primary criterion, P-value is frequently applied. A high p-value refers to the probability value at the extrema of a statistical distribution and it shows that a good fit as the null hypothesis prescribes. The null hypothesis means that no difference between the data set and the fitted statistical distribution function exists. Example 5.6.2 is utilized to demonstrate the application of statistical distribution identification to the fatigue life (cycles) of automotive latches.

EXAMPLE 5.6.2

Nine automotive latch systems went through a fatigue-life test [SAE J839 2019]. They failed at 70,560, 72,252, 73,267, 74,575, 75,697, 76,587, 77,777, 79,786, and 81,187 cycles [Ayub et al. 1997], respectively. How much are the life expectancies at (a) R = 90% and C = 95% and (b) R = 95% and C = 95%, respectively?

Solution:

The best fit for these nine data in this case study is the normal distribution, because its p-value = 98.6% that is higher than other statistical distributions tried. The average and standard deviation of the normal distribution function are, respectively:

$$Y_{ave} = 75742 \text{ cycles} \qquad \text{(Average)}$$

$$\text{and} \quad \sigma_Y = 3488 \text{ cycles} \qquad \text{(Sample standard Deviation)}$$

A quantile-quantile plot of cumulative failure distribution (in percentage) versus life cycles (i.e., the middle line) as shown in **Figure 5.6.3** is namely the fatigue-life curve. This middle line corresponds to the middle confidence level, i.e., C = 50%, and its related reliability R = 90% means that 90% of the latches in operation will exceed 73,000 cycles denoted by the solid triangle in **Figure 5.6.3** (using Minitab). This is often paradoxically referred to as B10 (or B_{10}) by engineers without addressing the confidence level.

Since the upper and lower limits (two-sided bootstrap curves) are defined for C = 90%, the lower-limit curve can be used as C = 95% for one-sided comparison because of the symmetry of a normal probability density function. The life cycles (denoted by $L_{R, C}$) are then

$$\text{(a)} \quad L_{R-90\%,\, C-90\%} = 69641 \text{ cycles (Solid square in Figure 5.6.3)}$$

$$\text{and} \quad \text{(b)} \quad L_{R=95\%,\, C=90\%} = 66999 \text{ cycles (Solid diamond in Figure 5.6.3)}$$

It means that we have a confidence level of 90%, that 90% of the latches in operation will exceed 69,641 cycles, and a confidence level of 90%, that 95% of the latches in operation will exceed 66,999 cycles.

FIGURE 5.6.3 Quantile plot of fatigue-life distribution of nine latch systems: C = 90% for two-sided limits.

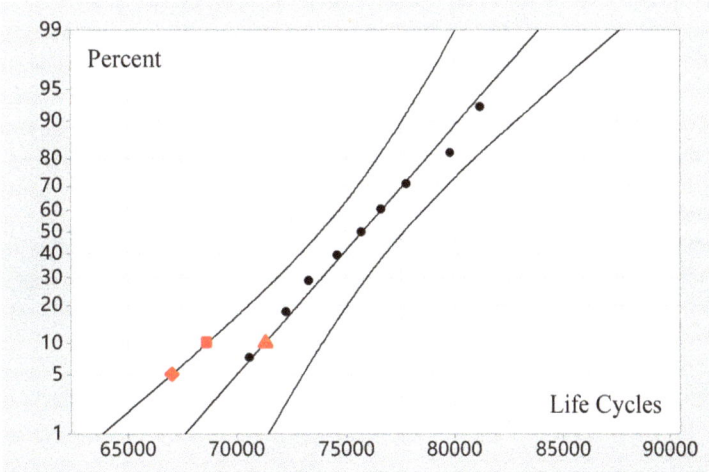

5.7. **Nested Design**

When designs with replicates are nested within at least a treatment, it is called nested DOE. A factor is nested within another factor when every level of one factor co-occurs with only one level of the other, as shown in Table 5.7.1, which demonstrates a two-factor nested design with six replicated tests each, having factor B nested within factor A. The "nested relationship" is an extension of one-way (one-factor) ANOVA and data collection and analysis are intended to give estimates of the sample means and for comparison of the means between treatments. Variation is partitioned among hierarchical levels and thus there is no reason to estimate interaction terms when dealing with conditional probability errors. Basic assumptions of ANOVA, including normality of residuals and constant variance, must hold.

5.7.1. **Two-Stage Nested Design: B Nested in A**

A factor is nested within another factor when each category of the first factor co-occurs with only one category of the other. In other words, an observation has to be within one category of factor B in order to have a specific category of factor A. Assume that an automotive maker wants to investigate the tensile strength of a part from different suppliers. The production parts come from two different suppliers. Three batches are sent for inspection by each supplier, while each batch contains five parts. Then it is a two-factor nested factorial design- factor B (batch) is nested in levels of factor A (supplier) as the levels of factor (B) are not identical to each other at different levels of factor (A). In other words, batches cannot be randomly allocated to different suppliers because they are particular to each individual supplier. For a two-stage nested DOE with factor B being nested within factor A as demonstrated in Table 5.7.1, an observation can be written as:

$$y_{ijn} = \bar{Y} + A_i + B_{j(i)} + \varepsilon_{n(ij)} \tag{5.7.1}$$

where

y_{ijn} is the observation

μ is the mean

A_i is the main effect of factor A at the ith level

$B_{j(i)}$ is the main effect of factor B at the jth level is nested under the ith level of factor A

i, j, and n are the indices, of which $1 \leq i \leq I$, $1 \leq j \leq J$, and $1 \leq n \leq N$

I, J are the total numbers of levels for factors A and B, respectively

N is the total number of replicates within A_i and B_j

$\varepsilon_{n(ij)}$ is the residues or errors, as replicates being nested at the run (i^{th} level of A and j^{th} level of B)

If one factor is nested within the other, there is no interaction between them because it is not feasible to have every combination of one factor along with every combination of the other. For example, I = 3, J = 3, and N = 6 for the case shown in Table 5.7.1. It is a balanced nested design because of an equal number of levels of B within each level of A and an equal number of replicates. When the levels of a factor are unique to the levels of one or more other factors, there is a nested factor. The case of experimental design given in Table 5.7.1 is a two-stage nested design and surely not a two-factor factorial design, since products in batch B_1 resourced from A_1 (Lab) is not the same as those in batch B_1 resourced from A_2 (Incoming Control Area) or those in batch B_1 resourced from A_3 (Warehouse).

TABLE 5.7.1 Two-stage nested design example with six replications.

Factor A (3 levels)	Lab area (A₁)			Control area (A₂)			Warehouse (A₃)		
Factor B (3-levels)	$B_1(A_1)$	$B_2(A_1)$	$B_3(A_1)$	$B_1(A_2)$	$B_2(A_2)$	$B_3(A_2)$	$B_1(A_2)$	$B_2(A_2)$	$B_3(A_2)$
	y_{111}	y_{121}	y_{131}	y_{211}	y_{221}	y_{231}	y_{311}	y_{321}	y_{331}
	y_{112}	y_{122}	y_{132}	y_{212}	y_{222}	y_{232}	y_{312}	y_{322}	y_{332}
	y_{113}	y_{123}	y_{133}	y_{213}	y_{223}	y_{233}	y_{313}	y_{323}	y_{333}
	y_{114}	y_{124}	y_{134}	y_{214}	y_{224}	y_{234}	y_{314}	y_{324}	y_{334}
	y_{115}	y_{125}	y_{135}	y_{215}	y_{225}	y_{235}	y_{315}	y_{325}	y_{335}
	y_{116}	y_{126}	y_{136}	y_{216}	y_{226}	y_{236}	y_{316}	y_{326}	y_{336}

© SAE International.

Note that six distinct parts are inspected in each inspection assignment.

Because each level of factor B is only present to one level of factor A, interaction AB cannot be separated from the main effect of either individual factor (i.e., A or B) in the two-factor nested design. However, a nested design de-confounds subsamples from true replicates. The null hypothesis (H_o) for a two-factor nested design is interpreted as:

H_o: Factor A (fixed): $\mu_A = 0$ (No difference among means of factor A, i.e. suppliers)
Factor B (fixed): $\mu_B = 0$ (No difference among means of factor B, i.e. batches)

H_o: Factor A (fixed): $\mu_A = 0$ (No difference among means of factor A, i.e. suppliers)
Factor B (random): $\sigma_B^2 = 0$ (No variability of factor B, i.e. batches)

or H_o: Factor A (random): $\sigma_A^2 = 0$ (No variability of factor A, i.e. suppliers)
Factor B (random): $\sigma_B^2 = 0$ (No variability of factor B, i.e. batches)

When a factor has a fixed effect, its levels applied in the study represent the specific levels of interest. On the other hand, when a factor has a random effect, its effect has been randomly specified as representatives of a larger population of levels of that factor, e.g., blocking effects.

Note that factor B usually has a random effect for a two-stage nested design.

At a minimum, the experimenter requires at least two measurements (i.e., N = 2) so that the experimenter can estimate the variability among measurements, σ_E^2. Of course, more than three replicates are recommended for meeting the statistical need—one DOF for the average, another for the standard deviation, and the third one is for the purpose of significance comparison. At least two batches per supplier (J = 2) have to be assayed in order to evaluate the variability among batches, $\sigma_{B|A}^2$.

The variation introduced at each hierarchy layer is inherently assessed relative to the layer below it when a nested-design approach is made. Noise limits the ability to detect effects, but known noise sources can be mitigated if used as blocking factors. The relative noise contribution of each layer can be utilized to optimally allocate experimental resources by means of nested ANOVA [Krzywinski et al. 2014] that can be obtained from relevant sums of squares. For calculating sums of squares of individual components as sampled in the DOE, Equation (5.7.1) can be rewritten into the following format:

$$y_{ijn} = \bar{Y} + A_i + B_{j(i)} + \varepsilon_{n(ij)}$$
$$y_{ijn} = \bar{Y} + (\bar{Y}_{A,i} - \bar{Y}) + (\bar{Y}_{AB,ij} - \bar{Y}_{A,i}) + (y_{ijn} - \bar{Y}_{AB,ij})$$

(5.7.2)

or $$y_{ijn} - \bar{Y} = (\bar{Y}_{A,i} - \bar{Y}) + (\bar{Y}_{AB,ij} - \bar{Y}_{A,i}) + (y_{ijn} - \bar{Y}_{AB,ij})$$

(5.7.3)

i.e., Total deviation = Main effect A + Main effect B within i^{th} level of factor A + Residual.

Similar to ANOVA table for two independent factors, the ANOVA table for the nested design—factor B nested in factor A can be derived and it is shown in Table 5.7.2. The nested design calls for the balanced replication at each level of the hierarchy, thus distributing the DOFs unequally so that the factor at the top of the hierarchy has relatively few DOFs. Sample sums of squares of the three items on the right side of Equation (5.7.3) are formulated as follows:

$$SS_A = J N \sum_{i=1}^{I} (\bar{Y}_{A,i} - \bar{Y})^2$$

(5.7.4)

$$SS_{B|A} = N \sum_{i=1}^{I} \sum_{j=1}^{J} (\bar{Y}_{AB,ij} - \bar{Y}_{A,i})^2$$

(5.7.5)

and

$$SS_E = \sum_{i=1}^{I} \sum_{j=1}^{J} \sum_{n=1}^{N} (Y_{ijn} - \bar{Y}_{AB,ij})^2 \qquad (5.7.6)$$

TABLE 5.7.2 Two-stage nested ANOVA with both random and fixed effects.

(a) A (Random) and B (Random):

Source of variance	SS	DOF	MS	$F_{u,v}$
A	SS_A	$I - 1$	$MS_A = SS_A/(I - 1)$	$F_{I-1,\ I(J-1)} = MS_A/MS_{B|A}$
B|A	$SS_{B|A}$	$I(J - 1)$	$MS_{B|A} = SS_{B|A}/[I(J - 1)]$	$F_{I(J-1),\ IJ(N-1)} = MS_{B|A}/MS_E$
Error	SS_E	$IJ(N - 1)$	$MS_E = SS_E/[IJ(N - 1)]$	—
Subtotal	SS_T	$IJ(N - 1)$	—	—
Grand ave.	SS_{ave}	1	—	—
Grand total	—	I J N		

(b) A (Fixed) and B (Random):

Source of variance	SS	DOF	MS	$F_{u,v}$
A	SS_A	$I - 1$	$MS_A = SS_A/(I - 1)$	$F_{I-1,\ I(J-1)} = MS_A/MS_{B|A}$
B|A	$SS_{B|A}$	$I(J - 1)$	$MS_{B|A} = SS_{B|A}/[I(J - 1)]$	$F_{I(J-1),\ IJ(N-1)} = MS_{B|A}/MS_E$
Error	SS_E	$IJ(N - 1)$	$MS_E = SS_E/[IJ(N - 1)]$	—
Subtotal	SS_T	$IJ(N - 1)$	—	—
Grand ave.	SS_{ave}	1	—	—
Grand total	—	I J N		

(c) A (Fixed) and B (Fixed):

Source of variance	SS	DOF	MS	$F_{u,v}$
A	SS_A	$I - 1$	$MS_A = SS_A/(I - 1)$	$F_{I-1,\ I(J-1)} = MS_A/MS_E$
B|A	$SS_{B|A}$	$I(J - 1)$	$MS_{B|A} = SS_{B|A}/[I(J - 1)]$	$F_{I(J-1),\ IJ(N-1)} = MS_{B|A}/MS_E$
Error	SS_E	$IJ(N - 1)$	$MS_E = SS_E/[IJ(N - 1)]$	—
Subtotal	SS_T	$IJ(N - 1)$	—	—
Grand ave.	SS_{ave}	1	—	—
Grand total	—	IJN		

© SAE International.

Notes:

I and J: Levels of factor A and B, respectively
N: Number of repeated runs to produce the error message
B\|A: Factor B on the given condition of factor A
SS: Sum of squares
DOF: Degree of freedom
MS: Mean square
$F_{u,v}$: F-distribution statistic of (u, v) DOFs

The subtotal of sum of squares is the combination of all the variances given above is:

$$SS_T = SS_A + SS_{B|A} + SS_E \qquad (5.7.7)$$

where

$$SS_T = \sum_{i=1}^{I} \sum_{j=1}^{J} \sum_{n=1}^{N} (Y_{ijn} - \bar{Y})^2 \tag{5.7.8}$$

Since factor B is nested within factor A, there can be no true interaction AB that can be calculated directly. Nevertheless, $SS_{B|A}$ can be also obtained from the following equation:

$$SS_{B|A} = SS_B + SS_{AB} \tag{5.7.9}$$

Note that the grand sum of squares (SS) is:

$$SS = SS_{Ave} + SS_T \tag{5.7.10}$$

where

$$SS_{Ave} = I J N \bar{Y}^2 \tag{5.7.11}$$

Equations derived above for calculating effects and doing diagnostic checking for two-factor factorial designs are still valid here with minor modifications. The ANOVA table has to be modified to accommodate the nested design structure as shown in Table 5.7.2, of which factor B is generally a conditional random variable nested in a level of factor A as the replicates for factor A. Since factor B is a random variable, the MS of the residuals is taken for granted to be the random sample error (S_E), as:

$$\sigma_E^2 = S_E^2 = MS_E \tag{5.7.12}$$

On the other hand, the error term for factor A is $MS_{B|A}$ instead of MS_E, as long as factor B is random whether factor A has a random effect or not. The error term for B|A is defined as factor B prescribed on the condition that factor A is given. The variances of factors A and B can be estimated by dividing the corrected MS by the number of data available for the corrected mean squares. Since there are N data for factor B and JN for factor A, the variances of factors A and B are, respectively:

$$\sigma_B^2 = (MS_{B|A} - MS_E) / N \tag{5.7.13}$$

and

$$\sigma_A^2 = (MS_A - MS_{B|A}) / (J N) \tag{5.7.14}$$

Note that the sum of square and DOF related to the nested design given in Table 5.7.3 can be also obtained directly from the full-factorial design of two independent factors as follows:

$$SS_{B|A} = SS_B + SS_{AB} \tag{5.7.15}$$

with

$$DOF_{B|A} = DOF_B + DOF_{AB} = (J - 1) + (I - 1) (J - 1) = I (J - 1) \tag{5.7.16}$$

In summary, the mean squares of a two-stage nested design can be calculated from experimental data and their expected mean squares are derived theoretically as shown in Table 5.7.3.

TABLE 5.7.3 Expected mean squares for two-stage nested design.

E(MS)	A: Fixed and B: Fixed	A: Fixed and B: Random	A: Random and B: Random
$E(MS_A)$	$\sigma_E^2 + SS_A/(I-1)$	$\sigma_E^2 + N\sigma_B^2 + SS_A/(I-1)$	$\sigma_E^2 + N\sigma_B^2 + JN\sigma_A^2$
$E(MS_{B|A})$	$\sigma_E + SS_{B|A}/[I(J-1)]$	$\sigma_E^2 + N\sigma_B^2$	$\sigma_E^2 + N\sigma_B^2$
$E(MS_E)$	σ_E^2	σ_E^2	σ_E^2

Note: SS_A and $SS_{B|A}$: Equations (5.7.4) and (5.7.5), respectively.

As a special application to the qualification of measurement, a two-stage nested design can be used to estimate gage repeatability and reproducibility of a designated experiment. It is demonstrated here using Example 5.7.1.

EXAMPLE 5.7.1

Three batches are randomly collected from each supplier (suppliers A_1 and A_2), respectively. Five steel parts are assayed from each batch. Material strength (MPa) data resulting from tensile tests are listed as follows:

Supplier	A_1			A_2		
Batch	$B_1(A_1)$	$B_2(A_1)$	$B_3(A_1)$	$B_1(A_2)$	$B_2(A_2)$	$B_3(A_2)$
	503	464	510	505	546	490
	510	473	515	496	515	495
	525	482	520	512	518	486
	498	595	508	512	518	486
	505	506	514	505	511	507

Solution:

Since design levels of both factors A and B are taken randomly, Table 5.7.2(a) is applied:

Source of variance	SS	DOF	DOF	$F_{u,v}$	$\alpha = 10\%$
Factor A	1.825	$I-1=1$	1.825 (=1.825/1)	$F_{1,4} = 0.161$	$F_{1,4;\alpha} = 4.55$
Factor B|A	45.41	$I(J-1)=4$	11.35 (=45.41/4)	$F_{4,24} = 9.39$	$F_{4,24;\alpha} = 2.2$
Error	29.02	$IJ(N-1)=24$	1.209 (=29.02/24)	—	—
Subtotal	76.25	$IJ(N-1)=29$			
Grand Ave.	—	1			
Grand total	—	$IJN = 30$			

Based on the p-values ($\alpha = 10\%$) given in the table, one jumps to the conclusion that there is no difference between the two suppliers (factor A) statistically but the variation from batch to batch (factor B) is extraordinary significant. Note that the higher the F-ratio, the more significant the corresponding factor.

Both factor B and sample error can contribute to the variation from batch to batch. Therefore, the "gage R&R" of the system has to be examined. The variance of measurement repeatability is the random sample error,

$$\sigma_{Repeatability}^2 = MS_E = 1.209$$

Note that there are five measurements (N = 5) in each batch. According to Equation (5.7.15), the sample variance from batch to batch at each measurement is:

$$\sigma_{Reproducibility}^2 = \sigma_B^2 = (MS_{B|A} - MS_E) / N = (11.35 - 1.209) / 5 = 2.03$$

The batch-to-batch variability (reproducibility) at each measurement as a percentage of the total variability (i.e., batch-to-batch variation + random sample error) is thus:

$$\sigma_{Reproducibility}^2 / (\sigma_{Reproducibility}^2 + \sigma_{Repeatability}^2) = 2.03 / (2.03 + 1.209) = 63\%$$

It is thus concluded that reduction in the reproducibility (batch-to-batch variability) has cut the product variability significantly.

5.7.2. Three-Stage Nested Design: B Nested in A and C Nested in B

Consider the three-stage fully nested design, in which factor C is assumed to be a conditional random variable nested in factor B as the replicates for factor B, while factor B is another conditional random variable nested in a predetermined design level of factor A as the replicates for factor A. Factor A can also have a random effect. Nevertheless, factor A may also have certain fixed design levels, which are selected arbitrarily, to the ultimate end. An observation (y_{ijmn}) for a three-factorial-nested (denoted as factors A, B, and C) DOE can be written as:

$$y_{ijmn} = \mu + A_i + B_{j(i)} + C_{m(ji)} + \varepsilon_{n(mji)} \tag{5.7.17}$$

where

μ is the mean

A_i is the main effect of factor A at the ith level

$B_{j(i)}$ is the main effect of factor B at the jth level, given that factor A is at the ith level

$C_{m(ji)}$ is the main effect of factor C at the mth level, given that B is at the jth level, which is in turn nested in the ith level of factor A

$\varepsilon_{n(mji)}$ is the residues or errors.

For calculating sums of squares of individual components as sampled in the DOE, each outcome corresponding to Equation (5.7.17) can be rewritten into the following format:

$$y_{ijmn} = \bar{Y} + (\bar{Y}_{A,i} - \bar{Y}) + (\bar{Y}_{AB,ij} - \bar{Y}_{A,i}) + (\bar{Y}_{ABC,ijm} - \bar{Y}_{AB,ij}) + (y_{ijmn} - \bar{Y}_{ABC,ijm}) \tag{5.7.18}$$

or $\quad y_{ijmn} - \bar{Y} = (\bar{Y}_{A,i} - \bar{Y}) + (\bar{Y}_{AB,ij} - \bar{Y}_{A,i}) + (\bar{Y}_{ABC,ijm} - \bar{Y}_{AB,ij}) + (y_{ijmn} - \bar{Y}_{ABC,ijm}) \tag{5.7.19}$

i.e., Total deviation = Main effect A + Main effect B in i^{th} level of A + Main effect C in j^{th} level of B and i^{th} level of A + Residuals.

By taking squares on both sides of Equation (5.7.19), one can derive the ANOVA table as given in Table 5.7.4. Sums of squares of the related terms in the table are summarized without derivations as follows:

$$SS_A = J\,M\,N \sum_{i=1}^{I} (\bar{Y}_{A,i} - \bar{Y})^2 \tag{5.7.20}$$

$$SS_{B|A} = M \, N \sum_{i=1}^{I} \sum_{j=1}^{J} (\bar{Y}_{AB,ij} - \bar{Y}_{A,i})^2 \qquad (5.7.21)$$

$$SS_{C|B} = N \sum_{i=1}^{I} \sum_{j=1}^{J} \sum_{m=1}^{M} (\bar{Y}_{ABC,ijm} - \bar{Y}_{AB,ij})^2 \qquad (5.7.22)$$

and

$$SS_{E} = \sum_{i=1}^{I} \sum_{j=1}^{J} \sum_{m=1}^{M} \sum_{n=1}^{N} (Y_{ijmn} - \bar{Y}_{ABC,ijm})^2 \qquad (5.7.23)$$

Formulae for $\bar{Y}$, $\bar{Y}_{A,I}$, $\bar{Y}_{AB,ij}$, and $\bar{Y}_{ABC,ijm}$ are given in Section 4.2. The number of repetitions, i.e., N, need not be large because information is being gathered on several check standards. The subtotal of sum of squares is the combination of all the variances as:

$$SS_{T} = SS_{A} + SS_{C|B} + SS_{B|A} + SS_{E} \qquad (5.7.24)$$

where

$$SS_{T} = \sum_{i=1}^{I} \sum_{j=1}^{J} \sum_{m=1}^{M} \sum_{n=1}^{N} (Y_{ijmn} - \bar{Y})^2 \qquad (5.7.25)$$

The grand total sum of squares is

$$SS = SS_{Ave} + SS_{T} \qquad (5.7.26)$$

where

$$SS_{Ave} = I \, J \, M \, N \, \bar{Y}^2 \qquad (5.7.27)$$

Note that a nested design involves at least one random factor.

TABLE 5.7.4 Three-stage nested DOE with fixed factors A and B and random factor C by analysis of variance.

Source of variance	SS	DOF	MS	$F_{u,v}$
Grand average	SS_{ave}	1	—	—
Factor A	SS_{A}	I − 1	$MS_{A} = SS_{A}/(I-1)$	$F_{I-1,\,I(J-1)} = MS_{A}/MS_{B|A}$
Factor B|A	$SS_{B|A}$	I(J − 1)	$MS_{B|A} = SS_{B|A}/[I(J-1)]$	$F_{I(J-1),\,IJ(M-1)} = MS_{B|A}/MS_{C|B}$
Factor C|B	$SS_{C|B}$	IJ(M − 1)	$MS_{C|B} = SS_{C|B}/[IJ(M-1)]$	$F_{IJ(M-1),\,IJM(N-1)} = MS_{B|A}/MS_{E}$
Error	SS_{E}	IJM(N − 1)	$MS_{E} = SS_{E}/[IJM(N-1)]$	—
Subtotal	SS_{T}	IJMN − 1	—	—
Grand average	SS_{Ave}	1	—	—
Total	—	IJMN	—	—

Notes:
I, J, M: Levels of factors A, B, and C, respectively
N: Number of repeated runs to produce the error message
B|A: Factor B on the given condition of factor A
C|B: Factor C on the given condition of factor B
DOF: Degrees of freedom
SS: Sum of squares
MS: Mean square
$F_{u,v}$: F-distribution statistic of (u, v) DOFs

Because each level of factor B is only present to one level of factor A, interaction AB cannot be separated from the main effect of either individual factor (i.e., A or B) in the three-factor nested design. Similar confounding pattern exists for factor C. The null hypothesis (H_o) for a three-factor nested design is interpreted as:

H_o: Factor A (fixed): $\mu_A = 0$ (No difference among means of factor A)
Factor B (fixed): $\mu_B = 0$ (No difference among means of factor B)
Factor C (fixed): $\mu_C = 0$ (No difference among means of factor C)

H_o: Factor A (fixed): $\mu_A = 0$ (No difference among means of factor A)
Factor B (random): $\sigma_B^2 = 0$ (No variability of factor B)
Factor C (random): $\sigma_B^2 = 0$ (No variability of factor C)

or H_o: Factor A (random): $\sigma_A^2 = 0$ (No variability of factor A)
Factor B (random): $\sigma_B^2 = 0$ (No variability of factor B)
Factor C (random): $\sigma_C^2 = 0$ (No variability of factor C)

When a factor has a fixed effect, its levels applied in the study represent the specific levels of interest. On the other hand, when a factor has a random effect, its effect has been randomly specified as representatives of a larger population of levels of factor such as blocking effects.

Note that factor C usually has a random effect for a three-stage nested design. Since the tests are conducted within factor C randomly, the MS of the residuals is taken for granted to be the variance of the random sample error:

$$\sigma_E^2 = MS_E \tag{5.7.28}$$

On the other hand, the error term for factor A is $MS_{B|A}$ instead of MS_E, as long as factor C is random no matter whether factor A has a random effect or not. The error term for B|A is defined as factor B prescribed on the condition that factor A is given.

In light of Equation (5.7.19), when all three factors have random effects the sample variance of experimental outcome Y can be partitioned as:

$$\sigma_Y^2 = \sigma_A^2 + \sigma_B^2 + \sigma_C^2 + \sigma_E^2 \tag{5.7.29}$$

The variances of factors A, B, and C can be estimated by dividing the corresponding corrected MS by the number of data available for the corrected mean squares. Since there are N data for factor C, MN for factor B, and JMN data for factor C, the variances of factors A and B are, respectively:

$$\sigma_C^2 = (MS_{B|A} - MS_E) / N \tag{5.7.30}$$

$$\sigma_B^2 = (MS_B - MS_{C|B}) / (M\,N) \tag{5.7.31}$$

and $\quad \sigma_A^2 = (MS_A - MS_{A|B}) / (J\,M\,N) \tag{5.7.32}$

If factor A has a fixed effect, $\sigma_A^2 \sim 0$, so does for factor B or C. Note that the sum of square and DOF related to the nested design given in Table 5.7.5 can be also obtained directly from the full-factorial design of three independent factors as follows:

$$SS_{B|A} = SS_B + SS_{AB} \tag{5.7.33}$$

$$\text{with} \quad \text{DOF}_{B|A} = \text{DOF}_B + \text{DOF}_{AB} = (J - 1) + (I - 1)(J - 1) = I(J - 1) \tag{5.7.34}$$

$$\text{and} \quad \text{SS}_{C|B} = \text{SS}_{C|A} + \text{SS}_{CB|A} = (\text{SS}_C + \text{SS}_{CA}) + (\text{SS}_{CB} + \text{SS}_{CBA}) \tag{5.7.35}$$

$$\text{with} \quad \text{DOF}_{C|B} = (\text{DOF}_C + \text{DOF}_{CA}) + (\text{DOF}_{CB} + \text{DOF}_{CBA})$$

$$= [(M - 1) + (M - 1)(I - 1)] + [(M - 1)(J - 1) + (M - 1)(J - 1)(I - 1)] \tag{5.7.36}$$

$$= I\,J\,(M - 1)$$

TABLE 5.7.5 Expected mean squares for three-stage nested design with random C.

E(MS)	A: Fixed and B: Fixed	A: Fixed and B: Random	A: Random and B: Random
$E(MS_A)$	$\sigma_E + N\sigma_C^2 + SS_A/(I-1)$	$\sigma_E^2 + N\sigma_B^2 + SS_A/(I-1)$	$\sigma_E^2 + N\sigma_C^2 + JN\sigma_B^2 + IJN\sigma_A^2$
$E(MS_{B|A})$	$\sigma_E + N\sigma_C^2 + SS_{B|A}/[I(J-1)]$	$\sigma_E^2 + N\sigma_C^2 + JN\sigma_B^2$	$\sigma_E^2 + N\sigma_C^2 + JN\sigma_B^2$
$E(MS_{C|B})$	$\sigma_E^2 + N\sigma_C^2$	$\sigma_E^2 + N\sigma_C^2$	$\sigma_E^2 + N\sigma_C^2$
$E(MS_E)$	σ_E^2	σ_E^2	σ_E^2

where SS_A and $SS_{B|A}$ obtained from Equations (5.7.4) and (5.7.5), respectively.

For application, a three-level nested design is recommended for a measurement system (e.g., new acquired material hardness tester) where sources of error are not well understood and have not previously been studied. The three levels may separate by time duration as:

- Level A: Measurements taken over a short-time to capture the precision of the gauge
- Level B: Measurements taken over runs separated by weeks
- Level C: Measurements taken over runs separated by months

By the same token, an observation (y_{ijmpn}) for a balanced nested DOE with four factors can be written as:

$$y_{ijmpn} = \mu + A_i + B_{j(i)} + C_{m(ji)} + D_{p(mji)} + \varepsilon_{n(pmji)} \tag{5.7.37}$$

where
μ is the mean
A_i is the main effect of factor A at the ith level
$B_{j(i)}$ is the main effect of factor B at the jth level, given that factor A is at the ith level
$C_{m(ji)}$ is the main effect of factor C at the mth level, given that B is at the jth level, which is in turn nested in the ith level of factor A
$D_{p(mji)}$ is the main effect of factor D at the pth level, given that C is at the mth level
B is at the jth level, which is in turn nested in the ith level of factor A
$\varepsilon_{n(pmji)}$ is the residues or errors

When a nesting design exists but neglected, one may face the following three problems:

(a) Wrongly attributing a main effect to an interaction effect when, in fact, no interaction exists
(b) Dividing by wrong DOFs when determining the MS and F statistics
(c) Assuming that a main effect has a smaller effect size because the sum of squares for that effect is being partly attributed to the interaction.

Many experimental designs in the behavioral sciences do qualify as a nested design. Response surface designs within split-plot structures can be assessed [Vining et al. 2005]. Nested errors for designing fractional two-level experiments are discussed in Schoen [1999].

5.7.3. Predictive Equation for Nested Designs

The regression model for a nested experimental design is a straightforward extension of the standard multiple linear regression model, as formulated in Section 4.3.

5.8. Crossed Design

Two factors are crossed when every category of one factor co-occurs in the design with every category of the other factor. If two factors are crossed, one can calculate an interaction between them. The difference between nested design and crossed design can be told easily using Figure 5.8.1:

(a) Nested design: Some parts examined by operator A_1 and other parts by operator A_2.

(b) Crossed design: Assume that all parts are to be examined by both operators.

A frequently observed application of crossed DOE is for Gage R&R study (Example 4.6.1), the other is the operational arbitration by either Kappa coefficient or common consent (Chapter 18, Volume IV). Crossed-nested design is to be discussed in Chapter 16 (Volume IV).

FIGURE 5.8.1 Difference between crossed design and nested design—two operators and eight parts.

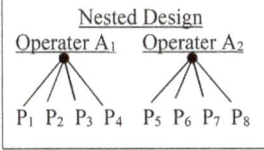
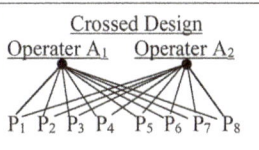

©SAE International.

References

Anderson-Cook, C.M., "When Should You Consider A Split-Plot Design?" *Quality Progress* 40, no. 10 (2007): 57-59.

Andrisano, A.O. et al. (2011), "Design of Simulation Experiments Method for Injection Molding Process Optimization," in *Proceedings of the IMProVe 2011 Int'l Conference on Innovative Methods in Product Design*, Venice, Italy, June 15–17, 2011.

Ankenman, B., "Design of Experiments with Two-Level and Four-Level Factors," *Journal of Quality Technology* 31 (1999): 363-375.

Ayub, M., Lee, D.J., Chiang, Y.J., and Barkczynski, D. (1997), "Robustness Study of Cables and Rods for Automotive Door Latch Systems," in *Proceedings of Total Product Development Symposium*, Dearborn, MI, November 5–6, 1997.

Bertsimas, D. et al., "The Power of Optimization Over Randomization in Designing Experiments Involving Small Samples," *Operations Research* 63, no. 4 (2015): 868-876.

Box, G.E.P. and Behnken, D.W., "Some New Three Level Designs for the Study of Quantitative Variables," *Technometrics* 2, no. 4 (1960): 455-495.

Box, G.E.P. and Draper, N.R., *Response Surfaces, Mixtures, and Ridge Analyses*, 2nd ed. (New York: Wiley, 2007).

Buckland, S.T., Burnham, K.P., and Augustin, N.H., "Model Selection: An Integral Part of Inference," *Biometrics* 53 (1997): 603-618.

Chiang, K.T. and Chang, F.P., "Analysis of Shrinkage and Warpage in an Injection-Molded Part with a Thin Shell Feature Using the Response Surface Methodology," *Int'l Journal of Advanced Manufacturing Technology* 35 (2007): 468-479.

Dean, A., Voss, D., and Draguljic, D., *Design and Analysis of Experiments*, 2nd ed. (Cham, Switzerland: Springer, 2017), ISBN:978-3319522487.

Efron, B. and Gong, G. (February 1983), "A Leisurely Look at the Bootstrap, the Jackknife, and Cross Validation," *The American Statistician*, 37(1), pp. 36-48.

Efron, B., "Censored Data and the Bootstrap," *Journal of the American Statistical Association* 76, no. 374 (1981): 312-319.

Fletcher, D. and Dillingham, P.W., "Model-Averaged Confidence Intervals for Factorial Experiments," *Computational Statistics and Data Analysis* 55 (2010): 3041-3048.

Franklin, L.A. and Wasserman, G.S., "Bootstrap Lower Confidence Limits for Capability Indices," *Journal of Quality Technology* 24 (1992): 196-210.

Guan, W., "From the Help Desk: Bootstrapped Standard Errors," *The Stata Journal* 3 (2003): 71-80.

Hartley, H.O., "Smallest Composite Designs for Quadratic Response Surfaces," *Biometrics* 15 (1959): 611-624.

Jones, B. and Nachtsheim, C.J., "Split-Plot Designs: What, Why and How," *Journal of Quality Technology* 41, no. 4 (2009): 340-361.

Kim, M.S. and Heo, S.J., "Conservative Quadratic RSM Combined with Incomplete Small Composite Design and Conservative Least Squares Fitting," *KSME International Journal* 17, no. 5 (2003): 698-707.

Krzywinski, M. et al., "Nested Design," *Nature Methods* 11 (2014): 977-978.

Lasky, R.C., Santos, D., and Cloyd, J. (2008), "An Effective Design of Experiment Strategy to Optimize SMT Processes," Dartmouth College, Indium Corporation of America, Utica, NY; Corpus ID: 27002088.

Long, J.S., *Regression Models for Categorical and Limited Dependent Variables* (Thousand Oaks, CA: SAGE Publications, 1997).

Mangal, S. and Kumar, A., "Geometric Parameter Optimization of Magneto-Rheological Damper Using Design of Experiments Technique," *Int'l Journal of Mechanical and Materials Engineering* 10, no. 4 (2015): 1-9.

Monohar, M. et al., "Application of Box Behnken Design to Optimize the Parameters for Turning Inconel 718 Using Coated Carbide Tools," *Int'l Journal of Scientific & Engineering Research* 4, no. 4 (2013): 620-642.

Montgomery, D.C. (2019), *Design and Analysis of Experiments*, 10th Edition, John Wiley & Sons, Hoboken, NJ, 688 pages; ISBN: 978-1-119-49244-3.

Myers, R.H., Montgomery, D.C., and Anderson-Cook, C.M., *Response Surface Methodology: Process and Product Optimization Using Designed Experiments*, 3rd ed. (New York: Wiley, 2009).

NIST (2022), "1.3.3.4. Bootstrap Plot," Engineering Statistics Handbook Department of Commerce, National Institute of Standards and Technology, Washington, DC, Retrieved November 5, 2022.

Nwanya, J. and Dozie, K., "Optimal Prediction Variance Capabilities of Inscribed Spherical Composite Design," *European Journal of Statistics and Probability* 8, no. 2 (2020): 41-48.

Oehlert, G.W., *Design and Analysis of Experiments: Response Surface Design* (New York: W. H. Freeman and Company, 2000).

Ohaegbulem, E.U. and Chigbu, P.E., "A Measure of Orthogonality for the Central Composite Designs," *Communications in Statistics-Theory and Methods* 51, no. 9 (2022): 2710-2724.

Oyejola, B.A. and Nwanya, J.C., "Selecting the Right Central Composite Design," *Int'l Journal of Statistics and Applications* 5, no. 1 (2015): 21-30.

Perincek, O. and Colak, M., "Use of Experimental Box-Behnken Design for the Estimation of Interactions between Harmonic Currents Produced by Single Phase Loads," *Int'l Journal of Engineering Research and Applications* 3, no. 2 (2013): 158-165.

Ross, P.J., *Taguchi's Techniques for Quality Engineering* (New York: McGraw-Hills, 1988).

SAE International (2019), "Passenger Car Side Door Latch Systems (STABILIZED Oct 2019)," SAE Standard J3839, Revised October 2019.

Sanchez, S. and Sanchez, P., "Very Large Fractional Factorials and Central Composite Designs," *ACM Transactions on Modeling and Computer Simulation* 15, no. 4 (2005): 362-377.

Schoen, E.D., "Designing Fractional Two-Level Experiments with Nested Error Structures," *Journal of Applied Statistics* 26 (1999): 495-508.

Sousa, A.M.F. et al., "Design of Experimental Design as a Tool for the Processing and Characterization of HDPE Composites with Sponge-Gourds (Luffa-Cylindrica) Agrofiber Residue," *Journal of Sustainable Development* 6, no. 4 (2013): 106-117.

Tang, S. et al., "The Use of Taguchi Method in the Design of Plastic Injection Mold for Reducing Warpage," *Journal of Materials Processing Technology* 182 (2007): 418-426.

Vieira, H. et al., (December 2011), "Improved Efficient, Nearly Efficient, Nearly Orthogonal, Nearly Balanced Mixed Designs," *Conference Paper in Proceedings-Winter Simulation Conference*, December 2011.

Vining, G.G., Kowalski, S.M., and Montgomery, D.C., "Response Surface Designs within a Split-Plot Structure," *Journal of Quality Technology* 37 (2005): 115-129.

Westlake, W.J., "Composite Design Based on Irregular Fractions of Factorials," *Biometrics* 21 (1965): 324-336.

Xu, H., "Algorithmic Construction of Efficient Fractional Factorial Designs with Large Run Sizes," *Technometrics* 51, no. 3 (2009): 262-277.

Xu, H. et al., "Combining Two-Level and Three-Level Orthogonal Arrays for Factor Screening and Response Surface Exploration," *Statistica Sinica* 24 (2014): 269-289.

Problems

P5.1: Continued from Problem P3.4. In order to make it a Box-Behnken design, five additional runs are replicated at the center point. Test results are listed below [Sousa et al. 2013]:

Treatment	a	b	c	d	σ_{uts} (MPa)	E (GPa)	I_{zod} (J/m)
17	0	0	0	0	17.71	1.079	71.73
18	0	0	0	0	17.30	1.091	71.60
19	0	0	0	0	17.21	1.111	69.56
20	0	0	0	0	17.82	1.125	74.96
21	0	0	0	0	18.00	1.085	81.51

Please formulate the predictive equation in light of response surface (up to quadratic terms and interactions only) for each mechanical property. Compare the predictive values for both cases: (1) using full-factorial design 2^4 and (2) the BBD.

P5.2: The Box-Behnken design was applied to optimizing the cutting parameters for turning Inconel 718 using coated carbide tools [Monohar et al. 2013]. The obtained data of surface roughness are listed as follows:

Run	Cutting speed (m/min)	Feed speed (mm/rev)	Depth of cut (mm)	R_a (Roughness) (µm)
1	50	0.2	2.0	3.13
2	40	0.2	1.5	3.15
3	60	0.25	2.0	3.28
4	40	0.3	1.5	3.71
5	50	0.25	1.5	3.25
6	60	0.3	1.5	3.60
7	50	0.3	1.0	3.56

Run	Cutting speed (m/min)	Feed speed (mm/rev)	Depth of cut (mm)	R$_a$ (Roughness) (μm)
8	50	0.2	1.0	2.98
9	50	0.3	2.0	3.75
10	60	0.25	1.0	3.15
11	40	0.25	2.0	3.42
12	50	0.25	1.5	3.24
13	50	0.25	1.5	3.23
14	60	0.2	1.5	3.01
15	40	0.25	1.0	3.24

The data given above are obtained from Table 3 of Monohar et al. [2013]. Please formulate the predictive equation. If possible, which composite design should be used instead of Box-Behnken design, and why?

6

Optimal Designs

W hen there are constrained mixture experiments that cannot be accomplished using balanced orthogonal designs or too many treatments to be completed in time or beyond human effort, one may resort to optimal designs. An optimal design is justified by its associated optimality criterion with a single number that summarizes how good a design is. The optimality is achieved by maximization, minimization, or target-orientation. D-optimal designs and A-optimal designs are set up to minimize the variance of parameter estimates, while I-optimal designs and G-optimal designs are expected to minimize the prediction variance. All of them belong to information-based optimality. Other algorithms for optimal designs include distance-based optimality and time-based optimality. If there are multiobjectives, Pareto optimality and/or gray relational analysis can be applied.

6.1. Information-Based Optimality

Ideally, balanced orthogonal designs are certainly preferred when applying the DOE. However, when only an "incomplete design of experiments" is available, optimization algorithms can be used to reduce the number of treatments as an alternative. The approach often gears toward exploring the following four optimization algorithms:

- D-optimality: Minimizing the trace of information matrix, i.e., variance-covariance matrix
- A-optimality: Minimizing the average variance of parameter estimates
- I-optimality (i.e., V-Optimality): Minimizing the average prediction variance
- G-optimality: Minimizing the maximum prediction variance

D-optimality and A-optimality are guided by the general variance (i.e., variance-covariance) matrix formed on parameter estimates, while I-optimality and G-optimality are guided by the prediction variance.

6.1.1. Information Matrix and Maximum Likelihood Method

D-optimality, A-optimality, I-optimality, and G-optimality are formulated using information matrix. Assume that f(t) is a continuous function such as a probability density function or polynomial equation for simulating the DOE test runs. The information can be drawn from the continuous function against a specific parameter, e.g., parameter θ, can be calculated as [Fisher 1966]:

$$I(\theta) = \int \left(\frac{\partial\{\ln[f(t\mid\theta)]\}}{\partial\theta}\right)^2 f(t\mid\theta)\, dt \tag{6.1.1}$$

$$\text{or} \quad I(\theta) = -E\left[\frac{d^2\{\ln[f(t\mid\theta)]\}}{d\theta^2}\right] \tag{6.1.2}$$

The derivations of the equations given above are detailed in Section 8.1 (Chapter 8, Volume II).

When there are two parameters of interest, namely θ_1 and θ_2, the off-diagonal elements in the information matrix are formed as

$$\text{or} \quad I(\theta_1, \theta_2) = -E\left[\frac{\partial^2\{\ln[f(t\mid\theta_1 \text{ and } \theta_2)]\}}{\partial\theta_1\, \partial\theta_2}\right] \tag{6.1.3}$$

It is a matrix of second cross-moments. Thus, an information matrix has both variances and covariances for multiple variables of interest.

Fisher information matrix can measure how much information one parameter (e.g., θ, as an input) carries off another value (e.g., $f(t\mid\theta)$, as an output), and thus it can be used to test whether an optimal DOE design or regression model is appropriate. In other words, Fisher information matrix techniques are widely used in DOE to forecast the precision of future experiments while they are still in the design phase. This is called optimality.

As shown in Example 6.1.1, the application of Equations (6.1.1)–(6.1.3) to the normal distribution is employed here to illustrate how an information matrix is formed.

EXAMPLE 6.1.1

Find the Fisher information matrix of variable t that follows a normal distribution.

Solution:

1. There are two unknown parameters of interest in a normal distribution function, i.e., mean μ and variance σ^2, the information matrix can be written as

$$[\,I\,] = \begin{bmatrix} I_{11} & I_{12} \\ I_{21} & I_{22} \end{bmatrix} \tag{6.1.4}$$

Diagonal elements I_{11} and I_{22} are variances of mean μ and variance σ^2, respectively, and they can be derived using Equation (6.1.2). Off-diagonal terms I_{12} and I_{21} that can be derived using Equation (6.1.3) are covariances between mean μ and variance σ^2.

2. When mean μ (as a parameter) is the unknown of interest:

Given that $t \sim N(\mu, \sigma^2)$, the probability density function is:

$$f(t) = \frac{1}{(2\pi)^{1/2}\,\sigma} \exp\left[-\tfrac{1}{2}\left(\frac{t-\mu}{\sigma}\right)^2\right]$$

where mean μ given in the above equation is the variable θ_1 given in Equation (6.1.1). Taking a natural-logarithmic transformation of the above equation,

$$\ln[f(t)] = -\ln[(2\pi)^{1/2}\,\sigma] - \left[\frac{(t-\mu)^2}{2\,\sigma^2}\right]$$

$$\frac{\partial\{\ln[f(t\mid\mu)]\}}{\partial\mu} = \frac{t-\mu}{\sigma^2}$$

and

$$\frac{\partial^2\{\ln[f(t\mid\mu)]\}}{\partial\mu^2} = \frac{-1}{\sigma^2} < 0$$

According to Equation (6.1.2), Fisher information I_{11} is:

$$I_{11} = I(\mu) = -E\left[\frac{\partial^2\{\ln[f(t\mid\mu)]}{\partial\mu^2}\right] = \frac{1}{\sigma^2} \qquad (6.1.5)$$

It shows that Fisher information of μ (mean value) is the inverse of the variance of a normal distribution function, and vice versa. This yields how well one can measure the location parameter (μ) of a normal distribution using its variance σ^2 with the available test data. It can be seen that standard error σ decreases with increasing Fisher information $I(\mu)$. The smaller the standard deviation, the better the fit to narrow down the range in which the true mean lies.

3. When variance σ^2 (working as another parameter) is the unknown of interest:

Given that $t \sim N(\mu, \sigma^2)$, the probability density function can be rewritten as:

$$f(t) = \frac{1}{(2\pi\,\sigma^2)^{1/2}} \exp\left[\frac{-(t-\mu)^2}{2\,\sigma^2}\right]$$

where variance σ^2 given in the above equation is another parameter θ_2 given in Equations (6.1.1)–(6.1.3). Taking a natural-logarithmic transformation of the above equation,

$$\ln[f(t)] = -\ln[(2\pi)^{1/2}] - \ln[(\sigma^2)^{1/2}] - \left[\frac{(t-\mu)^2}{2\,\sigma^2}\right]$$

$$\frac{\partial\{\ln[f(t\mid\mu)]\}}{\partial(\sigma^2)} = \frac{-1}{2\,\sigma^2} + \frac{(t-\mu)^2}{2\,(\sigma^2)^2}$$

and

$$\frac{\partial^2\{\ln[f(t\mid\mu)]\}}{\partial(\sigma^2)^2} = \frac{1}{2\,(\sigma^2)^2} - \frac{(t-\mu)^2}{(\sigma^2)^3}$$

According to Equation (6.1.2) and $E[(t - \mu)^2] = \sigma^2$, Fisher information I_{22} is:

$$I_{22} = I(\sigma^2) = -E\left[\frac{\partial^2\{\ln[f(t \mid \mu)]\}}{\partial(\sigma^2)^2}\right] = \frac{-1}{2(\sigma^2)^2} + \frac{1}{(\sigma^2)^2} = \frac{1}{2\sigma^4} \quad (6.1.6)$$

4. Off-diagonal terms:

$$I_{12} = -E\left[\frac{\partial^2\{\ln[f(t \mid \mu)]\}}{\partial(\sigma^2)\,\partial\mu}\right] = -E\left[\frac{2(t - \mu)}{2\sigma^4}\right] = 0 \quad (6.1.7)$$

since $E[(t - \mu)] = 0$. Similarly, $I_{21} = 0$.

5. Combining steps (1)–(4) leads to the following information matrix for a normal distribution:

$$[I] = \begin{bmatrix} 1/\sigma^2 & 0 \\ 0 & 1/(2\sigma^4) \end{bmatrix} \quad (6.1.8)$$

It can be seen that the information matrix consists of both variances of individual parameters as diagonal elements and covariances between the two parameters of interest as nondiagonal elements. In other words, the information matrix is a covariance matrix, which is also called the variance-covariance matrix or generalized variance matrix. The covariance matrix of the maximum likelihood estimator of $\mu(\theta_1)$ and $\sigma^2(\theta_2)$ is approximately equal to the inverse of the information matrix.

6. When the sample of size N is made of independent and identically distributed (IID) data:

The Fisher information of mean μ and variance σ^2 can be obtained as, respectively:

$$I_{t1, t2, \dots, tN}(\mu) = \frac{1}{\sigma^2} \approx \frac{1}{S^2} \quad (6.1.9)$$

$$\text{and} \quad I_{t1, t2, \dots, tN}(\sigma^2) = \frac{N}{2\sigma^4} \approx \frac{N}{2S^4} \quad (6.1.10)$$

of which S^2 is the sample variance. Equation (6.1.10) can be obtained using the additivity property of Fisher information. Thus, Fisher information matrix can be used to answer a common question among DOE practitioners and data analysts on how accurately the parameters of a distribution can be estimated.

6.1.2. **Opportunities to Use Optimal Designs**

When applying DOE, there are three opportunities for an experimenter to use optimal designs [Cook and Nachtsheim 1980]:

(a) Constrained mixture experiment [Goldfarb et al. 2004]: Constraints are imposed on factorial settings in the design space.

(b) Too many treatments to run: Each full factorial design or fractional factorial design demands a fixed number of treatments that may amount to a good number of resources and/or too much time required for carrying out all the tests.

(c) Treatments beyond human effort.

An optimal design is frequently selected by means of reductions in variances and/or covariances for special cases given above based on the Fisher information matrix, when complete design matrices (presented in Chapters 2–5) are not available. Optimal designs are mostly based on incomplete design matrices and often used in science and engineering. Marketing study based on optimal designs is often called conjoint analysis—how people value different features of a product or service.

6.1.3. **How to Select the Optimal Design**

An experimenter who wants to utilize an optimal design needs to first specify a metamodel for the design and the total number of treatments allowed. Then, select a candidate set of treatments for the design matrix, which usually consists of all possible combinations of the assigned factorial levels. Let the computing algorithm choose the optimal set of treatments for the design matrix from the specified candidate set of design matrix [Cauffriez et al. 2013]. The computing algorithm often resorts to the Fisher information matrix, which is applied for prioritizing candidate design matrices by measuring the amount of information that it contains about the true population value of the parameters (e.g., coefficients of a polynomial).

For example, given that k factors are of interest and a metamodel of p parameters (including coefficients and intercept if it is a polynomial) are selected, one may use the following as a guide to constructing the initial design matrix [Anderson and Whitcomb 2014a, 2014b]:

(a) Modeling: Building the metamodel using p points based on the selected optimal design.

(b) Lack of Fit: Checking the lack of fit by a certain number of treatments (e.g., 6) based on distance; if not fit, escalating to a higher-order model. Replications will lend the experimenter to assess residuals, which can be used for assuring diagnostic checking of the metamodel.

When higher trueness is desired, one is encouraged to rebuild the design with additional treatments. This practice also applies to selecting the experimental design from general factorial designs (L^{K-P}) and composite designs [Oyejola and Nwanya 2015].

6.1.4. **Comparison of Different Optimal Designs**

There are three general methods to compare different experimental designs, including DOE-t, DOE-F, DOE-W, DOE-E, DOE-LN, composite designs, and optimal designs. They are

1. Design efficiency: Parameter estimates computed from an experimental design with a higher design efficiency are more acceptable. Among all the computing algorithms, D-optimal design that is an information matrix-based optimality, is frequently applied because of its reasonable design efficiency.

2. Fraction of design space plot: The variance (or standard deviation) versus the fraction of design space of coded variables (or the applied range of an individual design factor) is called fraction of design space (FDS) plot. FDS plots can be used for examining the relative robustness among different optimality criteria [Ozol-Godfrey et al. 2005]. One may evaluate the accuracy of potential optimal designs via FDS plots, of which each is a plot of the prediction variance versus the applied range of an individual design factor. Of course, the smaller the variance, the better the fit.

3. Statistical power: Power in statistics is the probability that a hypothesis test can detect an effect in a sample when it exists in the population. High statistical power occurs when a hypothesis test is likely to find an effect that exists in the population. For example, when statistical power is 90%, a hypothesis test has a 90% chance of detecting an effect that actually exists.

Experimenters use these three comparative methods because it is rare to absolutely measure the corresponding population. Hypothesis tests applied in DOE enable people to draw conclusions over the entire population from the given finite number of sample data. In other words, experimenters are stuck with samples. In general, D-optimality leads to more accurate parameter estimations (e.g., coefficients and intercept of a polynomial) for screening and mechanistic models, while I-optimality makes more accurate predictions for empirical RSM.

6.1.5. Residual Maximum Likelihood (REML)

The average information residual maximum likelihood (AI-REML) algorithm was proposed for efficient variance parameter estimation for linear mixed models by [Patterson and Thompson 1971, Gilmour et al. 1995]. The expectation-optimization parameter-expanded algorithm (Section 6.1.1) based on the maximum likelihood method requires the specification of complete data (i.e., incomplete plus censored data), but the AI-REML algorithm can be used for system identifications with incomplete data [Diffey et al. 2017].

6.2. D-Optimality

D-optimality criterion is based on the trace of information matrix formulated for parameter estimates, and it aims to maximize the determinant of the information matrix. The information matrix is the inverse of the variance-covariance matrix of regression coefficients. D-optimality criterion is the most used one among all optimal designs. This has been in use as the first ever applied information-based optimality and is available in most statistical codes. For example, a user can specify the model, and then Minitab selects design points that optimize the D-optimal criterion from a set of candidate points.

6.2.1. D-Optimal Design for Linear Regression Model

For example, consider the constrained mixture experiment for estimating the impact of three factors on electric resistivity Y of a modified acrylonitrile powder [Atkinson et al. 2007, Piepel et al. 2002]. The electric resistivity of the powder depends only on the relative proportions of the three ingredients, i.e., factors A, B, and C, of which the lowest and highest design levels are given in Table 6.2.1.

TABLE 6.2.1 Design factors for electric resistivity of modified acrylonitrile powder [Goos and Leemans 2004].

Factor	Level (−)	Level (+)
A: Copper sulfate ($CuSO_4$)	20%	80%
B: Sodium thiosulfate ($Na_2S_2O_3$)	20%	80%
C: Glyoxal $(CHO)_2$	0%	60%

Reprinted from "Teaching Optimal Design of Experiments Usinga Spreadsheet," Peter Goos, Herlinde Leemans, Journal of Statistics & Data Science Education, 12, 3 © 2004 Taylor & Francis Ltd, http://www.tandfonline.com.

It is here assumed that factors A, B, and C are related to response Y as follows:

$$Y_q = \theta_a\, A_q + \theta_b\, B_q + \theta_c\, C_q + \varepsilon_q \qquad (6.2.1)$$

Note that q = 1, 2, 3, ..., and Q, where Q is the total number of treatments. Assume that nine treatments without any replication are designated to do the parameter estimates with the following four constraints [Goos and Leemans 2004]:

$$20\% < A < 80\%$$
$$20\% < B < 80\%$$
$$0\% < C < 60\%$$
$$\text{and} \quad A + B + C = 100\%$$

Given that nine observations are obtained from carrying out nine experimental tests, the nine simultaneous equations to be solved for three parameters to be estimated, i.e., θ_a, θ_b, and θ_c, can be formulated as:

$$
\begin{bmatrix}
A_1 & B_1 & C_1 \\
A_2 & B_2 & C_2 \\
A_3 & B_3 & C_3 \\
A_4 & B_4 & C_4 \\
A_5 & B_5 & C_5 \\
A_6 & B_6 & C_6 \\
A_7 & B_7 & C_7 \\
A_8 & B_8 & C_8 \\
A_9 & B_9 & C_9
\end{bmatrix}
\begin{Bmatrix}
\theta_a \\ \theta_b \\ \theta_c
\end{Bmatrix}
=
\begin{Bmatrix}
Y_1 \\ Y_2 \\ Y_3 \\ Y_4 \\ Y_5 \\ Y_6 \\ Y_7 \\ Y_8 \\ Y_9
\end{Bmatrix}
\qquad (6.2.2)
$$

or in form of matrix as:

$$[D]_{9x3}\ \{\theta\}_{3x1} = \{Y\}_{9x1} \qquad (6.2.3)$$

where
  [D] is the design matrix of a DOE
  $\{\theta\}$ is the vector of unknown parameters (coefficients)
  $\{Y\}$ is the vector of responses

Assume that matrix $[X_{ij}]$ is referred to as the extended design matrix of the experiment, of which each element is derived from the partial differentiation of response Y_i with respect to individual parameter θ_j as suggested by the regression model. Hence,

$$X_{ij} = \partial Y_i\, /\, \partial \theta_j \qquad (6.2.4)$$

When the least-squares method applies, the extended design matrix with the unknown parameters is to be estimated from the following matrix operations:

$$\{\theta\}_{3x1} = [([X]^T\, [X])^{-1}]_{3x3}\, [X]^T_{3x9}\, \{Y\}_{9x1} \qquad (6.2.5)$$

The associated variance-covariance matrix of estimated $\{\theta\}$ is then:

$$\text{Cov}(\{\theta\}) = \text{Var}(\{\theta\} = \sigma_\varepsilon^2 \,[([X]^T\,[X])^{-1}]_{3\times3} \qquad (6.2.6)$$

where

$\text{Cov}(\{\theta\})$ is the variance-covariance matrix, also called generalized variance $\text{Var}(\{\theta\})$
σ_ε^2 is the variance of errors ε_q, $q = 1, 2, \ldots,$ and 9

The diagonal elements of the variance-covariance matrix are the variances of the ordinary least-squares estimators for the slopes of θ_a, θ_b, and θ_c, respectively.

The mission for identifying a suitable DOE is now to figure out the "optimal" extended design matrix [X] that would actually lead to the more realistic relationship between the response vector $\{Y\}$ and factor vector $\{\theta\}$ via a certain criterion such as D-optimality. Design matrix [X] is to be derived from the differentiations of the model equation with respect to all parameters in the regression model, i.e., Equation (6.2.1),

$$[X] = \begin{bmatrix} \partial Y_1/\partial\theta_a & \partial Y_1/\partial\theta_b & \partial Y_1/\partial\theta_c \\ \partial Y_2/\partial\theta_a & \partial Y_2/\partial\theta_b & \partial Y_2/\partial\theta_c \\ \partial Y_3/\partial\theta_a & \partial Y_3/\partial\theta_b & \partial Y_3/\partial\theta_c \\ \partial Y_4/\partial\theta_a & \partial Y_4/\partial\theta_b & \partial Y_4/\partial\theta_c \\ \partial Y_5/\partial\theta_a & \partial Y_5/\partial\theta_b & \partial Y_5/\partial\theta_c \\ \partial Y_6/\partial\theta_a & \partial Y_6/\partial\theta_b & \partial Y_6/\partial\theta_c \\ \partial Y_6/\partial\theta_a & \partial Y_6/\partial\theta_b & \partial Y_6/\partial\theta_c \\ \partial Y_6/\partial\theta_a & \partial Y_6/\partial\theta_b & \partial Y_6/\partial\theta_c \\ \partial Y_6/\partial\theta_a & \partial Y_6/\partial\theta_b & \partial Y_6/\partial\theta_c \end{bmatrix}_{9\times3} = \begin{bmatrix} A_1 & B_1 & C_1 \\ A_2 & B_2 & C_2 \\ A_3 & B_3 & C_3 \\ A_4 & B_4 & C_4 \\ A_5 & B_5 & C_5 \\ A_6 & B_6 & C_6 \\ A_7 & B_7 & C_7 \\ A_8 & B_8 & C_8 \\ A_9 & B_9 & C_9 \end{bmatrix}_{9\times3} \qquad (6.2.7)$$

It is shown above that if response Y_i is a linear function of input variables (θ_j, $j = 1, 2, 3$),

$$[X] = [D]$$

An optimal design belongs to a class of experimental designs that are optimal with respect to some statistical criterion such as design efficiency.

Consider a general case study in quest of optimality for extended design matrix $[X]_{Q\times G}$, of which subscript Q is the number of treatments (runs) and subscript G is the number of state/parameter estimates, e.g., $Q = 9$ and $G = 3$ for Equation (6.2.7). In search of parameter estimates, i.e., $\{\theta\}$, the resulting generalized variance in a least-squares analysis is embedded in the information matrix [Atkinson et al. 2007] defined as:

$$[I] = ([X]^T[X])^{-1} \qquad (6.2.8)$$

where

[X] is the extended design matrix
[I] is the information matrix

Therefore, the variance-covariance matrix upon due a polynomial regression by the least-squares method for parameter estimates is given by [Fisher 1966],

$$\text{Cov}(\{\theta\}) = \sigma_\varepsilon^2 \,[I] \qquad (6.2.9)$$

of which σ_ε^2 is the variance of random errors. For treatment q in the design matrix, the variance of a predicted response in the applied design space is:

$$\text{Var}(Y_q) = \sigma_\varepsilon^2 \{X_q\} [I] \{X_q\}^T \tag{6.2.10}$$

where

Y_q is the predicted value of Y on treatment q, where q = 1, 2, ..., Q.
$\{X_q\}$ is the row vector that contains treatment q conditions as vector elements; for example, $\{X_6\} = \{A_6 \, B_6 \, C_6\}$ for the case study described by Equation (6.2.7), which is a linear model
$\{X_q\}^T$ is the transpose of $\{X_q\}$

Efficient designs, which reduce the total number of experiments necessary to extract maximum understanding of the input and output spaces and the time required, can be achieved by maximizing the information value of each experiment.

Design efficiency is traditionally a quantified measure as a function of the generalized variance of $[I]^{-1}$ for determining parameter estimates. It is utilized to measure the goodness of a specified experimental design. Design efficiency may range from 0% to 100%. Since design efficiency is used as an expedient for the purpose of comparison in the selection of optimal design, it may be accepted even if the design efficiency is significantly lower than 100%.

On the D-optimality, where D stands for determinant, the criterion is measured by the geometric mean of the eigenvalues that is given by the trace of matrix $[I]^{1/G}$. When the D-optimal design is employed to deal mainly with a supersaturated design [Lin 1993], the associated design efficiency is

$$\text{D-efficiency} = \frac{1}{Q \, \|[I]\|^{1/G}} \tag{6.2.11}$$

$$\text{and} \quad \|[I]\| \equiv \text{Det}([I]) \tag{6.2.12}$$

where

Q is the number of treatments
G is the number of independent state/parameter estimates
Det([I]) is the determinant of matrix [I]

As shown above, the D-efficiency is defined as a function of the number of treatments and the number of independent-parameter estimates for prediction over the specified design space [Kuhfeld et al. 1994].

This means to minimize the generalized variance of information matrix, i.e., $\|[I]\|$, against the to-be-estimated DOE regression parameter estimates (coefficients and intercept) and statistical parameters (state estimates) just like solving maximum likelihood problems (detailed in Chapter 8, Volume II). Because there are Q equations that are available to solve for G unknowns, the relationship that $Q \geq G$ has to be guaranteed for generating a feasible optimal design.

Consider the electric resistivity of the modified acrylonitrile powder. The computer algorithm generally uses a stepping and exchanging process to select the "extended design matrix" that minimizes the corresponding $\|[I]\|$, while there is no guarantee that the next design the computer generates is actually the most "optimal."

Two distinct "extended design matrices" are given in Example 6.2.1 to demonstrate the difference in their determinants and design efficiencies [Goos and Leemans 2004]. It can be seen that option (2) is better than option (1). Nevertheless, the comparison has to keep going until the "top" design efficiency is found. Since some significant informational contents might be lost even in the final choice, design validation is strongly encouraged. In practice, an experimenter would like to add treatments to a completed experiment to learn more about the process and evaluate higher-ordered models.

EXAMPLE 6.2.1

Given that $Y = \theta_a A + \theta_b B + \theta_c C$, compare the design efficiencies of the two distinct design matrices selected according to the D-optimal design criterion given below [Goos and Leemans 2004]:

	Option 1				Option 2		
	0.2	0.4	0.4		0.2	0.2	0.6
	0.2	0.6	0.2		0.2	0.2	0.6
	0.3	0.35	0.35		0.2	0.2	0.6
	0.4	0.2	0.4		0.2	0.8	0.0
[D] =	[0.4	0.6	0.0]	[D] =	[0.2	0.8	0.0]
	0.45	0.45	0.1		0.2	0.8	0.0
	0.5	0.25	0.25		0.8	0.2	0.0
	0.6	0.4	0.4		0.8	0.2	0.0
	0.6	0.4	0.0		0.8	0.2	0.0

Solution:

In light of Equations (6.2.4), (6.2.11), and (6.2.12), one can reach the following table:

Equation	Option 1	Option 2
By Equation (6.2.4)	[X] = [D] for a linear model	[X] = [D] for a linear model
By Equation (6.2.12)	$\|[I]\| = Det(([X]T[X])^{-1})$ = 4.2409	$\|[I]\| = Det(([X]T[X])^{-1})$ = 0.2858
By Equation (6.2.11)	D-Efficiency = $1/\{Q \|[I]\|^{1/G}\}$ = $1/\{9 (4.2409)^{1/3}\}$ = 6.9%	D-Efficiency = $1/\{Q \|[I]\|^{1/G}\}$ = $1/\{9 (0.2858)^{1/3}\}$ = 16.9%

It can be seen that the design efficiency of Option-2 is higher.

For a linear model like the one described by Equation (6.2.2), the variance matrix is proportional to [I], and so the solution to such design optimality problems is straightforward—minimizing $\|[I]\|$, or maximizing $\|[X]^T[X]\|$. A D-optimal design is accordingly not affected by linear transformations of levels of the experimental factors. However, for a nonlinear model the variance-covariance matrix depends on the true values of the state/parameter estimates themselves. It means that the "optimality" of a given D-optimal design is very model-dependent [Das and Lin 2011].

Here is another example adopted from Figure 1 in [Kuhfeld et al. 1994] on how to select the desired set of treatments for the design matrix based on D-optimality in the regime of linear models with no constraint, as:

$$Y_q = Y_{ave} + \theta_a A_q + \theta_b B_q + \varepsilon_q$$

(6.2.13)

Assume that an experimenter wants to pick four treatments out of the full factorial design 3^2 shown in Example 6.2.2. There are 126 possible design sets, given that C(9, 4) = 9!/[4!(9 − 4)!] = 126. After nonorthogonal designs are excluded, nine different sets of four runs are able to form orthogonal designs, of which each set of four runs forms a rectangle or square. After the design efficiencies of all nine candidate sets of runs are compared using Equation (6.2.11), there is one and only one set of four treatments, i.e., treatments 1, 3, 7, and 9, which happens to spread out to cover the entire given experimental range for yielding D-optimal design efficiency.

EXAMPLE 6.2.2

Given that nine orthogonal designs with four treatments are available as extracted from full factorial design 3^2, as shown below [Kuhfeld et al. 1994]:

(a) Design Points

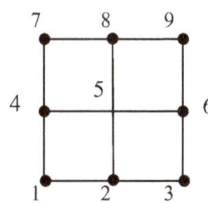

(b) Full Factorial Design:

Treatment	A	B
1	-1	-1
2	0	-1
3	1	-1
4	-1	0
5	0	0
6	1	0
7	-1	1
8	0	1
9	1	1

(c) D-Optimal Design:

Treatment	A	B
1	-1	-1
3	1	-1
7	-1	1
9	1	1

1. What would be the D-efficiency if the four corner points are selected to fit Equation (6.2.13)?

2. What would be the D-efficiency if the four center points are selected to fit Equation (6.2.13)?

3. Which one is the choice for D-optimal design?

Solution:

1. On fitting a linear model like Equation (6.2.13) for the D-optimal design with treatments 1, 3, 7, and 9, one has [X] = [D] according to Equation (6.2.4). Hence,

$$[X]^T [X] = [D]^T [D] = \begin{bmatrix} -1 & 1 & -1 & 1 \\ -1 & -1 & 1 & 1 \end{bmatrix}_{2\times4} \begin{bmatrix} -1 & -1 \\ 1 & -1 \\ -1 & 1 \\ 1 & 1 \end{bmatrix}_{4\times2} = \begin{bmatrix} 4 & 0 \\ 0 & 4 \end{bmatrix}_{2\times2}$$

Then $[I] = ([X]^T [X])^{-1} = (1/4) \begin{bmatrix} 1 & 0 \\ 0 & 1 \end{bmatrix} = \begin{bmatrix} 0.25 & 0 \\ 0 & 0.25 \end{bmatrix}$

Thus,   D-Efficiency = $1 / \{Q\,|[I]|^{1/G}\} = 1 / \{4\,|0.25|^{1/1}\}$ = 100%, in quest of intercept Y_{ave} only

and   D-Efficiency = $1 / \{Q\,|[I]|^{1/G}\} = 1 / \{4\,|0.25|^{1/4}\}$ = 35.4%, in quest of all four parameters, including intercept Y_{ave} and three coefficients for factor A, factor B and interaction AB.

It is always true that each diagonal element of [I] is equal to 1/Q, where Q is the number of total runs for an orthogonal design. Every off-diagonal term in matrix [I] for an orthogonal design is zero.

2. On fitting a linear model like Equation (6.2.13) for the D-optimal design with treatments 2, 4, 6, and 8, one has [X] = [D] according to Equation (6.2.4). Hence,

$$[X]^T [X] = [D]^T [D] = \left[\begin{matrix} 0 & -1 & 1 & 0 \\ -1 & 0 & 0 & 1 \end{matrix}\right]_{2\times4} \left[\begin{matrix} 0 & -1 \\ -1 & 0 \\ 1 & 0 \\ 0 & 1 \end{matrix}\right]_{4\times2} = \left[\begin{matrix} 2 & 0 \\ 0 & 2 \end{matrix}\right]_{2\times2}$$

Then $$[I] = ([X]^T [X])^{-1} = (1/2)\left[\begin{matrix} 1 & 0 \\ 0 & 1 \end{matrix}\right] = \left[\begin{matrix} 0.5 & 0 \\ 0 & 0.5 \end{matrix}\right]$$

Thus,   D-Efficiency = $1 / \{Q\,|[I]|^{1/G}\} = 1 / \{4\,|0.5|^{1/1}\} = 50\%$, in quest of intercept Y_{ave} only in Equation (6.2.13)

and   D-Efficiency = $1 / \{Q\,|[I]|^{1/G}\} = 1 / \{4\,|0.5|^{1/4}\} = 29.7\%$, in quest of all four parameters, including intercept Y_{ave} and three coefficients for factor A, factor B and interaction AB.

3. Similarly, the other seven orthogonal designs can be checked out one by one. In conclusion, D-optimality based on four corner points is the choice among all nine orthogonal designs in light of design efficiency.

6.2.2. D-Optimal Design for Quadratic Regression with Constraints

Here is one example adopted from [Goos and Leemans 2004] on how to select the desired set of treatments based on D-optimality in the regime of quadratic regression models with two factors,

$$Y_q = \theta_0 + \theta_a A_q + \theta_b B_q + \theta_{ab} A_q B_q + \theta_{aa} A_q^2 + \theta_{bb} B_q^2 + \varepsilon_q \tag{6.2.14}$$

where q = 1, 2, 3, ..., Q and Q is the total number of treatments. Each factor has three levels: $A_{-1} = 1$, $A_0 = 2$, $A_{+1} = 3$ and $B_{-1} = 1$, $B_0 = 2$, $B_{+1} = 3$. Nine observations are available for estimating the six unknowns of interest, i.e., $\theta_0, \theta_a, \theta_b, \theta_{ab}, \theta_{aa}, \theta_{bb}$. Nevertheless, there is one given constraint that:

$$A + B \le 5 \tag{6.2.15}$$

It means that high A and high B do not exist simultaneously. Although full factorial design 3^2 is optimal with a design efficiency of 100%, it is not applicable because of the constraint. How to derive the [I] matrix is demonstrated using an arbitrary design matrix given as follows:

1. List potential design matrix [D] with the given constraint

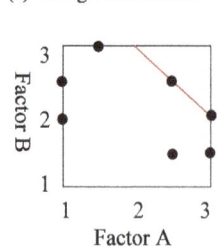

(a) Design Treatments

(b) Design Matrix

$$[D] = \left[\begin{matrix} 1 & 2 \\ 3 & 1.5 \\ 2.5 & 1.5 \\ 2.5 & 2.5 \\ 3 & 2 \\ 1 & 2.5 \\ 1.5 & 3 \\ 2.5 & 1.5 \\ 1 & 2 \end{matrix}\right]_{9\times2}$$

2. Calculate extended design matrix according to Equation (6.2.4), i.e., $[X] = [X_{ij}] = [\partial Y_i / \partial \theta_j]$

$$[X] = \begin{bmatrix} \partial Y/\partial\theta_0 & \partial Y/\partial\theta_a & \partial Y/\partial\theta_b & \partial Y/\partial\theta_{ab} & \partial Y/\partial\theta_{aa} & \partial Y/\partial\theta_{bb} \\ 1 & 1 & 2 & 2 & 1 & 4 \\ 1 & 3 & 1.5 & 4.5 & 9 & 2.25 \\ 1 & 2.5 & 1.5 & 3.75 & 6.25 & 2.25 \\ 1 & 2.5 & 2.5 & 6.25 & 6.25 & 6.25 \\ 1 & 3 & 2 & 6 & 9 & 4 \\ 1 & 1 & 2.5 & 2.5 & 1 & 6.25 \\ 1 & 1.5 & 3 & 4.5 & 2.25 & 9 \\ 1 & 2.5 & 1.5 & 3.75 & 6.25 & 2.25 \\ 1 & 1 & 2 & 2 & 1 & 4 \end{bmatrix}_{9x6} \tag{6.2.16}$$

3. Compute $[X]^T[X]$ and invert it to obtain $[I]_{sym}$

$$[I] = ([X]^T[X])^{-1} = \begin{bmatrix} 1263 & -435.6 & -825.8 & 99.08 & 57.83 & 144.9 \\ -435.6 & 159.5 & 279.1 & -34.38 & -22.32 & -48.8 \\ -825.8 & 279.1 & 545.0 & -64.34 & -36.58 & -96.11 \\ 99.08 & -34.38 & -64.34 & 8.597 & 4.180 & 10.90 \\ 57.83 & -22.32 & -36.58 & 4.180 & 3.473 & 6.581 \\ 144.9 & -48.8 & -96.11 & 10.90 & 6.581 & 17.18 \end{bmatrix}_{6x6} \tag{6.2.17}$$

4. Estimate Design Efficiency

$$\text{D-efficiency} = \frac{1}{Q\,|[I]|^{1/G}} = \frac{1}{9\,|[I]|^{1/3}} = 29.8\% \tag{6.2.18}$$

6.2.3. D-Optimal Design with Block Factors

An optimality may lead to an optimal split-plot design in mixture modeling within the presence of process variables. A mixture test is formed in this manner such that the descriptive variable and response rely only on the mixture's relative ratio in the mix but not on its composition. An example of applying the D-optimality to an electric circuit of a low-pass filter [Kovacs et al. 2013] is utilized to demonstrate how to do optimality with a block factor. It is an emulation test system used for checking application variances and parameter spread during post-silicon verification of automotive smart power products [Harrant et al. 2014]. There are three factors of concern and their corresponding design levels are listed in Table 6.2.2.

TABLE 6.2.2 Factors and their design levels for electric circuit of a low-pass filter of automotive smart power products. Data adapted from Table 3 in [Kovacs et al. 2013].

Factors	Factor type	Level (−)	Level (0)	Level (+)
A: Supply voltage (V)	Block	8	12	16
B: Filter resistance (kΩ)	Design (Noise)	9.5	10	10.5
C: Filter capacitance (μF)	Design (Noise)	9	10	11

© SAE International.

The supply voltage is considered as a block factor (blocking effect) since it is a physical operating parameter for the filter subjected to the application environment, while filter resistance and conductance are the two design parameters under control. The measured output is the maximum value of the current, which must be kept under a specified maximum value. Since factor A has three design levels while factors B and C have two design levels each, each response can be represented by the following metamodel of a nested design:

$$Y_{ijmn} = Y_0 + A_i + A_i^2 + B_{j(i)} + C_{m(i)} + B\,C_{jm(i)} + \varepsilon_{njm(i)} \tag{6.2.19}$$

where

A_i (V) is the supply voltage and subscript i = 1, 2, or 3
$B_{j(i)}$ (kΩ) is the filter resistance and subscript j = 1, 2, or 3
$C_{m(i)}$ (µF) is the filter capacitance and subscript m = 1, 2, or 3
n is the number of replicated runs for each treatment

It is assumed that the block factor does not interact with the treatments in the above equation. It is an over-saturated design, error term $\varepsilon_{njm(i)}$ may include fitting residuals and insignificant effects.

For such a case study on two design factors (i.e., B and C) and one hard-to-change block factor (i.e., A), one would like to find out the state/parameter estimates with less than or equal six runs.

Commercial codes from MATLAB that offer functions such as "cordexch, daugment," for creating or augmenting D-optimal experiments can be applied [Harrant et al. 2014]. The algorithm used in Minitab opts for selecting design treatments from the first given candidate set to obtain the initial design matrix [Kovacs et al. 2013]. Sequential selection means that all the points in the initial design are added in an order that provided the maximum increase in D-efficiency. Assume that the selected design matrix is listed in Table 6.2.3, in which each Y value is the averaged current from replicated measurements performed over a three-level factorial design. Note that replications are used for checking whether the noise is acceptable.

TABLE 6.2.3 The D-optimal design matrix for electric circuit of a low-pass filter of automotive smart power products [Kovacs et al. 2013].

Run	a	b	c	A	B	C	Y (Amp)	Y_p
1	−1	−1	−1	8	9.5	9	0.45177	0.45160
2	−1	−1	1	8	9.5	11	0.90704	0.90713
3	−1	1	−1	8	10.5	9	0.48715	0.48742
4	0	0	0	12	10	10	0.68591	0.68600
5	1	−1	1	16	9.5	11	0.82178	0.82211
6	1	1	1	16	10.5	11	0.90595	0.90587

© SAE International.

Since factors B and C are nested within block factor A, it is a resolution III design. Corresponding to Equation (6.2.19), the metamodel can be written as:

$$Y_p = \theta_0 + \theta_a + \theta_{aa}\,A^2 + \theta_b + \theta_c + \theta_{bc}\,B\,C \tag{6.2.20}$$

Regression is carried out using Minitab (Stat → Regression → Regression → Fit Regression Model). Only individual factors and the interaction between factors B and C are examined and the final ANOVA results are presented in Table 6.2.4 after weeding out the nonsignificant factors at a statistical significance level (p-value) of 10%. The predictive equation of the maximum current is:

$$Y_p = 0.04239 + 0.018245\ A - 0.001203\ A^2 - 0.179960\ B + 0.023975\ BC$$

$$(R = 100\%\ \text{and}\ R_{adj} = 100\%)$$

(6.2.21)

The effectiveness of the above equation can be easily justified using the model correlation and corrected model correlation given above. The predictive values of maximum currents for individual treatments (run conditions) based on Equation (6.2.21) are listed in column Y_p in Table 6.2.3. The adequacy of the model is guaranteed by the predictive correlation $R_{pred} = 100.00\%$, as calculated using Excel.

The denominator for evaluating the effective mean squares (EMS, i.e., Table 6.2.4: column MS_{adj}) of whole plot A and its quadratic term A^2 is different from those for B, C, and BC in a split design (Chapter 16, Volume IV). Nevertheless, the effectiveness of terms A and A^2 can be checked out using the concept of likelihood ratio (Section 8.5, Volume II). The p-value for constant term θ_0 set for different metamodels are listed as follows:

Metamodel	P-value of θ_0
$Y_p = \theta_0 + \theta_a\ A + \theta_{aa}\ A^2 + \theta_b\ B + \theta_{bc}\ B\ C$	8.1%
$Y_p = \theta_0 + \theta_a\ A + \theta_b\ B + \theta_{bc}\ B\ C$	23.1%
$Y_p = \theta_0 + \theta_{aa}\ A^2 + \theta_b\ B + \theta_{bc}\ B\ C$	20.8%
$Y_p = \theta_0 + \theta_b\ B + \theta_{bc}\ B\ C$	41.4%

Obviously, both terms A and A^2 are statistically significant because conventionally the acceptance of a confidence level is that p-value $\leq 10\%$.

TABLE 6.2.4 ANOVA for electric circuit of a low-pass filter of automotive smart power products.

Source of variance	SS_{adj}	DOF	MS_{adj}	$F_{u,v}$	P-value
Regression	0.206641	4	0.051660	290,999.65	0.1%
A	0.000121	1	—	—	—
A^2	0.000297	1	—	—	—
B	0.034199	1	0.034199	192,639.49	0.1%
BC	0.125467	1	0.125467	706,748.97	0.1%
Error	0.000000	1	0.000000		
Subtotal	0.206641	5			
Grand average	—	1			
Total	—	6			

According to Equation (6.2.21), the output electric current increases quadratically with an increasing voltage (factor A). With a given voltage, the output electric current decreases with an increasing filter resistance, while complicated by the interaction between the filter resistance (factor B) and filer capacitance (factor C). If the supply voltage (operating factor) is fixed at 12 V, the response surface plot of electric current as a function of filter resistance and filter capacitance (design factors) is plotted in Figure 6.2.1. The filter capacitance is the major contributor to the output electric current.

FIGURE 6.2.1 Electric current as a function of filter resistance (kΩ) and capacitance (μF).

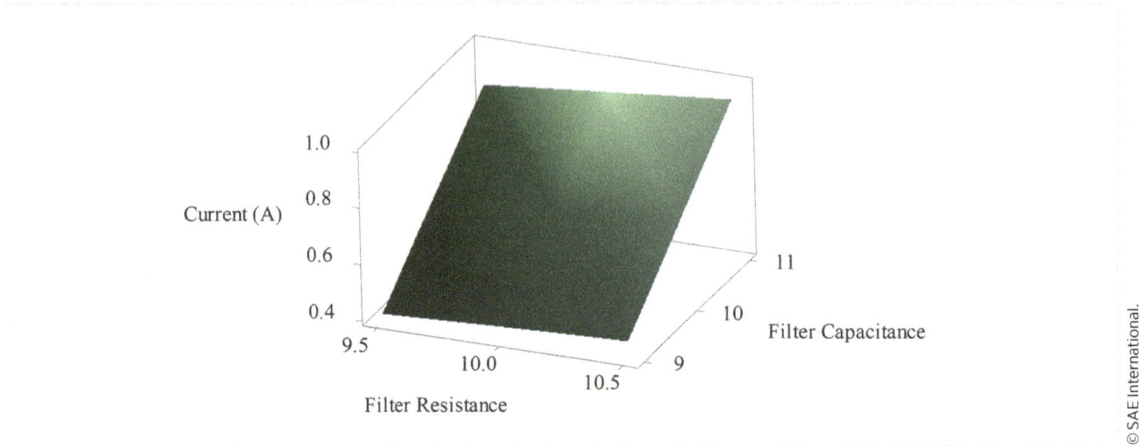

6.2.4. D-Optimal Design for Nonlinear Regression Models

Maximum likelihood estimates for product life prediction based on DOE exhibit the typical problem-solving process for nonlinear regression models, as addressed in Chapters 7–9 (Volume II) for traditional DOE, are also applicable to D-optimal design. The work procedure is similar to that presented in filter design for automotive smart power products in the last section (Section 6.2.3).

6.3. A-Optimality

Another popular optimal design option based on the information matrix [I] is A-optimality, where A refers to "average variance" for the regression coefficients/intercept of the fitted model. It seeks a set of experimental treatments that would minimize:

$$\sigma_{\theta^2 \text{ave}} = [\ \sum_{g=1}^{G} \text{var}(\theta_g)]\ /\ G \tag{6.3.1}$$

where G is the total number of independent regression parameters (coefficients and intercept) in the proposed metamodel.

A-optimality is based on the sum of the variances of the estimated parameters for the model, which is equivalent to the sum of the diagonal elements (i.e., the trace) of the corresponding information matrix [I]. The algorithm assumes that an overall measure of the average variance of the estimated parameters may lend it to the sum of the eigenvalues of $[I]^{-1}$. Mathematically,

$$\text{A-efficiency} = \frac{1}{Q\ \text{trace}([I])\ /\ G} \tag{6.3.2}$$

This is equivalent to using the arithmetic mean of the eigenvalues based on {trace of $[I]\}^{-1}$ to present the design efficiency, after Q (number of treatments) and G (number of parameter estimates) are determined. The A design that comes with a high A-efficiency is preferred to that with a low efficiency.

When incorporated with statistical distribution functions, A-optimality is to seek the design matrix that would minimize the average variance of the estimates of the regression coefficients/intercept and state variables (e.g., Weibull distribution function).

6.4. I-Optimal Design and FDS Plots

I-optimality, also called V-optimality, is based on how to reduce the average prediction variance in the design space. It seeks a set of experimental treatments that minimizes the average prediction variance over the design region:

$$\sigma_{Y\,ave}^2 = [\sum_{q=1}^{Q} Var(Y_q)] / Q$$

(6.4.1)

where
$Var(Y_q)$ is the generalized variance of predicted response Y_q, as defined by Equation (6.2.10)
Q is the total number of treatments in the design matrix [D]

As shown by Equations (6.2.10) and (6.2.14), the variance of each predicted response $Var(Y_q)$ is very model-dependent. Generally speaking, extended design matrix [X] is not equivalent to design matrix [D] for fitting a nonlinear polynomial, e.g., full quadratic model Equation (6.2.14). Furthermore, extended design matrix [X] also depends on which terms are effective in the polynomial, linear or nonlinear.

6.4.1. I-Efficiency

V-efficiency is defined as the average prediction variance with the former candidate set of design treatments divided by the average prediction variance with the latter candidate set of treatment as [Goos and Jones 2011]:

$$I\text{-efficiency} = \sigma_{Y\,aveL}^2 / \sigma_{Y\,aveF}^2$$

(6.4.2)

where
$\sigma_{Y\,aveL}^2$ is the average prediction variance of the latter candidate set of design treatments
$\sigma_{Y\,aveF}^2$ is the average prediction variance of the former candidate set of design treatments

When a I-efficiency, also called V-efficiency, is less than one, the former candidate set of design treatments is worse than the latter candidate set of design treatments. The comparison keeps going until the relatively "most robust experimental design" is identified.

6.4.2. FDS Plots

FDS plots may be used for examining the relative robustness between optimality criteria [Zahran et al. 2003], e.g., I-optimality versus D-optimality. In light of the prediction variance given by Equation (6.2.10), one may have the scaled prediction variance (SPV) calculated as [Ozol-Godfrey 2004]:

$$SPV(\{x_0\}) \equiv Var(Y_q) / \sigma_\varepsilon^2, \text{ where } q = 1, 2, 3. \ldots, \text{ and } Q$$

(6.4.3)

$$\text{Thus,}\quad SPV(\{x_o\}) \equiv \{x_0\}\,[I]\,\{x_0\}^T \tag{6.4.4}$$

where $\{x_0\}$ is the vector locating an arbitrary point of interest in the design space.

The plot of SPV versus individual coded variables in the design space is called the FDS plot, which reveals the profile of the relative size of prediction variance. An FDS plot is also the response surface plot of prediction variance if plotted against all coded variables simultaneously. How to construct an FDS plot is demonstrated in Example 6.4.1 [Ozol-Godfrey 2004].

EXAMPLE 6.4.1

Consider an experimental design that has "2^2 + center" points (5 treatments) and the fitting model is given as follows:

$$Y = \theta_0 + \theta_a\,a + \theta_b\,b + \theta_{ab}\,a\,b + \varepsilon \tag{6.4.5}$$

where a and b are coded variables for factors A and B, respectively. Note that $\theta_0 = Y_{ave}$. Please make an FDS plot to show the relative variance in the design space.

Solution:

1. Given that the experimental design has "2^2 + center" points, the design matrix is:

$$[D] = \begin{bmatrix} -1 & -1 \\ 1 & -1 \\ -1 & 1 \\ 1 & 1 \\ 0 & 0 \end{bmatrix}_{5\times2}$$

2. Extended design matrix [X] can be calculated using Equation (6.2.4), of which $\partial Y_a/\partial\theta_0 = 1$, $\partial Y_a/\partial\theta_a = a$, $\partial Y_a/\partial\theta_b = b$, $\partial Y_a/\partial\theta_{ab} = a\,b$. Plugging the five sets of (a, b) values in the design matrix into these partial differentials yields the extended design matrix:

$$[X] = \begin{bmatrix} \partial Y/\partial\theta_0 & \partial Y/\partial\theta_a & \partial Y/\partial\theta_b & \partial Y/\partial\theta_{ab} \\ 1 & -1 & -1 & 1 \\ 1 & -1 & 1 & -1 \\ 1 & -1 & 1 & -1 \\ 1 & 1 & 1 & 1 \\ 0 & 0 & 0 & 0 \end{bmatrix}_{5\times4}$$

3. According to Equation (6.2.8), the information matrix is calculated as:

$$[I] = ([X]^T\,[X])^{-1} = \begin{bmatrix} 1/5 & 0 & 0 & 0 \\ 0 & 1/4 & 0 & 0 \\ 0 & 0 & 1/4 & 0 \\ 0 & 0 & 0 & 1/4 \end{bmatrix}_{4\times4}$$

4. The SPV can be calculated using Equation (6.4.5),

$$SPV(\{x_o\}) = \{x_0\}\,[I]\,\{x_0\}^T = \{1, a, b, ab\}\,[I]_{4 \times 4}\,\{1, a, b, ab\}^T$$

$$= \{1 \ a \ b \ ab\}\begin{bmatrix} 1/5 & 0 & 0 & 0 \\ 0 & 1/4 & 0 & 0 \\ 0 & 0 & 1/4 & 0 \\ 0 & 0 & 0 & 1/4 \end{bmatrix}\{1 \ a \ b \ ab\}^T$$

$$= 0.2 + 0.25\,(a^2 + b^2 + a^2 b^2)$$

Note that $\{x_0\} = \{1\ a\ b\ ab\}$ is to be evaluated at any design point of interest in the design space. $SPV(\{x_0\})$ at some special locations are tabulated as follows:

a	b	$SPV(\{x_0\}) = 0.2 + 0.25\,(a^2 + b^2 + a^2 b^2)$
0	0	0.2
0.5	0.5	0.340625
0	1	0.45
1	0	0.45
−1	−1	0.95
1	−1	0.95
−1	1	0.95
−1	1	0.95

5. FDS plot: The response surface plot of prediction variance, i.e., $SPV(\{x_0\}) = 0.2 + 0.25\,(a^2 + b^2 + a^2 b^2)$, is given in Figure 6.4.1.

FIGURE 6.4.1 FDS plot for relative prediction variance profile (Example 6.4.1).

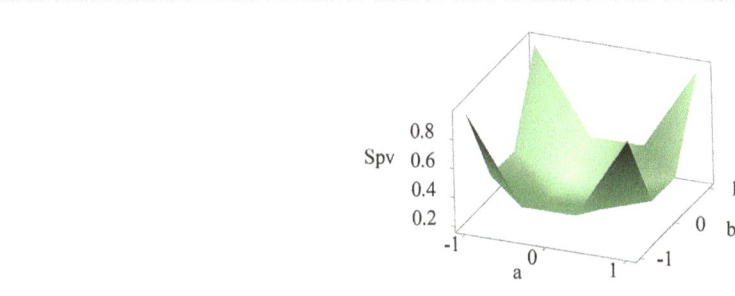

6.4.3. I-Optimality versus D-Optimality

The I-optimality is regarded as a natural choice because an experimental design can be constructed according to the setting under I-optimality, but instead adjusting the setting has to be adjusted to fit the experimental design under D-optimality. The difference between I-optimality and D-optimality in the context of experimental design with two factors will be shown here by two different applications: (1) linear model and (2) response surface model.

A comparison between D-optimality and I-optimality based on a linear model is demonstrated in Example 6.4.2. It is based on a completely randomized response surface experiment involving two categorial factors with three coded levels per factor for a linear model. Both design matrices were constructed using the JMP software [JMP 2021]. It is obvious that the advantage of D-optimality over I-optimality can be observed in light of either D-efficiency or SPV.

EXAMPLE 6.4.2

Two design matrices constructed with two distinct optimization algorithms are given below:

(a) I-optimality: Design matrix		(b) D-optimality: Design matrix	
0	0	0	0
1	0	1	0
−1	−1	−1	−1
0	0	1	−1
0	1	0	1
[D] = [−1	0]$_{12 \times 2}$	[D] = [−1	0]$_{12 \times 2}$
0	0	−1	−1
−1	1	−1	1
1	−1	1	−1
0	0	−1	1
0	−1	0	−1
1	1	1	1

Which one is better off in terms of D-efficiency and prediction variance for fitting the linear model that $Y = \theta_a \, a + \theta_b \, b$?

Solution:

1. Assume that the regression model is linear, then $[X] = [D]$.

2. For I-optimality,

$$[X]^T [X] = \begin{bmatrix} 6 & 0 \\ 0 & 6 \end{bmatrix} \text{ and } [I] = ([X]^T [X])^{-1} = 1/6 \begin{bmatrix} 1 & 0 \\ 0 & 1 \end{bmatrix}$$

D-Efficiency $= 1 / \{Q \, |[I]|^{1/G}\} = 1 / \{12 \, |1/6|^{1/4}\} = 13.04\%$, in quest of θ_a and θ_b.

3. For D-optimality,

$$[X]^T [X] = \begin{bmatrix} 9 & 0 \\ 0 & 9 \end{bmatrix} \text{ and } [I] = ([X]^T [X])^{-1} = 1/9 \begin{bmatrix} 1 & 0 \\ 0 & 1 \end{bmatrix}$$

D-Efficiency $= 1 / \{Q \, |[I]|^{1/G}\} = 1 / \{12 \, |1/9|^{1/4}\} = 14.4\%$, in quest of θ_a and θ_b.

4. According to the D-efficiency shown above in steps (2) and (3), D-optimality outperforms I-optimality in terms of model variance.

5. The SPV for the design matrix identified under I-optimality is:

$$SPV(\{x_o\}) = \{x_0\}\ [I]\ \{x_0\}^T = \{a, b\}\ [I]_{2x2}\ \{a, b\}^T$$

$$\text{i.e. } SPV(\{x_o\}) = \{a\ \ b\}\ \begin{bmatrix} 1/6 & 0 \\ 0 & 1/6 \end{bmatrix} \begin{Bmatrix} a \\ b \end{Bmatrix} = 1/6\ (a^2 + b^2)$$

6. The SPV for the design matrix identified under D-optimality is:

$$SPV(\{x_o\}) = \{a\ \ b\}\ \begin{bmatrix} 1/9 & 0 \\ 0 & 1/9 \end{bmatrix} \begin{Bmatrix} a \\ b \end{Bmatrix} = 1/9\ (a^2 + b^2)$$

7. According to the SPV shown above in steps (5) and (6), D-optimality also outperforms I-optimality in terms of scaled prediction variance. Note that $(a^2 + b^2) \geq 0$.

I-optimality is generally regarded more appropriate than the D-optimality criterion for generating RSM for prediction [JMP 2021]. Furthermore, I-optimal split-plot designs also provide substantial benefits in terms of improved prediction when compared with D-optimal split-plot designs [Jones and Goos 2012].

The advantage of I-optimality over D-optimality is demonstrated using the same design matrices given in Example 6.4.2, except that it is here to fit a quadratic model like Equation (6.2.14). Figures 5.41 and 5.43 given in [JMP 2021], focused on this case study, are combined here to generate a three-dimensional response surface plot of the SPV over both coded variables a and b as depicted in Figure 6.4.2, which is able to present the interactive effect between factors A and B. According to the two FDS plots (Figure 6.4.2), I-optimality outperforms D-optimality in the reduction of SPV for rendering a RSM.

FIGURE 6.4.2 FDS plots for comparing relative prediction variance profiles (Example 6.4.2).

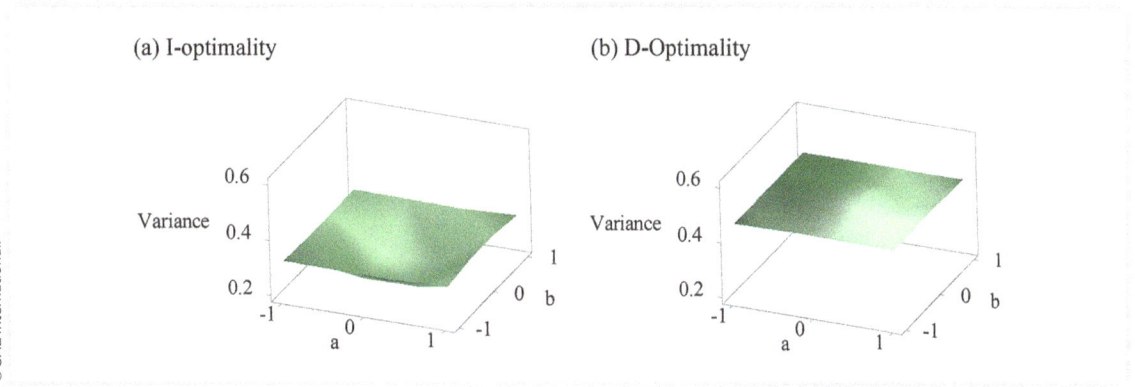

(a) I-optimality   (b) D-Optimality

6.5. **G-Optimal Design**

G-optimality refers to the reduction in the maximum prediction variance in the design space with a given candidate design matrix. In light of Equation (6.2.10), it means to minimize the maximum entry in the diagonal terms of matrix $[X]\,[I]\,[X]^T$ or $[X]([X]^T[X])^{-1}[X]^T$. Mathematically, the G-efficiency is defined as:

$$\text{G-efficiency} = \left[\frac{G/Q}{\sigma_Y^2{}_{max}}\right]^{1/2} \tag{6.5.1}$$

where

$\sigma_Y^2{}_{max}$ is the maximum prediction variance

Q is the total number of treatments

G is the total number of model parameters, e.g., coefficients and intercept for a polynomial

    I-optimality tends to place design points more uniformly in the design space than G-optimality. For the gain in a small fraction of the design space, the precision of G-optimality is sacrificed in the vast majority of the design space, when compared with I-optimality.

    There is an equivalence algorithm for measures between minimizing the maximum prediction variance for G-optimality and minimizing the generalized variance of information matrix for D-optimality [Kiefer and Wolfowitz 1959]. How to generate and assess exact G-optimal designs was addressed in [Rodriguez et al. 2010].

6.6. **Construction of Optimal Experimental Designs**

Each optimal design is an application-specific methodology. It may be employed to explore the cause-and-effect relationship instead of a traditional full or fractional factorial design. Neither D-efficiency nor I-efficiency is to be used as an absolute measure. Instead, they are supposedly to be used for relatively comparing one design with another in the same situation. It is a matter of how to choose the "best design" among various candidate design matrices?

6.6.1. Balance and Orthogonality of Optimal Experimental Designs

Design efficiency increases as the determinant of the information matrix "decreases," while a balanced orthogonal DOE yields the maximum design efficiency. Conversely, the more efficient an optimal design is, the more it moves in on balance and orthogonality [Kuhfeld et al. 1994]:

(a) Balanced and orthogonal design: When information matrix [I] of an optimal design is diagonal, it is a balanced and orthogonal design.

(b) Orthogonal designs: When the submatrix of information matrix [I], excluding the row and column for the intercept is diagonal, it is an orthogonal design. There may be off-diagonal nonzero for the intercept. Orthogonality usually implies that the coefficients will have a minimum variance.

(c) Balanced designs: A balanced design matrix yields that all off-diagonal elements in the intercept row and column are zero.

When orthogonal designs are not readily available particularly or the number of runs is limited, D-optimality is a great alternative. The D-efficiency of a standard fractional factorial design is 100%, but it is not feasible to achieve 100% D-efficiency with an unbalanced or nonorthogonal design matrix. If no proper orthogonal design can be detected, computer-generated nonorthogonal designs can be used instead.

As provided by computer algorithms, D-optimal, A-optimal, I-optimal, G-optimal design matrices and similar ones discussed above are usually nonorthogonal and estimated effects are correlated. Thus, no one of them is considerably a great exploratory DOE. On a second thought, they may well be more efficient than unbalanced orthogonal designs. Although nonorthogonal designs will never be more efficient than balanced orthogonal designs, good nonorthogonal designs may exist in the real world where no orthogonal design exists [Kuhfeld et al. 1994].

6.6.2. Construction Algorithms

Consider an example that an experimenter expects to select 18 treatments (Q = 18) from all the possible subsets of factorial design $2^2 3^3$ [Kuhfeld et al. 1994]. The selection involves

$$C(2x2x3x3x3, 18) = 108! / [18! (108 - 18)!] = 1.39 \times 10^{20}$$

(6.6.1)

possible candidate design sets. The central theme of computing algorithm is to search for "the best" set of 18 treatments out of 1.39×10^{20} candidate sets in light of their relative design efficiencies. Of course, this can be done in a straight forward procedure such as the random-coordinate selection rule [Nesterov 2012].

A simple way to construct an optimal design is to use the corresponding central composite design as the initial design.

An alternative rule is the cyclic coordinate-exchange algorithm proposed by [Meyer and Nachtsheim 1995] that can be utilized for constructing optimal designs in order to reduce the computing effort for identifying the truthful optimality. The algorithm uses a variant of the cyclic coordinate-descent algorithm within the k-exchange algorithm to achieve a substantive reduction in the required computing effort. However, an explicit randomization step of the assigned design points is recommended before conducting an experiment using the design produced by the cyclic coordinate-exchange algorithm [Stouwen and Goos 2019].

6.6.3. Generating an Optimal Design Using Commercial Codes

Most commercial codes in the area of DOE can be used for generating an optimal design. When using Minitab, one may

1. Choose Stat > DOE > Response Surface > Select Optimal Design
2. In "Number of points in optimal design," type the expected number of treatments (runs)
3. Click "Terms" to set up the metamodel of interest
4. Click OK in each dialog box

6.7. Statistical Power Evaluation of Experimental Designs

One approach to estimating the number of treatments and the number of replications required for a specific experimental design is to assess the statistical power. It is the probability of accepting the alternative hypothesis if it is true, and therefore can be applied for assessing the efficacy of each main or interactive effect. The higher the statistical power of a design factor, the lower the risk of making a Type II error. Calculation for statistical power is twofolds. One is to estimate the confidence an experimenter might have in the conclusions drawn from a completed experimental design and the other is to estimate the sample size required to detect an effect in an experimental design.

6.7.1. Operating Characteristic Curve

As defined for the acceptance of null hypothesis H_0, the related p-value is represented by the probability that H_0 will be accepted when the individual DOE parameter of interest, namely θ, is truly effective,

$$\alpha(\theta) \equiv \text{p-value} = P\{\text{Acceptance of } H_0 \mid H_0 \text{ is true}\} \tag{6.7.1}$$

The p-value is also termed the significance level of an experimental test. Statistically, the operating characteristic curve (OC curve) is the plot of $\alpha(\theta)$ versus t-statistic as:

$$t_N = N^{1/2} |\theta - \theta_0| / \sigma \tag{6.7.2}$$

where
  θ_o is the true value of variable θ
  N is the sample size
  σ is the standard deviation

As demonstrated in Chapters 2–5 for the traditional DOE, that p-value $\alpha(\theta) < 5\%$ (or 10%) is generally accepted to mitigate the risk of a false positive outcome, known as a Type I error. The OC may be used to determine how large the sample size (N) has to be in order to comply with a given specification against type I error. Note that $1 - \alpha$ is called confidence level or confidence.

6.7.2. Statistical Power

As a complementary function to the confidence level, the statistical power of an experiment is presented by the probability of correctly rejecting H_0 when H_0 is false, i.e., estimated θ is not the true θ_o,

$$1 - \beta(\theta) \equiv P\{\text{Rejection of } H_0 \mid H_0 \text{ is false}\}$$

$$= P\{\text{Acceptance of } H_1 \mid H_1 \text{ is true}\} \tag{6.7.3}$$

This is exhibited in Figure 6.7.1. The power function may be used for determining the sample size for how the corresponding experiment can comply with a given specification against Type-II error. How much is the statistical power needed for an experiment regarding the rejection of the null hypothesis, given that the null hypothesis is false?

As being the four primary quantities that determine the derivative properties of an experimental design, effect amount, confidence level, statistical power and sample size are closely related. For example, a larger sample size can make an effect easier to detect, and the statistical power can be increased in an experiment by increasing the significance level. In case the experimenter is confident that the null hypothesis will be rejected (e.g., $\alpha > 5\%$ or $\alpha > 10\%$) because the true mean is not the one used for the null hypothesis, the statistical power of the experiment must be acceptably large to warrant the Type II error is not significant statistically (e.g., $1 - \beta \geq 80\%$ conventionally).

FIGURE 6.7.1 Statistical power, significance level, and confidence for evaluating DOE effects.

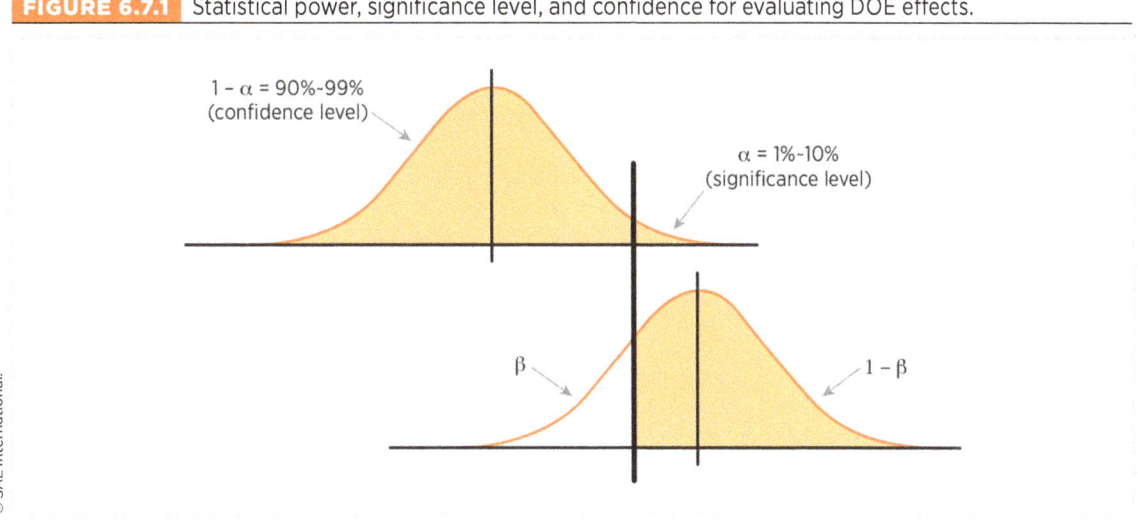

6.7.3. Statistical Power of Parametric Estimation in DOE

Higher variability undermines the capability to detect statistical significance. Statistical power and sample size estimates are as good as the estimate of variability. A truth table to determine if a calculated effect is statistically significant or not is given in Table 6.7.1. Both α (significance level) and $1 - \beta$ (power) have to be agreed upon to assure that the residues are really random errors. The practice based on significance level α (Figure 6.7.1) toward DOE has been addressed in Chapters 4–5. Simultaneously, statistical power is a sure warrant of success that the experimenter rejects H_0 when H_0 is false.

TABLE 6.7.1 Truth table for accepting or rejecting a calculated effect.

Accept	H_0 ($\alpha \leq 10\%$)	Reject H_0 ($\alpha > 10\%$)
If H_0 is true	OK	False alarm (Type I error)
If H_0 is false	Fail to detect that H_0 is false (Type-II error)	OK (Power $\geq 80\%$)

Prior to confirming the validity of an experimental design, one may undertake statistical power calculations to determine the sample size (i.e., number of treatments and the number of replications for DOE) required to generate a meaningful scientific effect with sufficient statistical power according to the structure of experimental design. It is unwise to construct a DOE that has a low statistical power for any main or interactive effect.

6.7.4. Noncentrality Parameter

The generalized F-ratio is derived from the ratio of two variance estimates, dividing the between-groups estimate by the within-group estimate as:

$$\text{F-ratio} \equiv (\sigma_\varepsilon^2 + \sigma_A^2) / \sigma_\varepsilon^2 = 1 + \sigma_A^2 / \sigma_\varepsilon^2 \tag{6.7.4}$$

where

σ_ε^2 is the estimate of population variance

σ_A^2 is the estimate of variability due to differences between group means

A is the factor of interest

When there is no difference between group means in the population, i.e., $\sigma_A^2 = 0$, and thus F-ratio = 1. This F-ratio follows the central F distribution, i.e., the F distribution based on the null hypothesis H_0, as presented in Chapter 4. If the null hypothesis is ideally true, i.e., the amount of an effect is 0, the test statistic has an F distribution (also called central F distribution) with that $\sigma_A^2/\sigma_\varepsilon^2 = 0$.

If there is a difference between group means in the population, $\sigma_A^2 \neq$ zero. Then, the F-ratio follows noncentral F distribution based on H_1, rejecting the null hypothesis H_0. An extra parameter, called noncentrality parameter

$$\delta_A = \sigma_A^2 / \sigma_\varepsilon^2 \tag{6.7.5}$$

is introduced to further characterize noncentral F distribution. As a generalized F distribution, the noncentral F distribution has three parameters, i.e., numerator degrees of freedom, denominator degrees of freedom, and noncentrality parameter.

The central F distribution presents how the F-statistic is distributed when H_0 is assumed to be true, but the noncentral F distribution instead shows how the F-statistic is distributed when H_0 is assumed to be false (i.e., H_1 is assumed to be true).

6.7.5. Noncentrality Parameters Based on ANOVA

The process for estimating noncentrality parameters for balanced orthogonal design 2^k is here demonstrated using the 2^4 design, of which each response obtained from different treatments can be expressed as:

$$Y_{ijmpn} = \mu + A_i + B_j + C_m + D_p + A_i B_j + A_i C_m + A_i D_p + B_j C_m + B_j D_p + C_m D_p$$
$$+ A_i B_j C_m + A_i B_j D_p + A_i C_m D_p + B_j C_m D_p + \varepsilon_{ijmpn} \tag{6.7.6}$$

of which A, B, C, and D are the four independent factors of interest. For a general factorial design such as L^{K-P} or composite design, the analysis of variance based on the sum of squares can be useful.

Noncentrality parameter is the ratio of "sum of squares between" to "mean square within" in the ANOVA table. For example, when R (number of replications) = 1, the noncentrality parameter of factor A can be written in terms of sum of squares as:

$$\delta_A^2 = SS_A / \sigma_\varepsilon^2 = J\,M\,P\,N\,[\ \sum_{i=1}^{I} (\bar{Y}_{A,i} - \bar{Y})^2]\ /\ \sigma_\varepsilon^2 \tag{6.7.7}$$

where

$$SS_A = J\,M\,P\,N\,[\ \sum_{i=1}^{I} (\bar{Y}_{A,i} - \bar{Y})^2] \tag{6.7.8}$$

and

$$\bar{Y}_{A,i} = [\ \sum_{j=1}^{J}\ \sum_{m=1}^{M}\ \sum_{n=1}^{N}\ \sum_{r=1}^{R}\ Y_{ijmpn}]\ /\ (J\ M\ P\ N) \tag{6.7.9}$$

where

$\quad$ $\bar{Y}$ is the grand average

$\quad$ SS_A is the sum of squares for factor A

$\quad$ $\bar{Y}_{A,i}$ is the average around factor A at the ith level, of which i ranges from -1 to I

$\quad$ I, J, M, and N are the design levels of factors A, B, C, and D, respectively

Similarly, other noncentrality parameters for other main effects and interactions such as $\delta_B{}^2$, $\delta_C{}^2$, $\delta_D{}^2$, $\delta_{AB}{}^2$, $\delta_{AC}{}^2$, $\delta_{AD}{}^2$, ..., and $\delta_{ABCD}{}^2$, can be derived accordingly. When the total number of treatments (runs) gets smaller or the effect is smaller, the experimental design has less statistical power [Draper and Lin 1990]. The statistical power of an effect increases with its amount of effect, increasing number of treatments (runs), and increasing number of replicates, as implicated by Equation (6.7.6).

Given that the null hypothesis is true, the amount of effect is zero, the noncentrality parameter is zero, and the test statistic has a F distribution (central F distribution). As for factor A, the goal of statistical power evaluation is to detect the difference between $\bar{Y}_A$ (estimated) and μ_A (true), written as Δ_A,

$$\Delta_A \equiv \bar{Y}_A - \mu_A \tag{6.7.10}$$

In light of Equation (6.7.7), Δ_A can be related to noncentrality parameter $\delta_A{}^2$ by the following equation:

$$\delta_A{}^2 = J\ M\ P\ N\ \Delta_A{}^2\ /\ (2\ \sigma_e{}^2) \tag{6.7.11}$$

Accordingly, all the main effects in Equation (6.7.6) have the same statistical power and all the 2-factor interactions have also the same statistical power, etc. In other words, parameters of the same order will have the same statistical power, as long as it is a balanced experimental design.

Again, the statistical power is the probability that the calculated amount of effect at a given significance level can be detected against Type-II error. The statistical power of an effect conventionally aims at 80% or above for justifying that Type-II error is statistically insignificant.

6.7.6. Noncentrality Parameters for a Two-Level Factorial Design

Equation (6.7.6) can be also expressed in terms of coded variables a, b, c, and d as:

$$Y_{ijmpn} = \gamma_0 + \gamma_a\ a_i + \gamma_b\ b_j + \gamma_c\ c_m + \gamma_d\ d_p$$

$$+\ \gamma_{ab}\ a_i\ b_j + \gamma_{ac}\ a_i\ c_m + \gamma_{ad}\ a_j\ d_p + \gamma_{bc}\ b_j\ c_m + \gamma_{bd}\ b_j\ d_p + \gamma_{cd}\ c_m\ d_p$$

$$+\ \gamma_{abc}\ a_i\ b_j\ c_m + \gamma_{abd}\ a_i\ b_j\ d_p + \gamma_{bcd}\ b_j\ c_m\ d_p \tag{6.7.12}$$

$$+\ \gamma_{abcd}\ a_i\ b_j\ c_m\ d_p + \varepsilon_{ijmpn}$$

where

γ_0 is the grand average referring to mean value μ

a_i, b_j, c_m, d_p are the coded variables for factors A, B, C, and D at levels i, j, m, and p, respectively

i, j, m, p are the indices for levels, ranging from 1 (for −1 level) to 2 (for +1 level)

n is the replication number

γ_0, γ_a, γ_b, γ_c, γ_d, γ_{ab}, γ_{ac}, γ_{ad}, γ_{bc}, etc., are the intercept and coefficients of the regression model

ε_{ijmpn} is the residual from the treatment at (a_i, b_j, c_m, d_p) level and replicate n

When a design factor is represented by its corresponding coded variables, ranging from −1 and +1, the design contrast of each main effect refers to the difference of its individual averages at the +1 and −1 sides (Chapter 2). By the same token, the interactive effects are derived. Each of the main and interactive effects is then obtained by dividing its corresponding design contrast by 2, i.e.,

$$\gamma_a = (\bar{Y}_A^{+1} - \bar{Y}_A^{-1}) / 2 \tag{6.7.13a}$$

$$\gamma_b = (\bar{Y}_B^{+1} - \bar{Y}_B^{-1}) / 2 \tag{6.7.13b}$$

$$\gamma_c = (\bar{Y}_C^{+1} - \bar{Y}_C^{-1}) / 2 \tag{6.7.13c}$$

$$\gamma_d = (\bar{Y}_D^{+1} - \bar{Y}_D^{-1}) / 2 \tag{6.7.13d}$$

$$\gamma_{ab} = (\bar{Y}_{AB}^{+1} - \bar{Y}_{AB}^{-1}) / 2 \tag{6.7.13ab}$$

and $$\gamma_{abcd} = (\bar{Y}_{ABCD}^{+1} - \bar{Y}_{ABCD}^{-1}) / 2 \tag{6.7.13abcd}$$

Consider factor A (corresponding to coded variable a). In light of Equation (6.7.5), its noncentrality parameter can be calculated as:

$$\delta_A = \gamma_a^2 / \sigma_{\gamma a}^2 / R \tag{6.7.14}$$

where

δ_A is the estimated noncentrality parameter of factor A

$\sigma_{\gamma a}^2$ is the variance of γ_a

R is the number of replicates

If the noncentrality parameter $\delta_A \neq 0$, it is recognized that there is a difference between the true value of effect γ_a and its value hypothesized, as measured by its corresponding population variance. By the same token, δ_B, δ_C, δ_D, δ_{AB}, δ_{AC}, δ_{AD}, δ_{ABC}, δ_{ABD}, δ_{ACD}, δ_{BCD}, and δ_{ABCD} can be calculated.

6.7.7. Statistical Power Evaluation for D-Optimal Design

Experimenters should check the alias relationships, orthogonality, and power for viable effects inherent with an experiment before conducting the tests. An inferior design cannot lead to a correct solution, but consuming time and money. If the statistical power is low for an effect of interest, the sample size of the experiment can be increased to reach higher statistical power.

The procedure for assessing the statistical power of main effects of a linear model is illustrated using a D-optimal design matrix with 13 runs and eight distinct treatments, as displayed in Example 6.7.1.

EXAMPLE 6.7.1

D-optimal design having 13 runs with eight distinct treatments [Nelson 2013].

Run	A	B	C	D
1	1	1	1	1
2	1	1	-1	-1
3	1	-1	1	-1
4	-1	1	1	-1
5	-1	1	-1	1
6	-1	-1	1	1
7	-1	-1	-1	-1
8	0	0	0	0
9	0	0	0	0
10	0	0	0	0
11	0	0	0	0
12	0	0	0	0
13	0	0	0	0

Assume the metamodel of interest is $Y = \gamma_0 + \gamma_a\, a + \gamma_b\, b + \gamma_c\, c + \gamma_d\, d$. Please estimate the statistical power for the intercept and main effects.

Solution:

1. Extended design matrix [X] and information matrix [I] can be derived from the design matrix [D] given, respectively, as:

$$[X]=[D]=\begin{bmatrix} 1 & 1 & 1 & 1 \\ 1 & 1 & -1 & -1 \\ 1 & -1 & 1 & -1 \\ -1 & 1 & 1 & -1 \\ -1 & 1 & -1 & 1 \\ -1 & -1 & 1 & 1 \\ -1 & -1 & -1 & -1 \\ 0 & 0 & 0 & 0 \\ 0 & 0 & 0 & 0 \\ 0 & 0 & 0 & 0 \\ 0 & 0 & 0 & 0 \\ 0 & 0 & 0 & 0 \\ 0 & 0 & 0 & 0 \end{bmatrix}_{13\times5}$$

and $[I]_{5\times5} = ([X]^T [X])^{-1}$

$$= \begin{bmatrix}
\gamma_0 & \gamma_a & \gamma_b & \gamma_c & \gamma_d \\
0.08333 & 0.02083 & -0.02083 & -0.02083 & 0.02083 \\
0.02083 & 0.16146 & -0.03646 & -0.03646 & 0.03646 \\
-0.02083 & -0.03646 & 0.16146 & 0.03646 & -0.03646 \\
-0.02083 & -0.03646 & 0.03646 & 0.16146 & -0.03646 \\
0.02083 & 0.03646 & -0.03646 & -0.03646 & 0.16146
\end{bmatrix}$$

2. The diagonal elements of information matrix [I] are sample variances for the intercept (γ_0) and coefficients γ_a, γ_b, γ_c, and γ_d, respectively, in light of Equation (6.2.9). Thus, the noncentrality parameter for the main effect of factor A can be calculated using Equation (6.7.13) as:

$$\delta_A = \frac{\gamma_a^2}{\sigma_{\gamma a}^2} = \frac{1}{0.16146} = 6.192$$

where

$$\gamma_a: (\bar{Y}_A^{+1} - \bar{Y}_A^{-1})/2 = \{[1 + 1 + 1 + 1]/4 - [(-1) + (-1) + (-1) + (-1)]/4\}/2 = 1$$

$\sigma_{\gamma a}^2$ is the second diagonal term in the information matrix [I]

Note that $\delta_A = \delta_B = \delta_C = \delta_D$. The first center point is mainly used to detect curvature effects, while five additional runs at (0, 0, 0, 0) are set up for evaluating lack of fit and used for increasing the statistical power somewhat.

3. Statistical significance by Type I error: Assume that the statistical significance level $\alpha = 5\%$ (Type I error) is taken for granted, then the F-value = 5.32. Then,

$$f_{critical} = F^{-1}{}_{1,8}(5\%) = 5.32$$

There are eight different treatments that make eight degrees of freedom for deriving $\sigma_{\gamma a}$, while the main effect γ_a instantly takes one degree of freedom.

4. Statistical significance by Type II error: The statistical power of main effect γ_a, as a measure of Type II error, is carried out via the corresponding conditional probability that:

$$\text{Power} = 1 - \beta_a = 1 - F_{1,8}(f_{critical} \mid \delta_A = 6.192) = 1 - F_{1,8}(5.32 \mid \delta_A = 6.192) = 58.3\%$$

Since the statistical power is below 80%, the number of treatments and/or the number of replications applied in the experimental design has to increase for improving the statistical power, i.e., reducing the Type-II error. Note that $1 - \beta_a = 1 - \beta_b = 1 - \beta_c = 1 - \beta_d$, since it is a balanced design.

5. For the purpose of comparison, assume that Type I error $\alpha = 10\%$ is acceptable. Then, one has the critical value from Table 4.1.1 as:

$$f_{critical} = F^{-1}{}_{1,8}(10\%) = 3.458$$

Thus, $$\text{Power} = 1 - \beta_a = 1 - \beta_b = 1 - \beta_c = 1 - \beta_d = 1 - F_{1,8}(3.458 \mid \delta_A = 6.192) = 72.4\%$$

Again, the statistical power is still below the desired value (i.e., 80%), though better. Still, the number of treatments and/or the number of replications applied in the experimental design has to increase for reducing the Type-II error.

6. Calculations done above are for balanced designs only. Due to unbalanced experimental designs, the statistical power of each effect will be different from each other, i.e.,

$$1 - \beta_a \neq 1 - \beta_b \neq 1 - \beta_c \neq 1 - \beta_d.$$

In case the design matrix based on the RSM do not exhibit orthogonality, the effects (including main effects and interactions) are correlated and the statistical power is degraded.

6.8. Definitive Screening Designs

An innovative class of three-level designs in the presence of quadratic effects, called definitive screening designs (DSD), was introduced by [Jones and Nachtsheim 2011, 2013]. The efficient definitive screening design strategy shows in the sense definitively that the estimates of all main effects can be unbiasedly de-aliased with active second-order effects (interactions and quadratic terms). The approach to forming DSD is here narrated an experimental design with five factors (k = 5) of interest. The full second-order polynomial model takes the following form:

$$Y = \gamma_0 + \sum_{i=1}^{5} \gamma_i x_i + \sum_{i=1}^{5} \sum_{j=1}^{5} \gamma_{ij} x_i x_j + \varepsilon \qquad (6.8.1)$$

As pointed out by [Polhemus 2018], there are 21 unknown parameters to be estimated in the above equation, i.e., one intercept (the constant), five main effects, ten 2-factor interactions, and five quadratic terms. It is intended that the resolution of a definitive screening design is IV. Main effects are not aliased with each other, so they can be estimated directly, but 2-factor interactions may be partly confounded with each other and quadratic terms.

6.8.1. Definitive Screening Design Matrices

The algorithm for generating definitive screening design matrices is presented in Table 6.8.1 and it only accommodates an experimental design with five factors or more. Thus, 13 (= 2 × 5 + 3 or 2 × 6 + 1) treatments is the minimum requirement. Nevertheless, a definitive screening design with 13 treatments can assess quadratic (curvilinear) effects when the model contains only intercepts, five main effects, and five quadratic effects. Only a subset of parameters in Equation (6.8.1) may be fit. It is essential to determine which 13 unknown parameters should be taken in prior to knowing their aliases in a definitive screening design matrix. For an experimental design with k factors:

(a) 2k+1 treatments are required for forming the design matrix, when k is an even number
(b) 2k+3 treatments are required for forming the design matrix, when k is an odd number

Dropping a factor still let the design retain all its properties.

As each definitive screening design matrix constitutes a single center point, it is encouraged to have a replicated run at the center point that would allow the experimenter to have access to the random error.

TABLE 6.8.1 Definitive screening design k factors [Jones and Nachtsheim 2011, 2013].

Fold-over pair	Treatment	x_1	x_2	x_3	...	x_k
1	1	0	±1	±1	...	±1
	2	0	−(±1)	−(±1)	...	−(±1)
2	3	±1	0	±1	...	±1
	4	−(±1)	0	−(±1)	...	−(±1)
3	5	±1	±1	0	...	±1
	6	−(±1)	−(±1)	0	...	−(±1)
...					...	...
					...	...
k	2k − 1	±1	±1	±1	...	0
	2k	−(±1)	−(±1)	−(±1)	...	0
Centerpoint	2k + 1	0	0	0	...	0

Reprinted from "Definitive Screening Designs with Added Two-Level Categorical Factors," Bradley Jones, Christopher J. Nachtsheim, Journal of Quality Technology © 2013 American Society for Quality, reprinted by permission of Taylor & Francis Ltd, http://www.tandfonline.com on behalf of American Society for Quality.

Given that there are six independent factors (k = 6), namely factors A, B, C, D, E and F, the generic design matrix can be organized as given in Table 6.8.1, in which a, b, c, d, e and f (i.e., x_1, x_2, x_3, x_4, x_5, and x_6) are coded variables for the six variables. One of the possible DSD for six factors is given as follows [Xiao et al. 2012]:

Run	a	b	c	d	e	f
1	0	1	1	1	1	1
2	1	0	−1	1	1	−1
3	1	−1	0	−1	1	1
4	1	1	−1	0	−1	1
5	1	−1	1	−1	0	−1
6	1	1	1	1	−1	0
7	−1	−1	−1	−1	1	0
8	−1	1	−1	1	0	1
9	−1	−1	1	0	1	−1
10	−1	1	0	1	−1	−1
11	−1	0	1	−1	−1	1
12	0	−1	−1	−1	−1	−1
13	0	0	0	0	0	0

Furthermore, one may generate a DSD for six factors; and drop the last column, should it be a case study on five factors only.

Another frequently DSD with 17 runs for eight factors as follows [Xiao et al. 2012]:

Run	a	b	c	d	e	f	g	h
1	0	1	1	1	1	1	1	1
2	1	0	−1	−1	1	−1	1	1
3	1	1	0	−1	−1	1	−1	1
4	1	1	1	0	−1	−1	1	−1
5	1	−1	1	1	0	−1	−1	1
6	1	1	−1	1	1	0	−1	−1
7	1	−1	1	−1	1	1	0	−1
8	1	−1	−1	1	−1	1	1	0
9	−1	1	1	−1	1	−1	−1	0
10	−1	1	−1	1	−1	−1	0	1
11	−1	−1	1	−1	−1	0	1	1
12	−1	1	−1	−1	0	1	1	−1
13	−1	−1	−1	0	1	1	−1	1
14	−1	−1	0	1	1	−1	1	−1
15	−1	0	1	1	−1	1	−1	−1
16	0	−1	−1	−1	−1	−1	−1	−1
17	0	0	0	0	0	0	0	0

Again, one may generate a DSD for eight factors; and drop the last column (factor h), should it be a case study on seven factors only.

6.8.2. Conference Matrix

As shown in Table 6.8.1, the design matrix of a DSD can be obtained using its corresponding conference matrix, also called C-matrix, i.e., denoted by square matrix [C] with 0 for elements on the diagonal and −1 and +1 (or sometime 0) for elements off the diagonal, such that $[C]^T[C]$ is a multiple of the identity matrix.

$$[C]_{kxk}{}^T [C]_{kxk} = [I]_{kxk} \tag{6.8.1a}$$

For example, the C-matrices for six and eight factors are the collection of the first six treatments in and the first eight treatments in Table 6.8.1, respectively [Xiao et al. 2012]:

$$[C]_{6x6} = \begin{bmatrix} 0 & 1 & 1 & 1 & 1 & 1 \\ 1 & 0 & -1 & 1 & 1 & -1 \\ 1 & -1 & 0 & -1 & 1 & 1 \\ 1 & 1 & -1 & 0 & -1 & 1 \\ 1 & -1 & 1 & -1 & 0 & -1 \\ 1 & 1 & 1 & 1 & -1 & 0 \end{bmatrix} \tag{6.8.2}$$

$$\text{and} \quad [C]_{8x8} = \begin{bmatrix} 0 & 1 & 1 & 1 & 1 & 1 & 1 & 1 \\ 1 & 0 & -1 & -1 & 1 & -1 & 1 & 1 \\ 1 & 1 & 0 & -1 & -1 & 1 & -1 & 1 \\ 1 & 1 & 1 & 0 & -1 & -1 & 1 & -1 \\ 1 & -1 & 1 & 1 & 0 & -1 & -1 & 1 \\ 1 & 1 & -1 & 1 & 1 & 0 & -1 & -1 \\ 1 & -1 & 1 & -1 & 1 & 1 & 0 & -1 \\ 1 & -1 & -1 & 1 & -1 & 1 & 1 & 0 \end{bmatrix} \tag{6.8.3}$$

Conference matrix $[C]_{kxk}$ is said to be normalized if all entries in its first row and first column are 1, except the (1, 1) entry, which is 0. The core of every normalized conference matrix of order k = 4t + 2(t = 1, 2, 3, …) is symmetric, such as $[C]_{6x6}$. On the other hand, the core of every normalized conference matrix of order k = 4t(t = 2, 3, …) is skew-symmetric, such as $[C]_{8x8}$. A systematic approach to constructing a conference matrix, namely $[C]_{kxk}$, with elements (0, ±1) having zero diagonal can be found in [Phoa and Lin 2015, Nguyen and Stylianous 2013].

After a conference matrix is obtained, its corresponding distinctive way of generating a DSD can be easily obtained by fold-over and then adding a center point test. As observed in Table 6.8.1, there are several different conference matrices available for generating DSD. One conference matrix may deliver a design matrix with a higher efficiency than the other, although both are DSD. The one with the highest design efficiency should be taken for granted. As pointed out by [Xiao et al. 2012], one $[C]_{12 \times 12}$ that leads to a DSD matrix that has the D-efficiency of 92.3% is given as follows:

$$[C]_{12x12} =$$

$$\begin{bmatrix} 0 & 1 & 1 & 1 & 1 & 1 & 1 & 1 & 1 & 1 & 1 & 1 \\ 1 & 0 & -1 & -1 & -1 & -1 & 1 & -1 & 1 & 1 & 1 & 1 \\ 1 & 1 & 0 & 1 & 1 & -1 & 1 & -1 & -1 & 1 & -1 & -1 \\ 1 & 1 & -1 & 0 & 1 & 1 & -1 & -1 & -1 & -1 & 1 & 1 \\ 1 & 1 & -1 & -1 & 0 & 1 & -1 & 1 & 1 & 1 & -1 & -1 \\ 1 & 1 & 1 & -1 & -1 & 0 & 1 & 1 & -1 & -1 & 1 & -1 \\ 1 & -1 & -1 & 1 & 1 & -1 & 0 & 1 & 1 & -1 & 1 & -1 \\ 1 & 1 & 1 & 1 & -1 & -1 & -1 & 0 & 1 & -1 & -1 & 1 \\ 1 & -1 & 1 & 1 & -1 & 1 & -1 & -1 & 0 & 1 & 1 & -1 \\ 1 & -1 & -1 & 1 & -1 & 1 & 1 & 1 & -1 & 0 & -1 & 1 \\ 1 & -1 & 1 & -1 & 1 & -1 & -1 & 1 & -1 & 1 & 0 & 1 \\ 1 & -1 & 1 & -1 & 1 & 1 & 1 & -1 & 1 & -1 & -1 & 0 \end{bmatrix} \tag{6.8.4}$$

6.8.3. **Room for Improvement**

Significant linear effects are more readily identified with traditional factorial screening designs such as fractional factorial designs with resolution IV, because they have somewhat higher statistical power for linear models than their corresponding similar-sized DSD. Compared with classical 2-level screening designs of the same size, there are some minor drawbacks with DSD [Donnelly 2016]:

   (a) Widened confidence intervals

   (b) Increased variance (standard error)

   (c) Reduced statistical power for main effects

   (d) Low statistical power for quadratic terms

One major fallout for applying the "DSD" algorithm is an experimenter's search for a follow-up factorial design to seek out design optimization. The experimenter has to resort retroactively back to traditional factorial designs [Weese et al. 2018].

6.8.4. **DSD with Added Two-Level Categorical Factors**

One application limitation of DSD is that all factors must be quantitative. Alternatively, a 2-level-factors augmented DSD with added 2-level categorical factors may lead to designs that have orthogonal linear main effects [Jones and Nachtsheim 2013]. It leads to an experimental design with resolution III, because some partial aliasing between main effects and interactions involving the categorical factors is present.

6.9. **Pareto Optimality**

Many engineering design problems in the real world involve the satisfaction of multiple conflicting objectives. For example, automotive manufacturers often need to meet demands concerning cost, quality, and delivery (CQD). A problem with vehicular design is typically characterized by the presence of a design space full of multiple variables and a criterion space filled with multiple objectives [Laumanns and Laumanns 2005]. For example, the development of a passenger vehicle depends on the delivery time, quality, and cost, along with other constraints.

Consider a multiobjective function that can be written as:

$$f(\{x\}) = \{f_1(\{x\}), f_2(\{x\}), ..., f_R(\{x\})\} \tag{6.9.1}$$

Then, a multiobjective decision problem can be represented in its general form as:
Minimize $f(\{x\})$ subject to

$$g_i(\{x\}) \le 0 \qquad \text{(Inequality constraint)} \tag{6.9.2}$$

$$h_j(\{x\}) = 0 \qquad \text{(Equality constraint)} \tag{6.9.3}$$

$$\text{and} \quad \{x\} \in \{x_1, x_2, x_3, ..., x_Q\} \tag{6.9.4}$$

where

  {x} is a vector of Q factors (design variables) in the design space
  f({x}) is the vector containing R objective functions, i.e., $f_1(\{x\})$, $f_2(\{x\})$, ..., and $f_R(\{x\})$
  $g_i(\{x\})$ is the inequality constraint i, where i = 1, 2, ..., I
  $h_j(\{x\})$ is the equality constraint j, where j = 1, 2, ..., J
  R is the total number of objective functions of interest
  I is the total number of inequality constraints
  J is the total number of equality constraints

Varying vector {x} in the design space may lead to a feasible solution in the criterion space. If there exist several solutions that would minimize all components of f(x) simultaneously, the problem-solving process may resort to the so-called Pareto optimality. The concept of Pareto optimality lends itself to many different areas including research, design, manufacturing, engineering, marketing, and economics. Recently, Pareto optimality has been a premise for multiobjective optimization in a feasible design space for decision-making in the automotive industry [Balta et al. 2014, Laumanns and Laumanns 2005].

6.9.1. Pareto Frontier

For the purpose of multiobjective optimization with Pareto optimality, the following three prerequisites must be met:

1. Definition of the design space.
2. Clear objective functions of interest.
3. Models that constitute mappings from the design space to the criterion space.

For example, the model for multiobjective design optimization of battery packs for electric vehicles constitutes a set of equations that are formulated to relate deliverable power, torque, and life cycles in the criterion space to input characteristics such as voltage, current, temperature, internal resistance, and cycle number in the design space.

Assume that there are two objective functions for the production of mounting bolts of batteries. One is the cost in dollars per lot and the other is the quality in terms of the number of defective parts per lot. Each is a function of the two independent factors,

$$Y_1 = Y_1(A, B) \qquad \text{(Dollars per lot)} \qquad (6.9.5)$$

$$\text{and} \quad Y_2 = Y_2(A, B) \qquad \text{(No. of defects per lot)} \qquad (6.9.6)$$

where

  Y_1 is the cost index in terms of dollars per lot in production
  Y_2 is the quality index in terms of number of defective parts per lot in production
  A is the factor A, e.g., manpower
  B is the factor B, e.g., hours of inspection

Here {x} = {A B} forms the design space and {y} = {Y_1, Y_2} forms the criterion space (decision space). Assume that the goal is to minimize both cost and number of defects. Three pieces of data resulting from an experimental design are available as follows:

Treatment	A	B	Y_1 ($ per lot)	Y_2 (defects per lot)
1	1	10	$7,000	21
2	5	4	$12,000	13
3	10	3	$21,500	16

The design space and criterion space of the cost-versus-quality problem are plotted in Figure 6.9.1. As exhibited in Figure 6.9.1(a), the feasible design space is limited to the upper right side of line 12 and the upper right side of line 23 in the first quadrant. The corresponding criterion space is the region inside the triangle as depicted in Figure 6.9.1(b).

FIGURE 6.9.1   Pareto frontier of two independent factors.

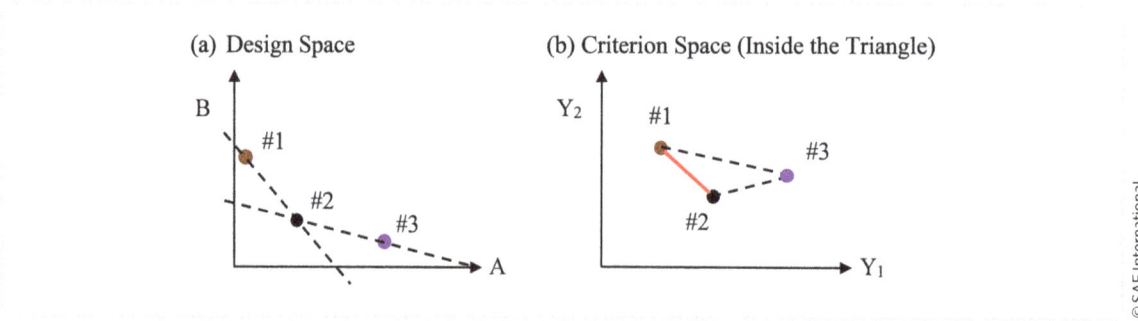

The Pareto frontier is the line segment between the point with lowest Y_1 (response with treatment 1) and the point with lowest Y_2 (response with treatment 2), as shown in Figure 6.9.1(b). Any point on Pareto frontier in the criterion space is "Pareto optimal." By moving the point along the line, which is a straight line here but mostly a curve for general applications, one could either cut down the cost at the sacrifice of product quality or reduce the number of defects at a higher cost, while both cannot be improved at the same time.

6.9.2. Pareto Efficiency

Pareto efficiency is a conceptual function in a situation where solutions can be provided to benefit some options while making other options worse, as used to measure Pareto optimality. Pareto efficiency will occur on a Pareto frontier. When an economy is operating on a simple production possibility frontier, e.g., at point 1, 2, or 3 shown in Figure 6.9.1(b), it is impossible to reduce the cost without increasing the defects. Any operating point that falls in the area of the triangle of Figure 6.9.1(b) but not on the Pareto frontier, is Pareto inefficient and thus not an optimal solution.

Profiling Pareto frontiers is important in multiobjective optimization. By yielding all the potentially optimal solutions in the feasible design space, a designer can focus on tradeoffs within this constrained set of parameters, rather than considering the full ranges of parameters. The focus of the embedded optimization algorithm is to combine model-based methods with statistical DOE, especially when dealing with RSM with conflicting objectives [Costa and Lourenco 2015].

The potential solution in a general criterion space consists of nondominated and dominated particles [Büyük et al. 2020], as illustrated in Figure 6.9.2. A particle (point) on the Pareto frontier is called nondominated particle (solids), since no other feasible design dominates. Other particles (open circles in the figure) are dominated particles that are Pareto inefficient. The profile of the objective functions whose nondominated particles are in the Pareto optimal set is called the Pareto frontier.

FIGURE 6.9.2 Conceptual nondominated particles (circles) and dominated particles (solids) in the criterion space.

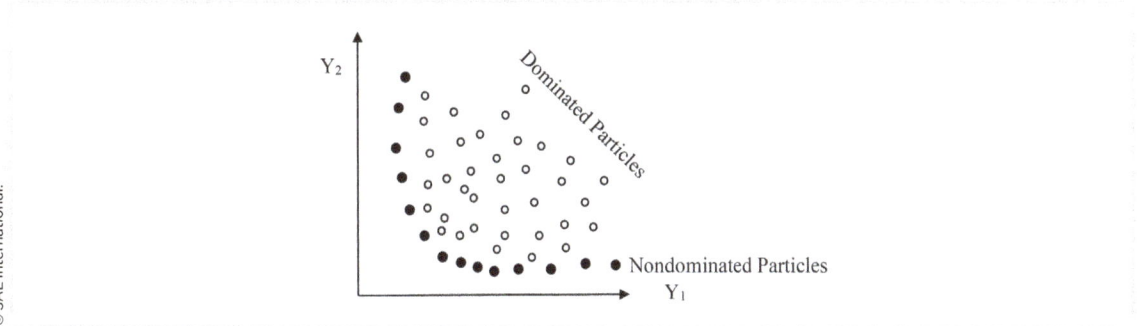

© SAE International.

When the conceptual design of Pareto optimality is extended to cover more than available test points (particles) to find a set of possible solutions, it is called Pareto dominance approach. Searching the Pareto frontier for multiobjective optimization problems usually involves the use of a population-based search algorithm or a deterministic method with a set of aggregate objective functions [Hu et al. 2013]. Two frequently applied searching techniques are given as follows:

(a) Multiobjective particle swarm optimization with Pareto optimality.

(b) Evolutionary multiobjective design with Pareto optimality.

Structural optimization of automotive parts is usually modeled using finite element methods and Pareto optimality based on one of the above searching algorithms.

6.9.3. Marginal Rate of Substitution and Utility Function

If a change to the allocation of resources is made such that certain objectives gain and no other objectives lose from the change, this change is regarded as moving toward Pareto efficiency. Such a change results in a Pareto improvement. This may be illustrated by means of utility function and marginal rate of substitution:

(a) Utility function: A function to quantify an individual's preferred utilization of goods and services.

(b) Marginal rate of substitution (MRS): The rate, at which a buyer replaces one good (or service) with another in order to maximize utility function.

Equations (6.9.1) and (6.9.2) can be generalized for a multiobjective optimization with multiple influential factors.

Assume that there are N different goods (e.g., car models available at a dealer site) and Q potential buyers. The overall utility function for mapping N goods to Q buyers can be written as:

$$\{Y\} = F([X_{qn}]) + \{\varepsilon\}$$
(6.9.7a)

where

$\{Y\}$ is the feasible set of criterion vector

$[X_{qn}]$ is the matrix of goods as a set of feasible decisions according to buyers

X_{qn} is the nth term of goods for the qth buyer; $n = 1, 2, 3, \ldots, N$ and $q = 1, 2, 3, \ldots, Q$

$\{\varepsilon\}$ is the random error

Similar to the context of DOE, where buyers are treatments, goods are factors, and matrix of goods is the design matrix. The utility function for buyer q, as compared to the response of treatment q, is then

$$Y_q = f(\{X_{qn}\}) + \varepsilon_q \tag{6.9.7b}$$

$$\text{and} \quad \{X_{qn}\} = \{X_{q1}, X_{q2}, X_{q3}, ..., X_{qN}\} \tag{6.9.8}$$

where

Y_q is the qth element of the feasible set of criterion vectors in answer to the qth buyer

$\{X_{qn}\}$ is the vector of goods as a set of feasible decisions for buyer q

The feasibility constraint is here assumed simply to be a linear combination of the goods

$$\sum_{q=1}^{Q} X_{qn} = C_n \quad (n = 1, 2, 3, ..., N) \tag{6.9.9}$$

The above equation means that the total quantity of the nth goods beloved by all Q buyers is limited to its availability.

The goods are to be put on allocation to reach out Pareto optimality. Buyer q may maximize the Lagrangian[1] according to:

$$L_q = f(\{X_{qn}\}) + \sum_{\substack{p=1 \\ (p \neq q)}}^{Q} \lambda_p [Y_p - f_p(\{X_{pn}\})] + \sum_{n=1}^{N} \mu_n (C_n - \sum_{p=1}^{Q} X_{pn}) \tag{6.9.10}$$

where

λ_p and μ_n is the Lagrangian multipliers

p is the dummy variable, p = 1, 2, 3, ..., Q but p ≠ q

The above equation is virtually a rewritten format of $L_q = f(\{X_{qn}\})$, except that Y_p is obtained from experimental data. Note that $Y_p - f_p(\{X_{pn}\}) = 0$ and $C_n - \Sigma X_{pn} = 0$, if there are random errors that are statistically insignificant. A sensitivity study on multiple influential factors would necessitate a feasibility space, also called design space. The sensitivity analysis suggests that:

$$\partial L_q / \partial X_{qn} = \partial f_q / \partial X_{qn} - \mu_n = 0 \qquad \text{(for goods)} \tag{6.9.11}$$

$$\text{and} \quad \partial L_q / \partial X_{pn} = -\lambda_p (\partial f_p / \partial X_{pn}) - \mu_n = 0 \quad \text{(for buyers)} \tag{6.9.12}$$

For goods, Equation (6.9.11) leads to $\partial f_q / \partial X_{nq} = \mu_n$ (for n = 1, 2, 3, ..., N), which reveals the constraint among goods. Given that n = s and n = t for any two different goods (s ≠ t), Pareto optimality is met when

$$\frac{\partial f_q / \partial X_{qs}}{\partial f_q / \partial X_{qt}} = \frac{\mu_s}{\mu_t} \tag{6.9.13}$$

---

[1] https://en.wikipedia.org/wiki/Lagrangian

The curve corresponding to Pareto optimality is called indifference curve or indifference schedule. The MRS_q for consumer q is the slope of the corresponding indifference curve at the point of interest, i.e.,

$$MRS_q = \frac{-\partial f_q / \partial X_{qs}}{\partial f_q / \partial X_{qt}}$$

(6.9.14)

In conclusion, the MRS (e.g., substituting good s for good t) is the same for buyer q when the allocation of goods follows Pareto optimality. The MRS can thus be also defined as the rate at which a consumer is willing to forgo good s for good t of the same utility function, when Pareto optimality is reached. How to calculate the MRS when moving from one treatment (one buyer's option) to another (another buyer's option) is illustrated using Example 6.9.1.

EXAMPLE 6.9.1

An industrial engineer would like to trade in the number of defects (factor A) for reduction of manufacturing cost (factor B) of a production part. The indifference schedule with four different combinations is given below:

Treatment	A (defects)	B ($)
1	0	1000
2	2	6000
3	4	4000
4	6	3000

What would be the MRS?

Solution:

The concept of MRS is an infinitesimal increment only. Based on Equation (6.9.14) with its partial differentiations being replaced by its corresponding finite differences, the marginal rates of substitution for treatment 2, 3, and 4, are, respectively

Treatment	MRS (Equation (6.9.14))
1	Baseline
2	$MRS_2 = -(\partial f_2/\partial X_{2s})/(\partial f_2/\partial X_{2t}) \approx -(6{,}000 - 10{,}000)/(2 - 0) = 2{,}000$
3	$MRS_3 = -(\partial f_3/\partial X_{3s})/(\partial f_3/\partial X_{3t}) \approx -(4{,}000 - 6{,}000)/(4 - 2) = 1{,}000$
4	$MRS_4 = -(\partial f_4/\partial X_{4s})/(\partial f_4/\partial X_{4t}) \approx -(3{,}000 - 4{,}000)/(6 - 4) = 500$

Note that the MRS begins with the second treatment. There could be two constraint equations that related to the total cost: (a) timely delivery and (b) innovative technology.

For the purpose of illustration, the idealization of MRS calculated as exhibited the indifference schedule is depicted in Figure 6.9.3, which happens to be a convex Pareto frontier with a diminishing MRS. If it is a case with an increasing MRS, Pareto frontier will be nonconvex.

FIGURE 6.9.3   Diminishing MRS—trading defects for production cost.

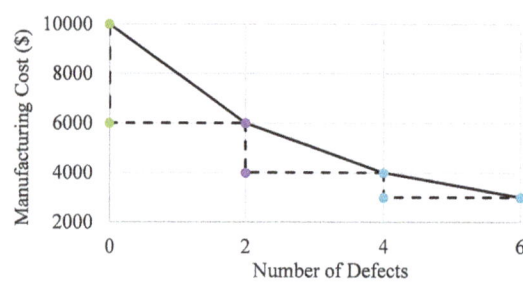

6.10. Multiobjective Optimization Based on DOE

Some optimization techniques have been used tremendously as alternative methods to conventional optimization techniques (e.g., least-squares and linear programming optimization) that could suffer from the dependence on the perceived initial model, weighting requirements, linearization problems, and traps in local optima. They are:

1. Particle swarm optimization [Eberhart and Kennedy 1995]
2. Evolutionary multiobjective optimization [Schaffer 1984]
3. Gray relational analysis [Deng 1989]

All these optimization techniques may resort to Pareto optimality. An example problem in the automotive area is given to demonstrate how the method of particle swarm optimization (PSO) is coupled with DOE for multiple-objective optimizations.

6.10.1. Multiobjective PSO with Pareto Optimality

PSO exploits a population of potential design points, which is called a swarm of particles, for feasible solutions by means of optimization algorithms. Swarm refers to the population of particles of interest. Because it is simple and effective, PSO has been successfully applied in many practical engineering applications.

Multiobjective PSO (MOPSO) with Pareto optimality allows an experimenter to obtain a joint optimal solution to a problem that has multiple objectives of interest. The determination of specific nondominated design points that are located on the Pareto frontier is the goal of multiobjective PSO. For the case of "the smaller the better," a Pareto-optimal solution to the aforementioned problem described by Equations (6.9.1)–(6.9.4) based on MOPSO is decided on the basis that there exists no {x} that makes the following Pareto optimality:

$$y_r(\{x\}^*) \le y_r(\{x\}) \text{ for all objective functions} \tag{6.10.1}$$

$$\text{and} \quad y_r(\{x\}^*) < y_r(\{x\}) \text{ for at least one i, of which } \{x\} \ne \{x\}^* \tag{6.10.2}$$

where

r is the subscript referring to the rth objective function, r = 1, 2, 3, ..., R
R is the total number of objective functions.
{x} is the design point, {x} = {A B C ...}, where A, B, C, ..., are design factors (variables)
{x}* is the design point whose projected particle in the criterion space is on the Pareto frontier

The projection of such an $\{x\}^*$ vector in the design space onto a $\{y\}^* = \{y_1(\{x\}^*) \; y_2(\{x\}^*) \; \dots \; y_R(\{x\}^*)\}$ vector in the criterion space makes a nondominated particle, also called a Pareto particle (Figure 6.9.2). Combining all $\{y\}^*$ particles makes the profile of Pareto frontier. Many computational algorithms for solving Equations (6.10.1) and (6.10.2) have been proposed to identify the profile of Pareto frontier [Baumgartner et al. 2004, Hu et al. 2013]. Given that "the smaller the better" for all objective functions, a particle moves to the next position according to the following equation [Eberhart and Kennedy 1995]:

$$\{v\}_i^{k+1} = w \, \{V\}_i^{k+1} + c_L \, \gamma_L \, (\{y_{pbest}\}_i - \{y\}_i^{k+1}) + c_G \, \gamma_G \, (\{y_{gbest}\}_i - \{y\}_i^{k+1}) \qquad (6.10.3)$$

$$\text{and} \qquad \{y\}_i^{k+1} = \{y\}_i^k + \{v\}_i^{k+1} \qquad (6.10.4)$$

where
  i is the ith particle
  k is the kth iteration
  $\{y\}_i^k$ is the position vector, where $\{y\}_i = \{y_1(\{x\}) \; y_2(\{x\}) \; \dots \; y_R(\{x\})\}_i$
  $\{v\}_i^k$ is the velocity vector, which determines the movement of a particle
  w is the inertia weight, relating the history of velocity to the current velocity
  c_L is the local learning constant, also called social parameter
  c_G is the global learning constant, also called cognitive parameter
  γ_L is the random number, uniformly distributed in [0,1]
  γ_G is the random number, uniformly distributed in [0,1]
  $\{y_{pbest}\}_i$ is the local best solution, i.e., the best particle in the neighborhood of a given particle
  $\{y_{gbest}\}_i$ is the global best solution, i.e., the best in the entire swarm

The coefficient w is called an inertia factor and regulates the exchange between the global exploration and local exploitation abilities of the swarm. It is typically chosen in the range of [0,1]. The PSO algorithm includes three steps, namely:

(a) Generating particles' positions and velocities
(b) Updating velocities
(c) Updating positions

The position vectors of particles can be initiated using the design matrix of influential factors in a complete factorial design 2^p for linear models and factorial design 3^p for nonlinear models, where the exponent p is the number of factors that are statistically significant, if DOE is applied. If no DOE is available, the initial position and velocity vectors for the particles are generated according to the following interpolating equations:

$$\{y\}_i^k = \{y\}_{min} + (\{y\}_{min} - \{y\}_{min}) \, \gamma_L \qquad (6.10.5)$$

$$\{v\}_i^k = \{v\}_{min} + (\{v\}_{min} - \{v\}_{min}) \, \gamma_G \qquad (6.10.6)$$

In practice, the nondominated solutions detected are stored in a repository. The search space is divided in many hypercubes. Each hypercube is assigned a fitness value that is inversely proportional to the number of particles it contains. Then, the classical roulette wheel selection is used to select a hypercube and a leader from it. The basic syntax for computer programming with PSO with Pareto optimality is shown in Figure 6.10.1.

FIGURE 6.10.1 Basic syntax for PSO with Pareto optimality.

```
for particle i in the swarm
        Initialize vector {y}ᵢ randomly within ({y}₁, {y}₂, ...., {y}total particles available)
        Initialize velocity vector {v}ᵢ with zero
        Assign {y}ᵢ to {y_pbest}ᵢ
end for
for particle i in the swarm do
        Evaluate objective function {y}ᵢ = {y₁({x}ᵢ) y₂({x}ᵢ) ... yᵣ({x}ᵢ)}
        Initialize a velocity vector {v}ᵢ = {v₁ v₂ ... vᵣ}ᵢ with zero
        if yᵣ({x}ᵢ) < yᵣ({x_pbest}ᵢ), r = 1, 2, ..., R
        Assign {x}ᵢ to {x_pbest}ᵢ
        end if
end for
Store non-dominated particles in an external archive
Select leaders
repeat
        for particle i in the swarm do
                Update velocity vector {v}ᵢ according to Equation (6.10.5)
                Update position vector {y}ᵢ according to Equation (6.10.4)
                Increase diversity
                Evaluate objective function {y}ᵢ = {y₁({x}ᵢ) y₂({x}ᵢ) ... yᵣ({x}ᵢ)}
                if yᵣ({x}ᵢ) < yᵣ({x_pbest}ᵢ), r = 1, 2, ..., R
                Assign {x}ᵢ to {x_pbest}ᵢ
                end if
        end for
        Update non-dominated particles in the external archive
        Select leaders
until maximum iterations are reached or minimum error criterion is not satisfied
```

An external archive file is used for maintaining the nondominated particles found using a MOPSO algorithm. Increasing the diversity means improving the numerical convergence. It requires the computation of all distances among particles at each iteration during the actual search in criterion space. This includes the selection of leaders and neighborhood topologies. A leader is a particle used to guide another particle toward a better region in the search space. Numerous nondominated particles on Pareto frontier indicate the presence of multiple leaders. An effective neighborhood topology would determine the set of particles that contribute to the calculation of the local best value of a given particle. This issue can be addressed by defining neighborhoods in the criterion space of particles' indices, assuming particles can be connected to each other based on different neighborhood topologies including ring, star, tree, and fully connected graphs, as detailed in [Reyes-Sierra and Coello 2006].

PSO with Pareto optimality for two or three objective functions can be solved directly with Excel and/or Minitab directly without tedious computer programming. Nevertheless, computer programming is essential for analyzing many objective functions at the same time.

The original PSO algorithm given above has undergone a number of modifications since it was first proposed. It is known that MOPSO algorithm is more efficient than generic algorithm (GA) at exploring the solution space, but it does not guarantee the global optimum as other evolutionary approaches. Most of these modifications are focused on the way of updating the velocity of a particle for the next iteration. For example,

1. An easy-to-understand alternative proposed by [Clerc and Kennedy 2002] is:

$$\{v\}_i^{k+1} = w\left[\{V\}_i^{k+1} + c_1\,\gamma_L\,(\{y_{pbest}\}_i - \{y\}_i^{k+1}) + c_2\,\gamma_G\,(\{y_{gbest}\}_i - \{y\}_i^{k+1})\right] \qquad (6.10.7)$$

where

w is the constriction factor

$c_1 = c_L/w$

$c_2 = c_G/w$

The change of constants from Equation (6.10.3) to Equation (6.10.7) seems to be off on another quest for numerical stability. Modeling with different combinations of χ, c_1 and c_2 in Equation (6.10.7) has been tried out by researchers, while one promising solution is $\chi = 0.729$ and $c_1 = c_2 = 2.05$ [Parsopoulos and Vrahatis 2008].

2. The inertia weight is to be characterized using random variables and Gaussian distribution function. The corresponding inertia weight is updated as [Yao et al. 2010]:

$$w = w_{min} + (w_{max} - w_{min})\, rand(0,1) + \sigma\, N(0,1) \tag{6.10.8}$$

where

w_{min} is the minimum value of inertia weight w

w_{max} is the maximum value of inertia weight w

rand(0, 1) is a random number falls between 0 and 1

N(0, 1) is a number derived from the standard normal distribution

σ is a constant

The above equation is designed to avoid falling into local optimal solutions and improve the global searching ability, as demonstrated by examples in [Li et al. 2020].

MOPSO has been a delightful way to solve engineering problems that have nondominated solutions to multiple objectives, such as the torque requirement and power need in operating an electric vehicle.

6.10.2. Example Problem for PSO with Pareto Optimality

Most practical multiple-objective optimization problems in the automotive industry have only two or three objective functions. An example, which is given by a study [Chen, S. et al. 2013] for demonstrating gray relational analysis, is rephrased here for illustrating the calculation procedure of carrying out PSO with Pareto optimality for two objective functions based on a multiple-factorial DOE. The study is focused on dampening of the automotive exterior noise with laminates based on physical tests. Each laminate is made of three layers, i.e., glass wool with aluminum foil, glass fiber in the middle, and PE foam. The layer thicknesses of these three layers are the design factors of concern and their design levels are listed in Table 6.10.1.

TABLE 6.10.1 Factors and their design levels for automotive noise reduction [Chen, S. et al. 2013].

Factor	Level (−1)	Level (0)	Level (+1)
A (mm): Glass wool/Al foil	10	13	16
B (mm): Glass fibers	7	10	13
C (mm): PE foam	4	7	10

As listed in Table 6.10.2, factorial design 3^{3-1} is employed to deploy the design matrix. Two objective functions (responses) of interest are:

1. S (dB): Noise level of sound (the smaller the better)
2. W (kg): Weight (the smaller the better)

TABLE 6.10.2 Factorial design 3^{3-1} for assessing the performance of a photovoltaic panel [Chen, S. et al. 2013].

Run	A	B	C	S (dB)	W (kg)
1	16	13	10	71.98	7.14
2	16	10	7	72.16	5.95
3	16	7	4	72.9	4.76
4	13	13	7	72.02	6.33
5	13	10	4	72.78	5.14
6	13	7	10	72.89	5.06
7	10	13	4	72.36	5.52
8	10	10	10	72.41	5.44
9	10	7	7	73.38	4.25

Reprinted from "Automotive Exterior Noise Optimization Using Grey Relational Analysis Coupled with Principal Component Analysis," Chen, Shuming; Wang, Dengfeng; Liu, Bo, Fluctuation and Noise Letters, 12, © 2013 World Scientific Publishing Co., Inc.

The data analysis for objective functions S (noise level of sound) and W (weight) is done with Minitab (Stat → Regression → Regression → Fit Regression Model). The "most insignificant" factor or interaction that comes with the largest p-value is weeded out one by one until all p-values (significance levels) are less than 10%, which is here defined as the parting significance level based on F distribution. In light of the ANOVA presented in Table 6.10.3, the final predictive equations are, respectively:

$$S_p = 81.425 - 0.3469 \ A - 0.8343 \ B - 0.3365 \ C + 0.02852 \ A \ B$$

$$+ \ 0.01537 \ B^2 + 0.01796 \ C^2$$

$$(R = 99.91\%, \ R_{adj} = 99.65\%, \ and \ R_{pred} = 96.80\%)$$

(6.10.9)

and $W_p = 0.006667 + 0.1467 \ A + 0.2733 \ B + 0.1233 \ C$

$$(R = 100\%, \ R_{adj} = 100\%, \ and \ R_{pred} = 100\%)$$

(6.10.10)

The effectiveness of each of the two equations given above can be validated by its respective high model correlation and adjusted model correlation, while the adequacy can be validated by the high predictive correlation. The following conclusions can be drawn from this study:

1. All three main effects are influential on the noise level.
2. Increasing the thickness of any lamina will reduce the noise level.
3. The individual impact of factor B or factor C on the noise level is nonlinear.

4. The interactive effect between factors A and B on the noise level is also statistically significant.

5. The weight of each laminate is simply an arithmetic mixture of individual layers. It is calculated, not from physical tests. No error message is available.

TABLE 6.10.3 ANOVA tables for assessing the performance of a photovoltaic panel.

(1) S:	Source	SS$_{adj}$	DOF	MS$_{adj}$	F$_{u,v}$	P-value
	Regression	1.78988	6	0.298314	194.20	0.5%
	A	0.18922	1	0.189219	123.18	0.8%
	B	0.19712	1	0.197116	128.32	0.8%
	C	0.06762	1	0.067619	44.02	2.2%
	B^2	0.03827	1	0.038272	24.92	3.8%
	C^2	0.04182	1	0.041818	27.22	3.5%
	AB	0.13176	1	0.131756	85.77	1.1%
	Error	0.00307	2	0.001536		
	Subtotal	1.79296	8			
	Grand ave.	—	1			
	Total	—	9			
(2) W:	Source	SS$_{adj}$	DOF	MS$_{adj}$	F$_{u,v}$	P-value
	Regression	6.01740	3	2.00580	*	*
	A	1.16160	1	1.16160	*	*
	B	4.03440	1	4.03440	*	*
	C	0.82140	1	0.82140	*	*
	Error	0.00000	5	0.00000		
	Total	1.79296	8			

A plot of noise level (dB) versus weight with the prescribed ranges of the design factors is exhibited in Figure 6.10.2. The red dashed line, which goes from the design point with minimum weight to the point with minimum noise level, forms the Pareto frontier. Since the noise level of 72 dB is usually accepted in the automotive industry, one may jump to the following conclusions:

1. Any design point on the Pareto frontier having a noise level below 72 dB can be selected if there is no weight limit. For example, the experimenter can have a design on the Pareto frontier with $S_p = 71.94$ dB and $W_p = 5.889$ kg.

2. Setting that $\partial S_p/\partial A = 0$, $\partial S_p/\partial B = 0$ and $\partial S_p/\partial C = 0$ leads to the minimum noise level $S_p = 70.53$ dB, with A = 16.14 mm, B = 12.16 mm, and C = 9.368 mm. Nevertheless, the weight is $W_p = 6.854$ kg. Note that $(S_p, W_p) = (70.53, 6.854)$ is not on the Pareto frontier.

The Pareto frontier with two objective functions can be figured out and observed easily using the regression models based on DOE, such as Equations (6.10.7)–(6.10.8), to generate design points (particles) using Excel or Minitab. The curve underlying nondominated particles, including the two minimum values corresponding to the two objective functions, is the Pareto frontier for a problem with two objective functions. Design point on the Pareto frontier can be read out in a Excel file or a Minitab file.

FIGURE 6.10.2   Noise level (SPL) versus laminate weight for dampening automotive exterior noise.

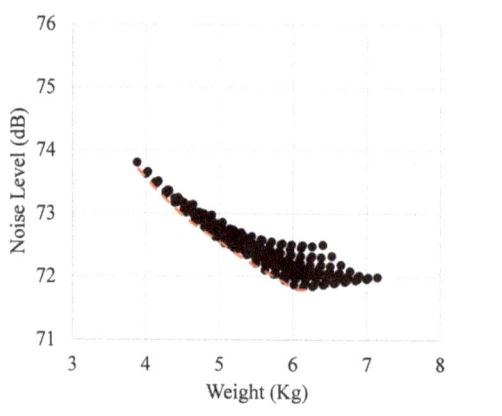

It is easier to read out the Pareto frontier when all the objective functions follow "the smaller the better" rule, as shown in Figure 6.10.2. If the optimization requests for an objective function, e.g., $X_1(A, B, C, …)$, which is "the larger the better," the experimenter may transform the objective function to make it "the smaller the better," e.g., $Y_1(A, B, C, …) = constant − X_1(A, B, C, …)$, in the graphical presentation. After figuring out the solution (nondominated particles) based on $Y_1(A, B, C, …)$, one may transform it back into the original function $X_1(A, B, C, …)$.

Assume that there is an additional objective function involved in the example problem given above. The predictive equation for the third objective function is given as:

$$M_p = 10 + 3\,A + 2\,B + 1\,C \qquad (6.10.11)$$

of which M_p is the predictive equation for manufacturing cost. So, the surface underlying nondominated particles including the three minimum values corresponding to these three objective functions forms the Pareto frontier in accordance with Equations (6.10.9)–(6.10.11). However, a contour plot of noise level versus weight and manufacturing cost, as exhibited in Figure 6.10.3, would reveal the feasible criterion space with a given acceptable noise level of 72 dB. The lower left edge denoted by black solid dots, as shown in the figure, becomes the Pareto frontier with the constraint of noise level. The weight and manufacturing cost of a design point corresponding to each of the solid dots in the figure can be identified in the Excel file.

FIGURE 6.10.3   Contour plot of noise level versus weight and manufacturing cost.

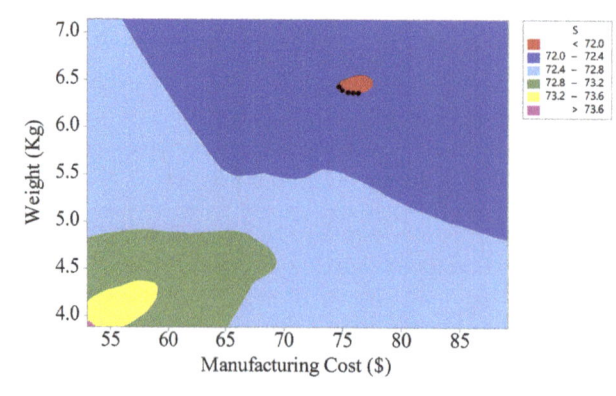

6.10.3. Evolutionary Multiobjective Design with Pareto Optimality

Evolutionary multiobjective optimization (EMOO) is another way to identify the Pareto frontier without using any single aggregated optimal solution for a multiobjective problem with a set of possible simultaneous solutions iteratively [Schaffer 1984]. Various evolutionary approaches to multiobjective optimization have been developed and they are capable of searching for multiple solutions concurrently in a single run. The genetic algorithm (GA) is a heuristic method inspired by the process of natural selection, as one original member of the larger class of evolutionary algorithms. Given a population of randomly selected candidate solutions, often called chromosomes, the steps for applying the basic GA include:

1. Selection: Fitness-based selection of candidate solutions
2. Genetic evolution: Applying genetic operations, such as crossover and mutation, to the previously selected solutions to generate new solutions
3. Updating population: Selected from old and new solutions

Repeat the three steps given above until one of the predetermined termination criteria is met. For example, when the population converges as indicated by the computed result that two successive iterations are sufficiently close (e.g., absolute difference smaller than 10^{-5}) or the number of iterations exceeded the predetermined number.

As extended from the basic generic algorithm, implemented versions targeting evolutionary optimization include SPEA, NSGA, NSGA-II, microGA, MOGA, ε-MOEA, MOPSO, and PAES [Coello and Zacatenco 2002]. A comparison of these computational algorithms is made by [Zitzler and Thiele 1999]. Detailed procedures of implementing these algorithms as published in the academic domain can be found in [Nedjah et al. 2018]. One important step is to do mutation that allows for exploration of the search space beyond the available solution information in the population and prevent the possibility of getting trapped in local optima [Nedjah et al. 2018].

Algorithms involved in evolutionary multiobjective design are less susceptible to problem-dependent characteristics, e.g., the profile of Pareto frontier that can be convex, concave, or even discontinuous. Many packages of computer software for the decision-making process involving EMOO have been developed [Emmerich and Deutz 2018]. The source code for NSGA-II (Nondominated Sorting Genetic Algorithm) is available at the KanGal (Kanpur Genetic Algorithms Laboratory) website. The NSGA (nondominated sorting GA) was first unveiled with a stochastic remainder by [Srinivas and Deb 1994].

6.10.4. Gray Relational Analysis

Gray relational analysis is a mathematical algorithm suitable for doing multiobjective optimization as a data-mining skill [Deng 2002]. It is often used to combine multiple-objective functions (e.g., performance characteristics) into a single representative objective function that can be used for solving a simple-minded optimization problem [Kumar et al. 2019]. Gray relational analysis aims to find the optimal combination of the desired performance characteristics. One can then examine the influence of each parameter of a multiparametric DOE on the representative objective function. Then, DOE skills or other ranking methods can be applied.

One major concern about gray relational analysis is how to deal with the representative objective function if the performance characteristics of interest are not of the same type of statistical distribution function; and they are usually not.

6.11. Distance-Based Optimality

There are two major categories of optimality criteria. One is information-based and the other is distance-based. Information-based optimality requires that the experimenter must have an idea of what terms to be included in the assigned regression model, but distance-based optimality may be used without specifying a model in advance [Minitab n.d., Vořechovský et al. 2019].

For distance-based criteria, the selection of design candidates is justified via the way how to compromise a subset of the cloud in p-dimensional Euclidean space, where p is the number of terms in the model. There are major versions of distance-based optimality:

1. U-optimality: Design points are uniformly spread over the design space. The basic tactic is to choose a subset of the cloud that virtually covers the entire cloud as uniformly as possible. In other words, it seeks to minimize the sum of the distances from each anchor point to the design points, i.e.,

$$\text{Minimize} \sum_{a \in A} \text{Distance}(a, \boldsymbol{D}) \tag{6.11.1}$$

where

    a is the one anchor point of interest
    $\boldsymbol{A}$ is a set of anchor points
    $\boldsymbol{D}$ is the design space of interest

The best set of experiments may be identified through a fast exchange algorithm where a substitution is selected to provide the maximum increase of the minimum distance between the now selected experiments in each cycle [Marengo and Todeschini 1992].

2. S-optimality: The basic tactic is to choose a subset of the cloud that virtually covers the entire cloud as broadly spread as possible. It seeks to minimize the averaged "1/harmonic mean distance" from each design point to all the other points in the design space, i.e.,

$$\text{Minimize} \left[\sum_{d \in \boldsymbol{D}} 1 / \text{Distance}(d, \boldsymbol{D} - d) \right] / N_D \tag{6.11.2}$$

where

    d is a design point in design space $\boldsymbol{D}$
    N_D is the total number of design points in design space $\boldsymbol{D}$

If there is no idea about which model to begin with, physical laws can be deduced to guide the distance-based optimality. U-optimality or S-optimality is then used for reducing the number of experiments [Marengo and Todeschini 1992], just like D-optimality.

6.11.1. U-Optimality in Euclidean Space

In the p-dimensional Euclidean space R^p, the distance-based criteria are based on the distance from an anchor (target) point {a} to a set of p-dimensional vector {d} in design space $\boldsymbol{D}$ and {d} $\in R^p$. The distance is defined as follows:

$$\text{distance}(\{a\}, \{d\}) = \min_{\{a\} \in A} \|\{a\} - \{d\}\| \tag{6.11.3}$$

$$\text{and} \quad \|\{a\} - \{d\}\| = [(a_1 - d_1)^2 + (a_2 - d_2)^2 + \ldots + (a_p - d_p)^2]^{1/2} \tag{6.11.4}$$

where

    $\|\{a\} - \{d\}\|$ is the p-dimensional Euclidean distance
    $\{a\}$ is the anchor point
    $\{d\}$ is the design point

In three-dimensional space, the above equation reduces to

$$\|\{a\} - \{d\}\| = [(a_1 - d_1)^2 + (a_2 - d_2)^2 + (a_3 - d_3)^2]^{1/2} \tag{6.11.5}$$

U-optimality refers to minimizing the sum of all the distances from anchor point $\{a\}$ to design point $\{d\}$ as:

$$\text{U-optimality} \rightarrow \min \Sigma \text{ distance}(\{a\}, \{d\}), \text{ where } \{d\} \in \boldsymbol{D} \tag{6.11.6}$$

6.11.2. Distance-Based Maximum Likelihood Estimation

U-optimality is here further illustrated using the example of nodal localization in wireless sensor networks in light of distance-based maximum likelihood estimation. Surveillance systems and indoor mobile robots, used for detecting and tracking moving targets, are profound applications of wireless sensor networks.

When a signal propagates between a transmitter and its corresponding receiver, the power loss or attenuation increases with increasing distance between transmitter and receiver. The distance measurement based on the received signal strength indicator (RSSI) is frequently adopted because of its intrinsic simplicity and independence of dedicated hardware. Many emerging sensor network applications rely on the efficient use of unattended sensors to detect, identify, and track targets during a find or update for retaining on-going availability and agility. Wireless sensor networks (WSNs) that are comprised of small and inexpensive anchor nodes with constrained computing power, limited memories, and short battery lifetime, can be used to monitor and collect data in a region of interest [Miao et al. 2020], such as autonomous vehicles used in warehouses or logistic centers. The location information is vital for these autonomous vehicles to accomplish tasks. Via a wireless sensor network, the autonomous vehicles estimate their distance to a specific node called a reference anchor node [Hacioglu and Sesli 2022].

The distance-based (or range-based) multilateral measurement localization method is frequently applied for node localization in WSNs [Augusto et al. 2008], of which reference anchor nodes are often equipped with GPS modules and can thus locate themselves. A target node with an unknown position in the space is then located by these anchor nodes that are then known as reference nodes, of which each has well-defined information about its location.

With a random error, unknown design node (x, y, z) of interest can be located using the following physical model for the measured distance in the 3-dimensional space by anchor node n, as:

$$d_n = \mu_n + \varepsilon_n = [(x_n - x)^2 + (y_n - y)^2 + (z_n - z)^2]^{1/2} + \varepsilon_n \tag{6.11.7}$$

$$\text{and} \quad \mu_n = [(x_n - x)^2 + (y_n - y)^2 + (z_n - z)^2]^{1/2} \tag{6.11.8}$$

where

    d_n is the measured distance
    μ_n is the mean of the measured distance
    ε_n is the random error
    (x, y, z) is the design node (x, y, z) with an unknown location
    n is the number of anchor nodes, used to measure point (x, y, z); n = 1, 2, 3, ..., N
    N is the total number of anchor nodes selected and $N \geq 4$ for a 3-dimensional space

An advance to accommodating random errors in Equation (6.11.7) means to include a statistical theory in the model that can be resolved using the maximum likelihood estimate for a population. Assume that the measured distance d_n by node n, where n = 1, 2, 3, ..., N, has a normal-distributed random error σ. Then,

$$d_n \sim N(\mu_n, \sigma) \tag{6.11.9}$$

which means that the distance-based optimality may respond to stochastics in nature [Xu et al. 2016].

Maximum likelihood estimation is a statistical method for estimating the parameters of a model. It lets the assumed model fit the observed data with the maximum likelihood. In sampling from a normal distribution with mean μ_n and variance σ^2, the likelihood function based on all the measured data can be written as [Freeman 2010]:

$$L(\mu_1, \mu_2, ..., \mu_N, \sigma^2) = f(d_1)\, f(d_2)\, ... \, f(d_N) = \prod_{n=1}^{N} f(d_n) \tag{6.11.10}$$

where $f(d_1)$, $f(d_2)$, ..., and $f(d_N)$ are probability density functions of d_1, d_2, ..., and d_N, respectively. Distance-based maximum likelihood estimation (DB-MLE) refers to minimizing likelihood function L given above. By statistical theory, maximizing $L(\mu_1, \mu_2, ..., \mu_N, \sigma^2)$ is equivalent to maximizing $\ln[L(\mu_1, \mu_2, ..., \mu_N, \sigma^2)]$. The log-likelihood function is derived by taking a natural log of the above equation. Assume that $f(d_1)$, $f(d_2)$, ..., and $f(d_N)$ are all normally distributed following Equation (2.8.1). Then,

$$\ln[L(\mu_1, \mu_2, ..., \mu_N, \sigma^2)] = -\tfrac{1}{2} N \ln(2\pi) - \tfrac{1}{2} N \ln(\sigma^2) - \tfrac{1}{2} \sum_{n=1}^{N} \left(\frac{d_n - \mu_n}{\sigma}\right)^2 \tag{6.11.11}$$

Substituting Equation (6.11.8) into the above equation leads to:

$$\ln[L(\mu_1, \mu_2, ..., \mu_N, \sigma^2)] = -\tfrac{1}{2} N \ln(2\pi) - \tfrac{1}{2} N \ln(\sigma^2)$$
$$-\tfrac{1}{2} \sum_{n=1}^{N} \left\{\frac{d_n - [(x_n - x)^2 + (y_n - y)^2 + (z_n - z)^2]^{1/2}}{\sigma}\right\}^2 \tag{6.11.12}$$

The likelihood equations for locating (x, y, z) can be obtained by setting the first derivatives of the above equation with respect to x, y, and z, respectively:

$$\frac{\partial\{\ln[L(\mu_1, \mu_2, ..., \mu_N, \sigma^2)]\}}{\partial x} = \sum_{n=1}^{N} (x_n - x)(1 - d_n / \mu_n) = 0 \tag{6.11.13}$$

$$\frac{\partial\{\ln[L(\mu_1, \mu_2, ..., \mu_N, \sigma^2)]\}}{\partial y} = \sum_{n=1}^{N} (y_n - y)(1 - d_n / \mu_n) = 0 \tag{6.11.14}$$

and
$$\frac{\partial\{\ln[L(\mu_1, \mu_2, ..., \mu_N, \sigma^2)]\}}{\partial z} = \sum_{n=1}^{N} (z_n - z)(1 - d_n / \mu_n) = 0 \tag{6.11.15}$$

Solving the above three equations reveals the maximum likelihood estimation value of the unknown node (x, y, z). If four signals from four different anchor node locations are obtained, one can merely measure the signal strength and make a ratio from those locations.

The mean value and variance can be resolved by setting the first derivatives of Equation (6.11.11) with respect to μ_n and σ, respectively:

$$\mu_n = \frac{1}{N} \left[\sum_{n=1}^{N} d_n \right] \tag{6.11.16}$$

and

$$\sigma^2 = \frac{1}{N} \left[\sum_{n=1}^{N} (d_n - \mu_n)^2 \right] \tag{6.11.17}$$

where

μ_n is the mean estimated according to anchor node n

σ^2 is the variance estimated from N samples

This method improves the accuracy of localization compared to the range-based multilateral measurement localization method and quick local search method [Xu et al. 2016], although the maximum likelihood estimate of σ^2 is biased in statistical theory.

6.11.3. **S-Optimality**

S-optimality refers to minimizing the sum of the inverse of harmonic mean distance from each design point {d}* to all the other points {d} in the design space.

$$\text{S-optimality} \rightarrow \min \left[\frac{\sum_{\{d\}^* \neq \{d\}} \text{distance}^{-1}(\{d\}^*, \{d\})}{N_A} \right] \tag{6.11.18}$$

As long as distance({d}*, {d}) is large for all design points of concern, the points in design space are spread out widely.

6.11.4. **Distance-Based Measurements for Autonomous Vehicles**

The distance and relative velocity between the surrounding objects and an autonomous vehicle itself should be known in advance to ensure safety. Distance-based optimality becomes an important study for autonomous vehicles. Distance estimation can be realized by using the following information:

1. RSS (Received Signal Strength)

2. TOA (Time of Arrival)

3. TDOA (Time Difference of Arrival)

Each has its own advantages and disadvantages, given that different sensing algorithms deliver different functional needs.

A variety of sensors can be integrated into an autonomous vehicle to improve the performance. It is imperative to realize how sensors help autonomous vehicles know where to go and recognize objects on the road so as to prevent accidents. These sensors include (but not limited to):

1. mmWave Radar (Millimeter wave radar): The radar is essentially a short-wavelength electromagnetic wave that works in the millimeter wave band (1–12.5 mm). Its working principle is to emit a chirp signal (mostly a frequency-modulated continuous-wave (FMCW)) to the measured object, then compare the received echo with the transmitted signal, and obtain the relevant information of the measured object, such as the distance. Millimeter wave radar is widely used in driver-assisting systems, such as adaptive cruise control (ACC), automatic emergency braking (AEB), active lane control (ALC), blind spot detection (BSD), front collision warning (FCW), lane change assistance (LCA), parking assistance, etc. These are driver support features classified as Level 1 and Level 2 driving automation [SAE 2021]. Three widely used frequencies are:

    (a)   24 GHz mmWave radar (wavelength = 12.5 mm): Working in the short range of 0.15 to 30 m, such as for parking assistance.

    (b)   77 GHz mmWave radar: Working in the range of 1 to 100 m, such as for BSD.

    (c)   79 GHz mmWave radar: Working in the range up to 250 m, such as for ACC and FCW.

    An electromagnetic wave-based mmWave radar is sensitive to the target material's dielectric constant. Material with a low dielectric tends not to reflect an electromagnetic wave well.

2. Ultrasonic radar: As ultrasound is used as the carrier wave, it is a radar operating in the frequency band of ultrasound. Using the time elapse, it takes for the sound burst to return to the sensor after being sent, the distance between the sensor and the substance being measured or the level of the substance is calculated. Volume, weight, or other similar measurements can also be calculated from the measured distance, and thus useful for presence detection and object profiling. Things like physical obstructions, excessive foam, heavy vapors, thick dust, and light powders may distort the signal. When a vehicle travels at a high speed, the signal received may be delayed owing to the slow speed of the ultrasonic signal.

3. Lidar (laser radar): It is a radar operating in optical frequency band, as laser is used as the carrier wave. Its working principle is to emit a laser beam to the measured object, then compare the received echo with the transmitted signal, and obtain the relevant information of the measured object, such as the distance and orientation of the measured object. Solid-state lidar-based distance estimation, along with the continuous evolution of autopilot, has become an essential sensor for partial driving automation (i.e., Level 3 and Level 4: automatic driving [SAE 2021]) and autopilot (i.e., Level 5: self-driving at all times under any condition [SAE 2021]), because of its unique three-dimensional modeling of the environment. Working as a multilane multiobject tracking device designed for intelligent transportation systems, it provides basic information of traffic flow such as traffic, lane share, model, queuing length, and event analysis based on multiobjective real-time tracking traces in 5D mode, including X-, Y-, Z-coordinates, one-dimensional speed, and angle. Lidar systems typically have no trouble when identifying objects during autopilot, even under hazardous conditions including fog, rain, or material with a low dielectric.

4. 360° camera: 3D cameras are applied for displaying images in detail. These image sensors detect objects, classify them, and determine the distances between the vehicle and surrounding objects such as other cars, pedestrians, cyclists, traffic signs and signals, road markings, bridges, and guardrails. Cameras mounted on autonomous vehicles are the most cost-efficient among all technologies

5. Infrared night vision camera: These cameras would clearly identify the surrounding objects in the roadway in poor environmental conditions, such as driving in the night, rain, fog, or snow.

6. High-speed information processing systems and automotive Ethernet networking: GPS and inertial measurement units (IMUs) along with 3D road maps can be used for communications such as vehicle-to-vehicle, vehicle-to-network, and vehicle-to-grid.

Autonomous cars are made possible thanks to these optical devices and the related computer software, fused in a sensor network that is usually made up of spatially distributed sensor nodes given above, which interact with one another independently.

6.12. Time-Based Optimality

Time-based management (TBM) allows people to analyze and control cost-incurring activities within spaces of time such as seconds, minutes, and hours. Time-based optimality provides an interpretation for a case study that aims at optimizing the time against a certain goal such as cost reduction.

6.12.1. Time-Based Optimality for Cost Function of Charging Plug-in Electric Vehicles

One interesting subject of time-based optimality is about charging plug-in electric vehicles at a commercial charging facility. Influx of plug-in electric vehicles (PEVs) creates a pressing need for careful time-based optimality of charging infrastructure. Note that time-based optimality can be model-dependent. The primary goal is to devise a closed-form expression for the PEV charging station capacity problem. Markovian queues are here utilized to assess the cost function at a charging station system in terms of service load, waiting time, and their interaction [Bayram et al. 2022]. Service durations at the charging lot are assumed to be an M/M/1 queuing system:

1. First **M** (Memoryless): Arrivals follow a Poisson distribution function.
2. Second **M** (Memoryless): Service times follow an exponential distribution function—first come first serve.
3. **1**: There is only one server, i.e., the charging lot.

The M/M/1 queue has independent interarrival times and independent service times. The example given here is one server only, but the theory allows several charging stations to work simultaneously, namely M/M/K queue. State transition occurs when there is an arrival, or when the server finishes the service, sending the vehicle out of the process.

Consider a charging lot located at a workplace dedicated for employees and visitors. This is an optimal source calculation for estimating the size of the total power demand feeding a certain number of vehicles to be charged. Assume that:

1. μ: Service rate is the variable to be decided, i.e., the mean number of PEVs per unit time charged for the M/M/1 queue. The service time is then $1/\mu$.
2. λ: Mean number of PEVs per unit time that arrive at the station, which is assumed to be known in advance in the study. The expected time between adjacent arrivals is $1/\lambda$.

Note that service rate μ and arrival rate λ are independent of each other. When $\mu > \lambda$, the utilization of the system is:

$$u = \lambda / \mu \tag{6.12.1}$$

The measure of an M/M/1 queue is the number of jobs (i.e., n vehicles) currently occupying it. Assume that p_n is the probability that the process is current in state n. Then, in the state n+1:

$$\lambda \, p_n = \mu \, p_{n+1} \tag{6.12.2}$$

Substituting Equation (6.12.1) into the above equation yields:

$$u = p_{n+1} / p_n \tag{6.12.3}$$

As the total probability is 100%,

$$\sum_{i=0}^{\infty} p_i = \sum_{i=0}^{\infty} u \, p_{i-1} = \sum_{i=0}^{\infty} u^i \, p_0 = p_0 \left(\sum_{i=0}^{\infty} u^i \right) = \quad p_0 / (1 - u) = 100\% \tag{6.12.4}$$

Thus, $p_0 = (1 - u)$

Equations (6.12.3) and (6.12.4) imply that:

$$p_1 = u \, p_0 = u \, (1 - u)$$
$$\cdots$$
$$\text{and} \quad p_n = u \, p_{n-1} = u^n \, (1 - u) \tag{6.12.5}$$

Given that the charge operation is based on the principle of first-in-first-out (FIFO) that follows exponential waiting time, the expected volume of vehicles, namely N, in the system is:

$$E[N] \equiv \sum_{n=0}^{\infty} n \, p_n = \sum_{n=0}^{\infty} n \, u^n \, (1 - u) = (1 - u) \sum_{n=0}^{\infty} n \, u^n = (1 - u) \, u \left[\sum_{n=0}^{\infty} n \, u^{n-1} \right]$$

$$= (1 - u) \, u \, [(1 - u)^{-2}] = \frac{u}{1 - u} = \frac{\lambda}{\mu - \lambda} \tag{6.12.6}$$

The above equation is established on the condition that $\mu > \lambda$. When $\mu \leq \lambda$, $E[N] = 0$

After E(N) is known, one may estimate the expected total time in the system using Littles theorem that:

$$\lambda \, E[T] = E[N]$$

$$\text{Thus,} \quad E[T] = E[N] / \lambda = 1 / (\mu - \lambda)$$

Note that $E[T] = E[\text{time in queue}] + E[\text{time in service}]$, where $E[\text{time in service}] = 1/\mu$ because the service fits an exponential distribution. Thus,

$$E[\text{time in queue}] = E[T] - E[\text{time in service}] = 1 / (\mu - \lambda) - 1 / \mu = \lambda / [(\mu - \lambda) \, \mu]$$

$$\text{and} \quad E[\text{number in queue}] = \lambda \, E[\text{time in queue}] = \lambda^2 / [(\mu - \lambda) \, \mu] = \rho^2 / (1 - \rho)$$

EXAMPLE 6.12.1

Consider a situation where the exponentially distributed service rate is 20 vehicles/day and the arrival rate is 16 vehicles/day for the charging lot in the parking ground of an assembly plant. The plant is open 24 hours/day. According to M/M/1 queue, what would be the expected volume of vehicles in the charging lot (system), the expected total time in the charging lot (system), expected time in service, and expected time in queue?

Solution:

$$E[N] = \lambda \ / \ (\mu - \lambda) = 16 \ / \ (20 - 16) = 4 \text{ vehicles in the system (charging lot)}$$

Given that the charging lot is open for business 24 hours/day, one has

$$E[T] = 1 \ / \ (\mu - \lambda) = 1 \ / \ (20 - 16) = 0.25 \text{ day} = 6 \text{ hours}$$

$$E[\text{time in service}] = 1 \ / \ \mu = 1 \ / \ 20 = 0.05 \text{ day} = 1.2 \text{ hours (time to charge a vehicle)}$$

and $\quad E[\text{time in queue}] = E[T] - E[\text{time in service}] = 0.25 - 0.05 = 0.2 \text{ day} = 4.8 \text{ hours}$

Whether the time in service (the time duration required for charging a vehicle), i.e., 1.2 hours, can be met or not depends on the facility (e.g., AC Level 1, AC Level 2, and DC Fast) acquired.

Given that $\mu > \lambda$, the objective to minimize the long-term average cost per unit of time can be represented as [Bayram et al. 2022]:

$$C(\mu) = c_s \, \mu + c_w \, E[N] \tag{6.12.7}$$

where
  C (\$) is the operating cost function
  $E[N]$ is the expected volume of PEVs in the parking lot
  c_s (\$/vehicle) is the service cost rate for charging a PEV, related to power drawn from the grid
  c_w (\$/vehicle) is the cost per PEV, when it waits at the charging lot

Assume that the Poisson process reigns [Adan and Resing 2002], i.e., charging is a process with Poisson input, and Poisson output. With a single charging lot, plugging Equation (6.12.6) into Equation (6.12.7) yields:

$$C(\mu) = c_s \, \mu + c_w \, \lambda \ / \ (\mu - \lambda) \tag{6.12.8}$$

Given that $c_s = 2.5$ \$/vehicle and $c_w = 4.5$ \$/vehicle, according to the above equation, the operating cost per unit time is plotted in Figure 6.12.1.

FIGURE 6.12.1 Cost as a function of charging rate at service and arrival rate.

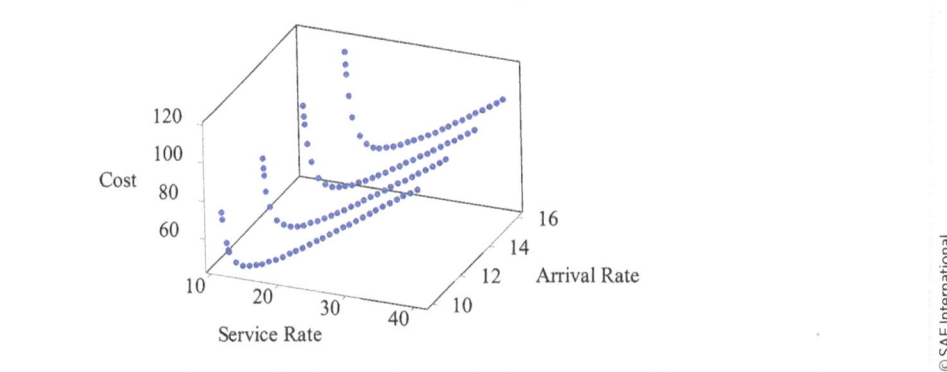

The minimum cost for ach curve plotted in Figure 6.12.1 can be found by observing the first and second differentiations with respect to variable μ, respectively:

$$\partial C(\mu) / \partial\mu = c_s - c_w \lambda / (\mu - \lambda)^2 \qquad (6.12.9)$$

and

$$\partial^2 C(\mu) / \partial\mu^2 = 2 c_w \lambda / (\mu - \lambda)^3 \qquad (6.12.10)$$

Since $\partial^2 C(\mu)/\partial\mu^2 > 0$, setting $\partial C(\mu)/\partial\mu = 0$ would unveil the incurrence of the minimum cost as estimated,

$$\mu^* = \lambda + [(c_w / c_s) \lambda]^{1/2} \qquad (6.12.11)$$

where μ* is the estimate of μ. Substituting the above equation back into Equation (6.12.4) leads to the optimal cost function as:

$$C(\mu^*) = c_s \lambda + 2 (c_s c_w \lambda)^{1/2} \qquad (6.12.12)$$

where

$c_s \lambda$ is the operating cost of supplying the minimal service level of a PEV charging service

$2(c_s c_w \lambda)^{1/2}$ is the operating cost incurred because of the 2-factor interaction, i.e., $c_s c_w$

Which yields the minimum value of operating cost with a given arrival rate (λ). To reduce cost, the service rate has to be higher than the arrival rate, as shown in Figure 6.12.2. For example, $\mu^* = 14.24$, if $\lambda = 10$. Thus, one way to bring down the operating cost is to speed up the service rate to assure that Equation (6.12.11) is satisfied. However, it may cause an increase in the fixed cost according to electric vehicle charger types, e.g., AC Level 1, AC Level 2, or DC Fast.

© SAE International.

FIGURE 6.12.2   Service rate versus arrival rate at least cost for a single PEV charging lot.

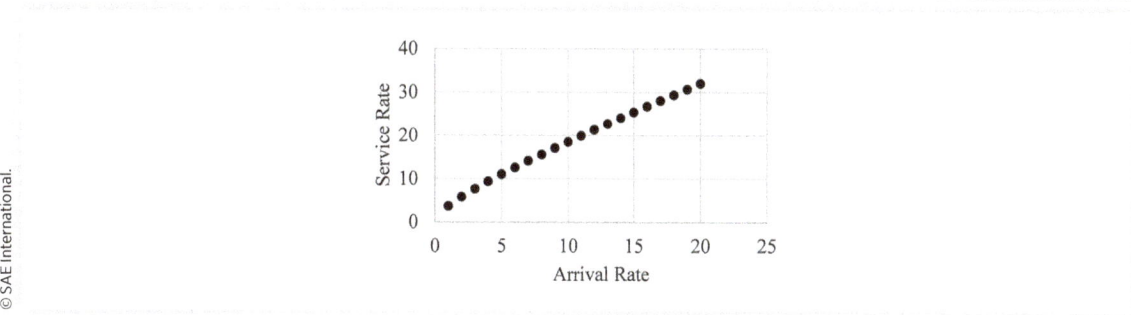

6.12.2. Time-Based Optimality for Profit Function of Charging PEVs

By the same token as Equation (6.12.4), the profit function can be obtained as:

$$P(\lambda) = c_p \, \lambda - c_w \, \lambda \, / \, (\mu - \lambda) \tag{6.12.13}$$

where

P ($) is the profit function

c_p ($/vehicle) is the profit rate, profit per charge, given that there are λ vehicles per unit time

Given that c_p = 5.0 $/vehicle and c_w = 4.5 $/vehicle, according to the above equation the operating profit per unit time is plotted in Figure 6.12.3.

FIGURE 6.12.3   Profit as a function of service rate and arrival rate.

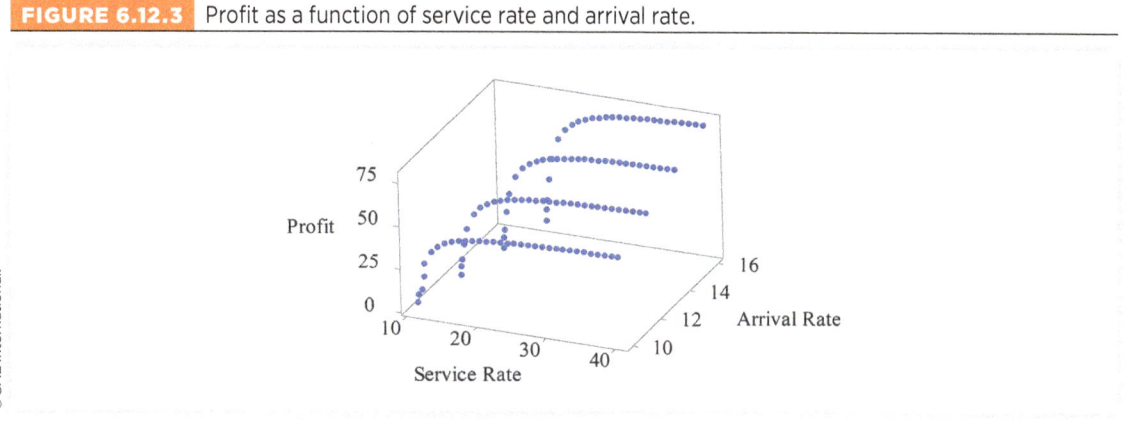

© SAE International.

The maximum profit for each curve plotted in Figure 6.12.3 can be found by observing the first and second differentiations with respect to variable μ, respectively:

$$\partial P(\lambda) \, / \, \partial \lambda = c_p - c_w \, \lambda \, / \, (\mu - \lambda)^2 \tag{6.12.14}$$

and
$$\partial^2 P(\lambda) \, / \, \partial \lambda^2 = -2 \, c_w \, \lambda \, / \, (\mu - \lambda)^3 \tag{6.12.15}$$

Given that $\mu > \lambda > 0$ for cost reduction as guided by Figure 6.12.2, $\partial^2 P(\lambda)/\partial\lambda^2 > 0$. Thus, the solution to $\partial P(\lambda)/\partial\lambda = 0$ would lead to the maximal profit at:

$$\mu^S = \lambda + (c_w \lambda / c_p)^{1/2} \tag{6.12.16}$$

Plugging the above equation into Equation (6.12.8) yields the maximum profit as:

$$P(\lambda) = c_p \lambda - c_w \lambda / (c_w \lambda / c_p)^{1/2} = c_p \lambda - (c_p c_w \lambda)^{1/2} \tag{6.12.17}$$

6.12.3. Just-in-Time Strategy for Production Workflow

The concept of just-in-time (JIT) is a workflow strategy aimed at reducing time flow, and consequently the cost, in a production facility and the distribution of materials. The philosophy of continuous improvement, known as Kaizen, that is associated with JIT strategy permeates most of the recent improvements realized in the time-based optimality. It is focused on customer responsiveness, time-based competition, and value stream analysis to eliminate all waste. It enables the first level schedule of all activities, including supply chain management and sales, which take place under the guidance of lean production and agile manufacturing [Mascitelli 2011]. Achieving operational objectives of JIT strategy requires the coordination of production planning, sourcing, and logistics. For example, how to schedule multiple servers to facilitate JIT part-supply in automobile assembly lines [Peng and Zhou 2018]. An advance to the accommodation of moving boundaries and/or time-varying boundaries in decision-making on the time-based optimality is presented by [Malhotra et al. 2018]. How to maximize production efficiency via JIT strategy is introduced in [Monden 2011].

6.12.4. Time-Efficient Optimal Torque Split Strategy for Electric Vehicles with Multiple Motors

To exploit the potential of energy saving of a dual-motored powertrain over a single-motored powertrain, three time-efficient optimal torque split strategies for the load distribution between the front and rear axles are examined [Zheng et al. 2020]:

1. Evenly distributed between the front axle and rear motor
2. Rule-based distribution
3. Adaptive nonlinear PSO (ANLPSO)

It is concluded that the dual-motored powertrain with the adaptive nonlinear PSO is able to save the energy consumption over the single-motored powertrain by 11.88% based on the NEDC (New European Driving Cycle) and 12.18% based on the WLTP (World Light Vehicle Test Procedure), respectively.

6.13. Topological Optimization

Topological optimization, also called shape optimization, indicates where nonvital material could be subtracted (eliminated) from the part and/or material in critical areas can be added (enhanced) to enhance the design. It plays a major role in structural optimization as the most common type of shape design tool. The part with a simple geometry is sketched as an initial conceptual design that is called block design. The detailed design configuration of the structure is then obtained from performing topological optimization techniques. As shown in Figure 6.13.1, a solid rectangular steel plate (1 m × 0.25 m × 0.1 m) as fabricated, may evolve into different finished parts by means of topological optimization at different loads with geometric nonlinearity using X-FEM and iso-values [Abdi 2015].

Traditional topological optimization algorithms for automotive parts include, but not limited to, the following [Broekaart 2015]:

(a) Maximizing stiffness with volume constraint

(b) Maximizing stiffness with frequency constraints

(c) Minimizing a displacement with volume constraint

(d) Minimizing a volume with displacement constraint

(e) Minimizing a reaction or internal force

(f) Maximizing first eigenfrequencies

(g) Maximizing the band gap around an eigenfrequency

(h) Compromising differences among multiobjective functions

These main/max strategies accelerate and improve the reliable design of lightweight, rigid, and durable components and systems.

FIGURE 6.13.1 Evolutionary design change by finite element analysis of a plate based on topological optimization [Abdi 2015].

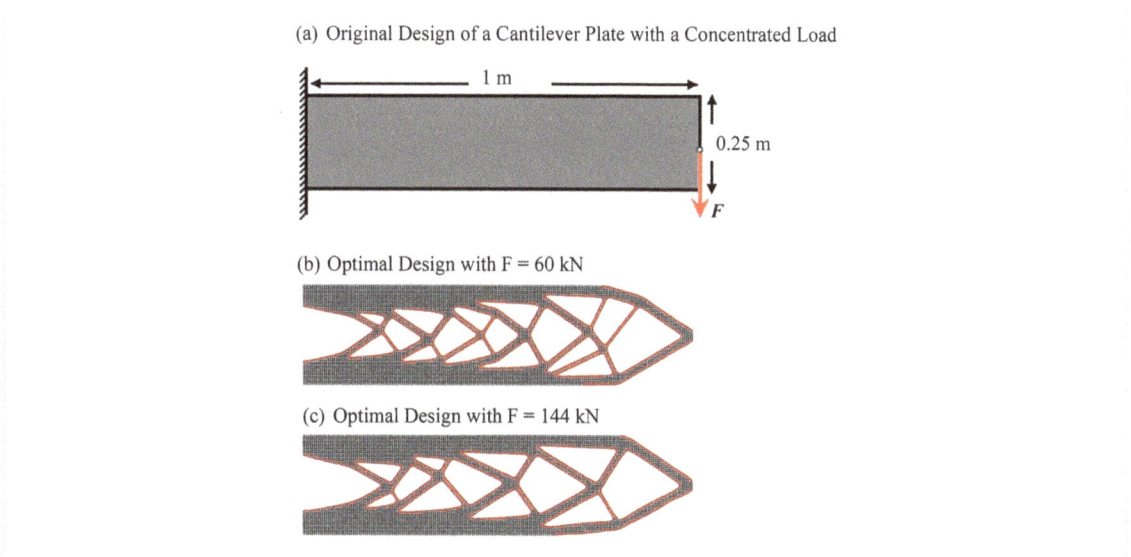

(a) Original Design of a Cantilever Plate with a Concentrated Load

1 m

0.25 m

F

(b) Optimal Design with F = 60 kN

(c) Optimal Design with F = 144 kN

Top software packages widely used for topological optimization of parts based on finite element methods are Altair Inspire/Solid Thinking, Ansys Mechanical/Discovery, CogniCAD (part of ParaMatters), Creo (Parametric Technology Corporation), DfAM (Sulis), Netfabb (Autodersk), nTopology (Headquartered in New York), Siemens NX/Solid Edge, Tosca Structure/Fluid (Dassault Systems), and 3DXpert. Enhanced with the power of cloud computing and machine learning (artificial intelligence), topological optimization brings product development with scalability and innovation at a cracking pace of design change.

In fact, there are often multiple objectives that undergo mutual conflicts. For example, the mass need and structural compliance with physics of failure are considered as conflicting objectives. They can be assessed in accordance with Pareto optimality, by which the corresponding curve of Pareto frontier is traced, instead of imposing additional constraints on the material properties. The strategy of multimaterial topological optimization (MMTO), which permeates the optimal design methodology by integrating material properties into the geometric design space, was promoted by [Thomsen 1992]. Implementation of this strategy using Pareto optimality for multiobjective optimization was organized by [Suresh 2010, Suresh and Takalloozadeh 2013].

6.13.1. Evolutionary Design Change by Finite Element Methods (FEM)

Assume that there are two independent objective functions. The first intuitive objective of an optimal design is to reduce the cost in terms of either mass reduction or total price, which is generally in demand as the first design goal. The second objective is to comply with physics of failure, by which it is expected to minimize unreliability such as shortened fatigue life of the stressed material in the geometric design space. The scenario of optimal design based on these two criteria can be translated into the following utility function of design factors of interest, as:

$$Y_q = f_q(\{X_q\}) + \varepsilon_q \qquad \text{where } q = 1 \text{ and } 2 \qquad (6.13.1)$$

$$\text{and} \qquad \{X_q\} = \{X_{1q}, X_{2q}, X_{3q}, \ldots, X_{Nq}\} \qquad (6.13.2)$$

where

Y_1 is the cost or mass

Y_2 is the fatigue damage

$X_{1q}, X_{2q}, X_{3q}, \ldots, X_{Nq}$ are N design factors of interest

The goal is to minimize both Y_1 (cost) and Y_2 (fatigue damage) with compromised solutions against the design factors of interest. In light of computer-aided design, the FEM is recommended for the stress-strain analysis and an inferential mechanism (e.g., fatigue life prediction) for estimating the material damage. Topological optimization is a generative design process.

Topological optimization involves addition or subtraction of elements in the finite element analysis. When an element is deleted from or added to the finite element mesh of a part, it leads to an incremental change in the stiffness matrix and displacement. Enlightened by the FEM based on the tangent-stiffness method in the regime of elastoplasticity, a set of simultaneous equations in equilibrium can be formulated and solved for optimizing the topological configuration in the global coordinate system. This can be written in the form of matrix as [Cook et al. 1998]:

$$[K_t]_{i-1} \{\Delta D\}_i = \{\Delta F\}_{i-1} \qquad (6.13.3)$$

$$\text{and} \qquad [K_t]_{i-1} = \sum_{n=1} [k_t]_{n,\ i-1} \qquad (6.13.4)$$

where

$[K_t]_{i-1}$ is the global tangent-stiffness matrix formed using the information available at step $i-1$

$[k_t]_{n,\ i-1}$ is the tangent-stiffness matrix of element n at step $i-1$

Σ is the assembly of elemental matrices, not summation

$\{\Delta F\}_{i-1}$ is the global vector of force increments applied at step $i-1$

$\{\Delta D\}_i$ is the global vector of displacement increments at step i

Subscript i is the current load increment and subscript $i-1$ is the previous load increment. The first load increment (i = 1) may be contrived to place only the highly stressed sampling point on the verge of yielding

[Cook et al. 1998]. Stiffness matrix $[k_t]$ may consist of orthotropic materials for composite structures [Chiang 1996]. Right after Equation (6.13.3) is solved, the global nodal displacements $\{\Delta D\}_i$ will be allocated to elemental nodal displacements $\{\Delta d\}_i$. The strain and stress components at the elemental level are then calculated using the following equations, respectively:

$$\{\Delta\varepsilon\}_i = [B]\,\{\Delta d\}_i \tag{6.13.5}$$

$$\text{and} \quad \{\Delta\sigma\}_i = [E]_i\,(\{\Delta\varepsilon\}_i - \{\Delta\varepsilon_p\}_i) \tag{6.13.6}$$

where
   $\{\Delta d\}_i$ is the vector of nodal-displacement increments of the element of interest at step i
   $[B]$ is the shape function matrix of the element of interest
   $\{\Delta\varepsilon\}_i$ is the vector of elastoplastic strain increments of the element of interest at step i
   $\{\Delta\varepsilon_p\}_i$ is the vector of plastic strain increments of the element of interest at step i
   $[E]_i$ is the stiffness matrix of the element of interest at step i

Both stress and strain increments of each element are evaluated individually at the elemental level. Shape function $[B]$ is an interpolation function that relates the strains at any point of the element to the strains at the nodal point of that element.

The global displacement vector and stress vector at the elemental level will be, respectively, updated as:

$$\{D\}_i = \{D\}_{i-1} + \{\Delta D\}_i \tag{6.13.7}$$

$$\text{and} \quad \{\sigma\}_i = \{\sigma\}_{i-1} + \{\Delta\sigma\}_i \tag{6.13.8}$$

Material degradation behaviors that result in potential material damage are also considered and updated in Equation (6.13.6). For example, the fatigue life of an automotive component because of cyclic loadings is expected to last at least 10 years or 240,000 km. After the potential material damage is figured out according to the life expectancy, the process continues with a new increment (i $\rightarrow$ i + 1) until the total applied load, i.e., $\Sigma\{\Delta F\}_i$, reaches for the following criterion:

$$\left| \sum_{i=1} \{\Delta F\}_i - \{F\} \right| < \text{Numerical tolerance} \tag{6.13.9}$$

The above nonlinear solution procedure has been implemented in most commercial finite element codes such as Abaqus, Ansys, and Nastran. In general, the following three different nonlinearities are taken into consideration in the numerical analysis:

(a) Material Nonlinearity: Nonlinear constitutive laws, e.g., nonlinear stress-strain curves.
(b) Geometric Nonlinearity: Large deformations and strains, e.g., large rotations.
(c) Boundary Nonlinearity: Variation of boundary conditions, e.g., contact/impact problems.

Topological optimization based on iso-surfaces of strain energy in 3-dimensional modes estimated by the finite element procedure given above and further refined by extended finite element method (XFEM) can be used as an evolutionary optimization method that enables the generation of optimized shape designs [Abdi et al. 2018]. Nevertheless, the final saying of the long-term reliability of a structural part depends on the fatigue life that cannot be expressed in a single term such as strain energy, strain, or stress. Topological optimization of structural parts often resorts to iso-surfaces of fatigue life, XFEM, and DOE [Chiang 1996].

6.13.2. Shape Design by Pareto Optimality Based on DOE

The shape optimization of an arm for a high-speed robot is implemented with Pareto optimality based on DOE. Experimental tests are conducted using commercial finite element codes (Ansys) and the overall finite element mesh in shown in Figure 6.13.2.

FIGURE 6.13.2   Finite element mesh for a robot in an assembly plant [Hsiao et al. 2020].

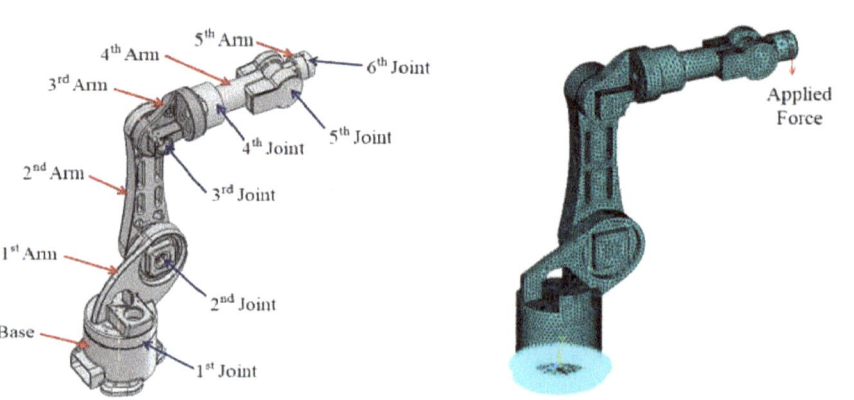

As exhibited in Figure 6.13.2, arm 2 is of great concern as the load is applied and its topology of baseline part is shown in Figure 6.13.2(a). Besides the arm thickness, four more design variables and their design ranges suggested for reducing the deflection at the tip [Figure 6.13.2(b)], where the force is applied for the finite element analysis (FEA), are listed on Table 6.13.1 as exhibited in Figure 6.13.3. Detailed dimensions and tolerances are given in Figure 2 in the paper presented by [Hsiao et al. 2020]. The goal is three folds: reducing the weight (Y_1), moment of inertia (Y_2), and deflection (Y_3), simultaneously. The FEA results based on Box-Behnken design matrix are given in Table 6.13.2.

FIGURE 6.13.3   Design parameters of the second arm (front view) selected for shape optimization [Hsiao et al. 2020].

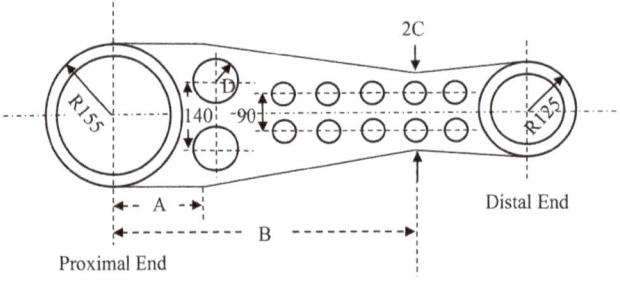

TABLE 6.13.1 Factors and their design levels of arm 2 on tip deflection [Hsiao et al. 2020].

Factor	Level (−1)	Level (0)	Level (+1)
A (mm): Arm length at maximum height	150	175	200
B (mm): Arm length at minimum height	450	500	550
C (mm): Half of minimum height	100	110	120
D (mm): Radius of weight-reduction hole	50	55	60
E (mm): Atm thickness	90	95	100

TABLE 6.13.2 Shape optimization based on FEM and DOE with Pareto optimality [Hsiao et al. 2020].

Run	A	B	C	D	E	Y_1	Y_2	Y_3
1	150	450	110	55	95	37.86	7.495	0.03921
2	150	550	110	55	95	38.39	7.623	0.03809
3	200	450	110	55	95	38.42	7.479	0.03700
4	200	550	110	55	95	39.24	7.696	0.03555
5	175	500	100	50	95	37.02	7.114	0.04079
6	175	500	100	60	95	36.09	7.066	0.04287
7	175	500	120	50	95	41.16	8.164	0.03313
8	175	500	120	60	95	40.22	8.115	0.03449
9	175	450	110	55	90	35.34	6.863	0.04671
10	175	450	110	55	100	41.68	8.274	0.0338
11	175	550	110	55	90	35.98	7.024	0.04520
12	175	550	110	55	100	42.38	8.452	0.03272
13	150	500	100	55	95	36.10	7.068	0.04390
14	150	500	120	55	95	40.44	8.127	0.03446
15	200	500	100	55	95	37.03	7.113	0.04001
16	200	500	120	55	95	40.97	8.154	0.03308
17	175	500	110	50	90	36.26	7.005	0.04481
18	175	500	110	50	100	42.64	8.429	0.03238
19	175	500	110	60	90	35.32	6.957	0.04663
20	175	500	110	60	100	41.71	8.380	0.03383
21	175	450	100	55	95	36.10	6.970	0.04304
22	175	450	120	55	95	40.21	8.012	0.03441
23	175	550	100	55	95	36.61	7.100	0.04158
24	175	550	120	55	95	41.02	8.219	0.03330
25	150	500	110	50	95	38.72	7.619	0.03744
26	150	500	110	60	95	37.79	7.571	0.03960
27	200	500	110	50	95	39.45	7.656	0.03535
28	200	500	110	60	95	38.52	7.607	0.03673
29	175	500	100	55	90	33.83	6.480	0.05110
30	175	500	100	55	100	40.01	7.854	0.03724
31	175	500	120	55	90	37.79	7.486	0.04164
32	175	500	120	55	100	44.37	8.959	0.03004
33	150	500	110	55	90	35.46	6.964	0.04696
34	150	500	110	55	100	41.80	8.386	0.03422
35	200	500	110	55	90	36.15	6.999	0.04450
36	200	500	110	55	100	42.57	8.425	0.03202
37	175	450	110	50	95	38.60	7.513	0.03721
38	175	450	110	60	95	37.67	7.464	0.03895
39	175	550	110	50	95	39.27	7.681	0.03599
40	175	550	110	60	95	38.33	7.633	0.03761
41	175	500	110	55	95	38.65	7.615	0.03700

where
Y_1 (kg): Weight
Y_2 (kg·m^2): Moment of inertia
Y_3 (μm): Tip deflection, i.e., deflection at the loading point (Figure 6.13.2)

The data analysis for objective functions weight Y_1, moment of inertia Y_2 and deflection Y_3 is done with Minitab (Stat → Regression → Regression → Fit Regression Model). Because data for Y_1, Y_2, and Y_3 in Table 6.13.2 are obtained numerically on the computer, there is no replicated errors. The "most insignificant" factor or interaction that comes with the smallest MS_{adj} is weeded out one by one until the model has all the correlations (i.e., R, R_{adj}, and R_{pred}) above 95% and the related lack of fit is above 80% based on F distribution. In light of the ANOVA given in Table 6.13.3, the final predictive equations are, respectively:

$$Y_1 = -17.16 + 0.01448\ A + 0.2087\ C + 0.003363\ E^2$$

$$(R = 98.63\%,\ R_{adj} = 98.53\%\ \text{and}\ R_{pred} = 98.38\%)$$

(6.13.10)

$$Y_2 = 26.07 - 0.592\ E + 0.003544\ E^2 + 0.000557\ CE$$

$$(R = 99.42\%,\ R_{adj} = 99.37\%\ \text{and}\ R_{pred} = 99.31\%)$$

(6.13.11)

and $$Y_3 = 927.0 - 0.04910\ A - 0.4124\ C - 16.32\ E + 0.0792\ E^2$$

$$(R = 98.75\%,\ R_{adj} = 98.61\%\ \text{and}\ R_{pred} = 98.42\%)$$

(6.13.12)

The effectiveness and adequacy of each of the three equations given above can be validated by its respective high model correlation, adjusted model correlation, and the strong insignificance of the lack of fit. These three equations are somehow different from those published by [Hsiao et al. 2020] where statistically insignificant terms are not weeded out.

TABLE 6.13.3 ANOVA on shape optimization based on FEM and DOE with Pareto optimality.

Y_1:	Source of variation	SS_{Adj}	DOF	MS_{Adj}	F-ratio	P-value
	Regression	235.117	3	78.372	439.8	0.000
	A	2.095	1	2.095	11.76	0.2%
	C	69.681	1	69.681	391.0	0.0%
	E^2	163.341	1	163.341	916.6	0.0%
	Error	6.594	37	0.178		
	Lack of fit	1.128	15	0.075	0.30	99.0%
	Pure error	5.466	22	0.248		
	Subtotal	241.711	40			
	Grand average	—	1			
	Total	—	41			
Y_2:	Source of variation	SS_{Adj}	DOF	MS_{Adj}	F-ratio	P-value
	Regression	12.6593	3	4.21977	1047	0.0%
	E	0.0592	1	0.05925	14.70	0.0%
	E^2	0.0766	1	0.07659	19.00	0.0%
	CE	4.4873	1	4.48728	1113	0.0%
	Error	0.1491	37	0.00403		
	Lack of fit	0.0018	5	0.00036	0.08	99.5%
	Pure error	0.1473	32	0.00460		
	Subtotal	12.8084	40			

(Continued)

TABLE 6.13.3 (Continued) ANOVA on shape optimization based on finite element methods and DOE with Pareto optimality.

	Grand average	—	1			
	Total	—	41			
Y_3:	**Source of variation**	**SS_{Adj}**	**DOF**	**MS_{Adj}**	**F-ratio**	**P-value**
	Regression	975.83	4	243.958	352.6	0.0%
	A	24.11	1	24.108	34.84	0.0%
	C	272.09	1	272.085	393.2	0.0%
	E	44.97	1	44.974	65.00	0.0%
	E^2	38.28	1	38.283	55.33	0.0%
	Error	24.91	36	0.692		
	Lack of fit	5.74	14	0.410	0.47	92.6%
	Pure error	19.17	12	0.871		
	Subtotal	1000.74	40			
	Grand average	—	1			
	Total	—	41			

The influences of individual main effects of factors A, C, and E on the tip deflection (Y_3) are statistically significant. Furthermore, the impact of factor E (arm thickness) on the tip deflection is nonlinear. The next step is to find out the potential design points with an allowable amount of tip deflection at the minimum weight and moment of inertia.

A contour plot of tip deflection (Y_3) versus weight (Y_1) and moment of inertia (Y_2), as exhibited in Figure 6.13.4, will reveal the feasible criterion space with a given acceptable tip deflection. For example, the maximum acceptable level of the desired deflection is 35 μm. The balance of weight and moment of inertia will provide the optimal solution. The lower left edge of "contour 35 μm," denoted by black solid dots shown in the figure, forms the Pareto frontier, of which the knee point (Figure 6.13.4) is a good choice. Note that the design point corresponding to each of the black solid dots at a given tip deflection can be found in the Excel file that was used for making the plot according to Equations (6.13.10)–(6.13.12).

FIGURE 6.13.4 Contour plot of deflection Y_3 (μm) versus weight Y_1 (kg) and moment of inertia Y_2 (kg·m²).

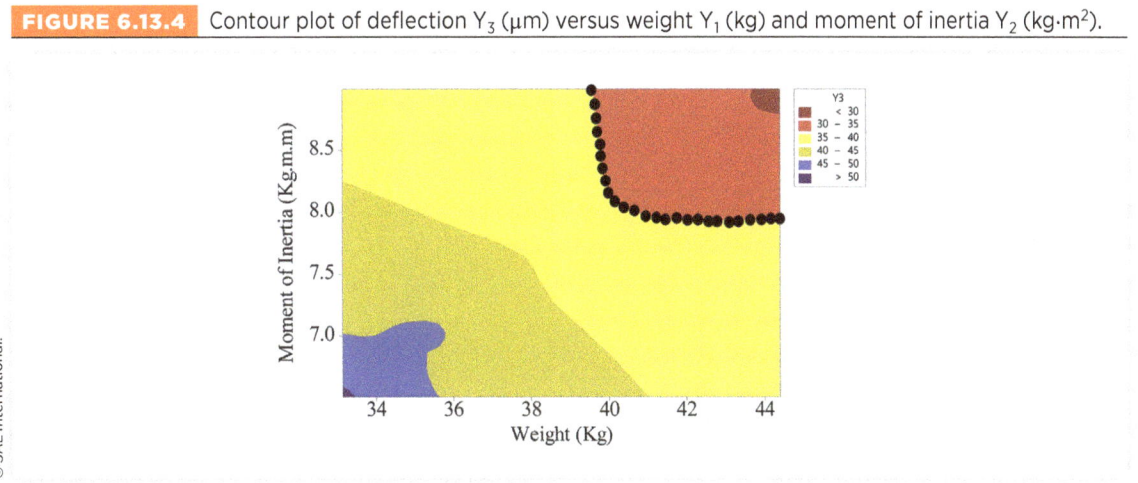

6.13.3. Robust Design of Iosipescu Shear Test Specimen for Composites

For improving the accuracy of the Iosipescu shear test method, researchers have been interested in analyzing the uniformity of shear stress distribution along the notched cross section for different design configurations of specimens made of various materials [Iosipescu 1967]. The statistical DOE based on finite element models is employed here to characterize impacts of main effects and interactions among various design factors on the uniformity of shear stress distribution along the notched cross section for unidirectional laminae using Epoxy/Gl laminae (epoxy reinforced with glass fibers). The material properties are given in Table 6.13.4. The orthotropy in modulus elasticity is moderate, with $E_{11}/E_{22} = 4.67$.

TABLE 6.13.4 Material properties of epoxy/Gf and Al/boron laminae [Chiang 1996].

Material	Young's modulus (GPa)	Shear modulus (GPa)	Poisson's ratio
Epoxy/Gf	$E_{11} = 38.6$	$G_{12} = 4.14$	$v_{12} = 0.26$
	$E_{22} = 8.27$	$G_{13} = 4.14$	$v_{13} = 0.26$
	$E_{33} = 8.27$	$G_{23} = 4.14$	$v_{23} = 0.26$
Al/Boron	$E_{11} = 227.1$	$G_{12} = 57.57$	$v_{12} = 0.237$
	$E_{22} = 139.3$	$G_{13} = 57.57$	$v_{13} = 0.237$
	$E_{33} = 139.3$	$G_{23} = 57.57$	$v_{23} = 0.237$

Relevant dimensions of the Iosipescu shear test specimen are shown in Figure 6.13.5, to which asymmetric boundary conditions (loadings and clamped areas) are applied. Note that only boundary conditions on the right side are shown in the figure. Statistical DOE is employed here to identify the main effects and multiple-factor interactions of the dimensions identified in Figure 6.13.5. Three design levels are identified for each dimension as given in Table 6.13.5: "−" means low level, "+" means high level, and "o" stands for the middle level between them.

FIGURE 6.13.5 Relevant dimensions of the Iosipescu shear test specimen with asymmetric loading [Chiang 1996].

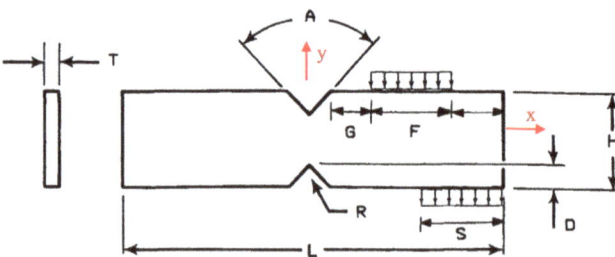

Reprinted with permission from Chiang, Y.J., "Robust Design of the Iosipescu Shear Tests Specimen for Composites," Journal of Testing and Evaluation, ASTM 24, no. 1 (1996): 1 - 11. © 1996 ASTM.

TABLE 6.13.5 Factors and their design levels for the DOE.

Factor	(−1)	(0)	(+1)	Coded variable
A: Notch angle (degrees)	90	105	120	a = (A − 105)/15
R: Notch radius (mm)	1.27	1.588	1.905	r = (B − 1.588)/0.3175
D: Notch depth (mm)	3.75	4.693	5.635	d = (D − 4.693)/0.9425
H: Specimen height (mm)	19.05	21.43	23.81	h = (H − 21.43)/2.38
T: Specimen thickness (mm)	1.041	1.3	1.56	t = (T − 1.3)/0.26
G: Marginal gap (mm)	3.75	5.625	7.50	g = (G − 5.625)/1.875
F: First contact length (mm)	3.75	5.625	7.50	f = (F − 5.625)/1.875
S: Second contact length (mm)	3.75	5.625	7.50	s = (S − 5.625)/1.875
L: Total length of specimen (mm)	−	76.2	−	−

Accuracy of the finite element calculations was validated by comparing simulations with the physical test results from [Ho et al. 1993], as shown in Figure 6.13.6 with the following dimensions: notch angle A = 110°, notch radius R = 1.27 mm, notch depth D = 3.81 mm, specimen width H = 19.05 mm, specimen thickness T = 1.0414 mm, marginal gap length G = 5.73 mm, first contact length F = 3.75 mm, and second contact length S = 3.75 mm. It can be shown that the shear stress is not uniformly distributed along the notched cross section. Let the average applied shear stress be:

$$\tau = \frac{1}{B} \int \tau_{12}\, dB \qquad (6.13.13)$$

where B is the notched cross-sectional area, i.e., (H − 2D) T, and τ_{12} is the shear stress along the cross section (along the y-axis). The integration of the shear stress in a differential area goes from one notch root to the other. The variation of shear stress uniformity is here defined as:

$$Z = \left[\frac{1}{B} \int \left(\frac{\tau_{12}}{\tau} - 1\right)^2 dB\right]^{1/2} \qquad (6.13.14)$$

where

τ_{12}/τ is the normalized stress

Z is the nonuniformity of shear stress along the shearing cross section; the smaller, the better

Reprinted with permission from Chiang, Y.J., "Robust Design of the Iosipescu Shear Tests Specimen for Composites," Journal of Testing and Evaluation, ASTM 24, no. 1 (1996): 1 - 11. © 1996 ASTM.

FIGURE 6.13.6 Validation of finite element accuracy with physical test data [Chiang 1996].

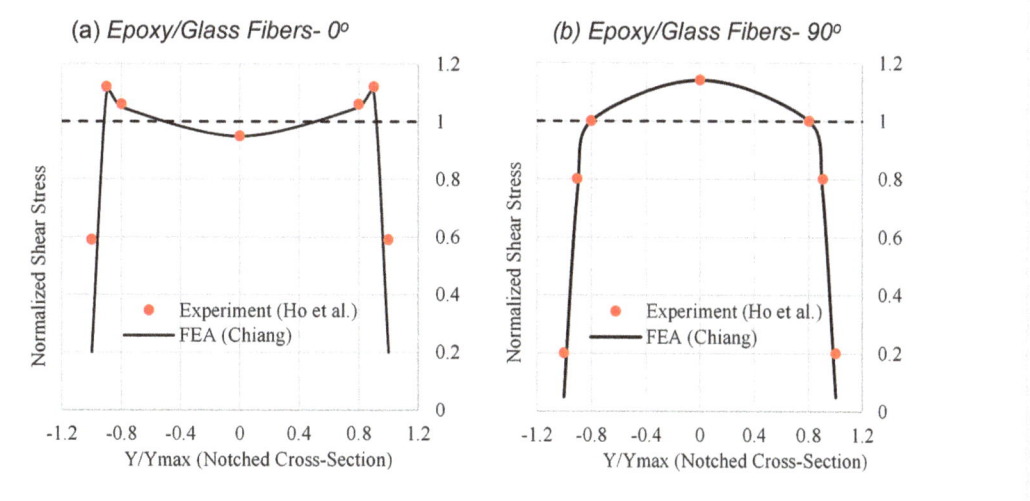

(a) *Epoxy/Glass Fibers- 0°* (b) *Epoxy/Glass Fibers- 90°*

TABLE 6.13.6 Design matrix and variation of shear stress uniformity along the notched cross section [Chiang 1996].

(i) 2^{8-4}_{IV} (T = RDH, G = ADH, F = ARD, and S = ARH) as screening design:											
Run	a	r	d	h	t	g	f	s	E/G_0	E/G_{90}	A/B_0	A/B_{90}
1	–	–	–	–	–	–	–	–	0.079	0.206	0.119	0.164
2	+	–	–	–	–	+	+	+	0.086	0.241	0.154	0.201
3	–	+	–	–	+	–	+	+	0.087	0.234	0.149	0.195
4	+	+	–	–	+	+	–	–	0.107	0.259	0.177	0.223
5	–	–	+	–	+	+	+	–	0.093	0.224	0.134	0.181
6	+	–	+	–	+	–	–	+	0.096	0.254	0.164	0.216
7	–	+	+	–	–	+	–	+	0.103	0.254	0.166	0.216
8	+	+	+	–	–	–	+	–	0.117	0.275	0.192	0.242
9	–	–	–	+	+	+	–	+	0.059	0.201	0.128	0.163
10	+	–	–	+	+	–	+	–	0.070	0.233	0.154	0.198
11	–	+	–	+	–	+	+	–	0.078	0.225	0.153	0.189
12	+	+	–	+	–	–	–	+	0.088	0.248	0.172	0.215
13	–	–	+	+	–	–	+	+	0.102	0.194	0.107	0.148
14	+	–	+	+	–	+	–	–	0.075	0.234	0.141	0.192
15	–	+	+	+	+	–	–	–	0.095	0.222	0.133	0.180
16	+	+	+	+	+	+	+	+	0.094	0.252	0.164	0.214
Average									0.089	0.235	0.150	0.196

(ii) Additional runs for epoxy/Gf-0°: 2^{5-1}_V (G = ARDH):								
Run	a	r	d	h	t	g	f	s	E/G_0
17	–	–	–	–	0	+	0	0	0.0695
18	+	–	–	–	0	–	0	0	0.0818
19	–	+	–	–	0	–	0	0	0.0873
20	+	+	–	–	0	+	0	0	0.1076
21	–	–	+	–	0	–	0	0	0.0963
22	+	–	+	–	0	+	0	0	0.0947
23	–	+	+	–	0	+	0	0	0.1036
24	+	+	+	–	0	–	0	0	0.1173
25	–	–	–	+	0	–	0	0	0.0807
26	+	–	–	+	0	+	0	0	0.0703
27	–	+	–	+	0	+	0	0	0.0783
28	+	+	–	+	0	–	0	0	0.0890
29	–	–	+	+	0	+	0	0	0.0881
30	+	–	+	+	0	–	0	0	0.0759
31	–	+	+	+	0	–	0	0	0.0943
32	+	+	+	+	0	+	0	0	0.0936

where
E/G_0: Nonuniformity of Epoxy/Gf-0°, of which fibers are aligned with x-axis
E/G_{90}: Nonuniformity of Epoxy/Gf-90°, of which fibers are perpendicular to x-axis
A/B_0: Nonuniformity of Al/Boron-0°, of which fibers are aligned with x-axis
A/B_{90}: Nonuniformity of Al/Boron-90°, of which fibers are perpendicular to x-axis

The data analysis is done with Minitab (Stat ➔ Regression ➔ Regression ➔ Fit Regression Model). The "most insignificant" factor or interaction that comes with the largest p-value is weeded out one by one until all p-values are less than 10% that is here defined as the parting significance level based on F distribution. Based on the analysis of variances, predictive equations for different operating modes are given as follows:

1. For Epoxy/Gf-90°: Based on the data obtained from factorial design 2^{8-4}_{IV} as listed in Table 6.13.6(i), the predictive equation derived from analysis of variance is:

$$\text{Epoxy/Gf-90°}_p = 0.1341 + 0.000980\ A + 0.03547\ R + 0.00407\ D$$

$$- 0.003629\ H \tag{6.13.15}$$

$$(R = 96.8\%,\ R_{adj} = 95.6\%,\ \text{and}\ R_{pred} = 93.1\%;\ \text{p-value for constant} = 0.0\%)$$

There is no complication from 2-factor interactions for Epoxy/Gf-90°, where the fibers of test laminae are perpendicular to the x-axis. The nonuniformly can be simply reduced by lowering the notch angle, downsizing the notch radius, decreasing the notch depth, and increasing the specimen height.

2. For Epoxy/Gf-0°: Based on the data obtained from factorial design 2^{8-4}_{IV} as listed in Table 6.13.6(i), the predictive equation for nonuniformity across the notched cross section is:

$$\text{Epoxy/Gf-0°}_p = -0.0588 + 0.002635\ A + 0.000216\ A\ R - 0.000194\ A\ D$$

$$- 0.000086\ A\ H - 0.000014\ A\ G + 0.001314\ D\ H \tag{6.13.16}$$

$$(R = 95.5\%,\ R_{adj} = 92.3\%,\ \text{and}\ R_{pred} = 80.5\%;\ \text{p-value for constant} = 15.4\%)$$

Since the p-value of constant for Epoxy/Gf-0° laminae is not statistically significant and confounded 2-factor interactions cannot be resolved, the above equation is not a valid model. Nevertheless, Equation (6.3.13) reveals that factors T, F, and S are statistically insignificant.

Thus, the second set of factorial design based on 2^{5-1}_V is subsequently set up for five factors to further refine the model for Epoxy/Gf-0°. Analysis of variance based on the data listed in Table 6.13.6(ii) leads to the predictive equation for nonuniformity across the notched cross section as:

$$\text{Epoxy/Gf-0°}_p = -0.2642 + 0.001527\ A + 0.04139\ D + 0.01659\ H - 0.00553\ G$$

$$+ 0.000749\ A\ R - 0.000149\ A\ D - 0.000100\ A\ H + 0.000047\ A\ G \tag{6.13.17}$$

$$- 0.002628\ R\ H - 0.000894\ D\ H$$

$$(R = 99.8\%,\ R_{adj} = 99.3\%,\ \text{and}\ R_{pred} = 97.5\%;\ \text{p-value for constant} = 0.1\%)$$

The nonuniformly can be reduced by lowering the notch angle (factor A), decreasing the notch depth (factor D), decreasing the specimen height (factor H), and increasing the marginal gap (factor G). Six 2-factor interactions are also statistically significant. Note that the 3-factor interactions are assumed to be negligible in this fractional factorial design, $2^{5-1}V$.

Note that the ratio of notch depth to specimen height, i.e., D/H, must be maintained above a certain level for a specific material to assure that shearing failure happens at the notched cross section rather than crush or buckling at the clamped-contact areas.

It can be seen that the shearing behavior of composite laminae varies according to the orthotropy ratio of the Iosipescu shear test specimen, which is defined as E_{xx}/E_{yy}, of which coordinates x and y are exhibited in Figure 6.13.6. Note that E_{11}/E_{22} is the material anisotropy of unidirectional laminae of interest. In addition to Epoxy/Gf-0° and Epoxy/Gf-90° laminae, the impact of orthotropy ratio E_{xx}/E_{yy} on the nonuniformity of shearing behavior in the notched cross section is further examined using Al/Boron-0° and Al/Boron-90°, of which the test data are listed in Table 6.13.6(i). In summary, the averages of nonuniformities of these four test cases based on factorial design 2^{8-4}_{IV} are given as follows:

(1) Epoxy/Gf-90° ($E_{xx}/E_{yy} = E_{22}/E_{11} = 0.214$): *23.5%*
(2) *Al/Boron-90°* ($E_{xx}/E_{yy} = E_{22}/E_{11} = 0.613$): *19.6%*
(3) *Al/Boron-0°* ($E_{xx}/E_{yy} = E_{11}/E_{22} = 1.63$): *15.0%*
(4) Epoxy/Gf-0° ($E_{xx}/E_{yy} = E_{11}/E_{22} = 4.67$): 8.9%

The plot of averaged nonuniformities corresponding to different orthotropy ratios of test specimens is plotted in Figure 6.13.7. It can be seen that the nonuniformity of shearing along the notched cross section decreases with an increasing orthotropy ratio E_{xx}/E_{yy} of the test specimen, and multifactorial interactions become statistically significant as comparing Equation (6.13.14) with Equation (6.13.12). In conclusion, the dimensions of each test specimen made of a distinct orthotropic lamina can be adjusted according to its material anisotropy to reduce the nonuniformity for achieving a robust Iosipescu shear test.

FIGURE 6.13.7   Nonuniformity of shearing along the notched cross section in Iosipescu shear tests as a function of orthotropy ratio E_{xx}/E_{yy}.

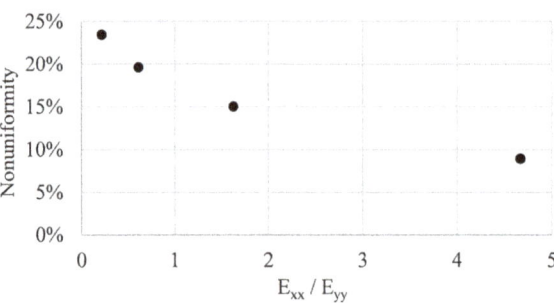

© SAE International.

6.14. Dynamic Optimization

Dynamic optimization involves topological parameters in the time domain. Among the approaches to solving the dynamic multiobjective optimization problems, DOE in combination with Pareto frontier is a promising versatile solution. Two major applications of dynamic optimization to automobiles are:

(a) Crashworthiness: The objective is to provide an automobile that can absorb the crash energy by controlled structural deformations and preserving enough space of the passenger compartment.

(b) Ride comfort and road holding capability: The capability to have a smooth ride and the road holding of the vehicle when subjected to road imperfections.

Explicit FEM with material constitutive equations with dynamic strain (stress) flow are frequently applied to such structural dynamics of automobiles. DOE for dynamic optimization is demonstrated using examples.

6.14.1. Explicit FEM and Material Behaviors with Dynamic Strain Flow

Thin-walled structures are widely used in automobile body and chassis, of which some components are designed to absorb more energy should a collision happen. Explicit FEM based on solid and shell elements in dynamic mode are applied for such analyses. Explicit FEM are utilized to calculate the state of a given system at a future time from the current time through iterations with a large time interval as compared with implicit FEM (Section 6.13.1). The explicit method should be used when the strain rate is significantly high, such as more than 10 units/second. Explicit analyses aim to solve for acceleration while implicit analyses are based on displacements. After accelerations are calculated at the nth step, velocities at $(n + 1/2)^{th}$ step and displacements at $(n + 1)^{th}$ step are calculated accordingly.

6.14.2. Dynamic Strain Flow

The constitutive behavior of thin shell elements or solid elements can be based on an elastoplastic material model with von Mises isotropic plasticity algorithm with piecewise linear plastic hardening. The strain-rate effect is the key to render a successful finite element model. One simple promising model is the Cowper-Symonds strain-hardening equation [Cowper and Symonds 1958],

$$\sigma = [A + B\,(\varepsilon_{eq}^{p})^{n}]\left[1 + \left(\frac{d\varepsilon_{eq}^{p}/dt}{D}\right)^{1/q}\right] \qquad (6.14.1)$$

where
  σ (MPa) is the flow stress, i.e., stress in motion
  ε is the strain
  ε_{eq}^{p} is the equivalent plastic strain, e.g., von Mises equivalent plastic strain
  A is the yield strength at a strain rate of 1 S^{-1} and room temperature (T_{room})
  B is the strain-hardening coefficient at a strain rate of 1 S^{-1} and room temperature (T_{room})
  D^{-1} is the strain-rate hardening coefficient at T_{room}
  t (s) is the time
  c and q are the coefficient and exponent for strain rate, respectively

which is often used to emphasize the material behavior at different strain rates ($d\varepsilon/dt$). The Cowper-Symonds strain-hardening equation has been successfully applied to crashworthiness simulations [Xie et al. 2008]. Coefficient D and exponent q assume different values at different reference stress levels, because each stress has its own sensitivity to strain rate and working temperature. For example, D = 40 S^{-1} and p = 5 for mild steel in general applications at the room temperature.

A more complete model is the Johnson-Cook constitutive equation for the nominal flow stress (σ) that is portrayed in terms of five material parameters, namely A, B, C, n, and m, as [Johnson and Cook 1983]:

$$\sigma = [A + B\,(\varepsilon_{eq}^{p})^{n}]\left\{1 + C\ln\left[\frac{d\varepsilon_{eq}^{p}/dt}{(d\varepsilon_{eq}^{p}/dt)_{0}}\right]\right\}\left[1 - \left(\frac{T - T_{room}}{T_{m} - T_{room}}\right)^{m}\right] \qquad (6.14.2)$$

where

T$_{room}$ is the room temperature

T$_m$ is the melting point (temperature) of the material

A is the yield strength at a strain rate of 1 S^{-1} and room temperature (T$_{room}$)

B is the strain-hardening coefficient at a strain rate of 1 S^{-1} and room temperature (T$_{room}$)

n is the strain-hardening exponent at T$_{room}$

C is the strain-rate hardening coefficient at T$_{room}$

m is the strain softening exponent with respect to temperature change

ε_{eq}^{p} is the equivalent plastic strain, e.g., von Mises equivalent plastic strain

$d\varepsilon_{eq}^{p}/dt$ is the equivalent plastic strain rate, e.g., von Mises equivalent plastic strain rate

$(d\varepsilon_{eq}^{p}/dt)_0$ is the reference equivalent plastic strain rate, usually at 1.0 S^{-1}

$(d\varepsilon_{eq}^{p}/dt)/(d\varepsilon_{eq}^{p}/dt)_0$ is the normalized equivalent plastic strain rate

in which strain, strain rate, and temperature are taken into consideration simultaneously.

The strain-rate-dependent parameters of both Johnson-Cook and Cowper-Symonds material models can be estimated using DOE [Škrlec and Klemenc 2016, 2020].

6.14.3. Crashworthiness of a Vehicular Structure

This is illustrated here by multiobjective Pareto swarm optimization for automobile crashworthiness, of which the data are adopted from [Yildiz and Solanki 2011] but the multiple linear regression using Minitab is applied instead. The components that have potential influences on the crash characteristics of a sedan are shown in Figure 6.14.1, based on which the factors of interest for crashworthiness and their design levels for tests are listed in Table 6.14.1. The test results according to the design matrix given in Table 4.7.8 based on fractional factorial design 3_{III}^{13-10} (L$_{27}$) are listed in Table 6.14.2. These crush-initiating treatments are carefully implemented virtually with nonlinear FEA based on LS-DYNA, and some of them are verified through physical crash tests.

FIGURE 6.14.1   Components having potential influences on sedan crashworthiness [Yildiz and Solanki 2011].

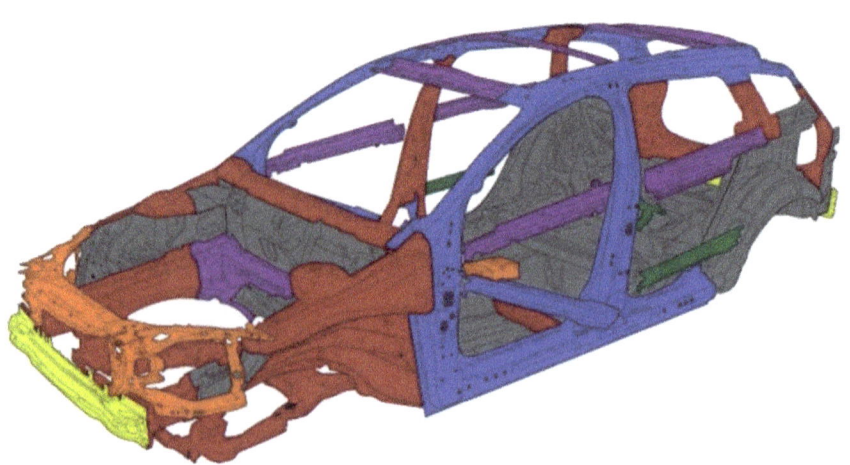

Reprinted with permission from "Multi-Objective Optimization of Vehicle Crashworthiness Using a New Particle Swarm Based Approach," Ali, R., Yildiz; Kiran, N., Solanki, 59 © 2011 Springer Nature BV.

TABLE 6.14.1 Factors and their design levels for crashworthiness tests [Yildiz and Solanki 2011].

Factor	(0)	(1)	(2)	Coded variable
A (mm): Left and right front doors	0.425	0.85	1.275	$a = (A - 0.85)/0.425 + 1$
B (mm): Left and right rear doors	0.415	0.83	1.245	$b = (B - 0.83)/0.415 + 1$
C (mm): Inner hood	0.325	0.65	0.975	$c = (C - 0.65)/0.325 + 1$
D (mm): Left and right outer B-pillars	0.805	1.61	2.415	$d = (D - 1.61)/0.805 + 1$
E (mm): Left and right middle B-pillar	0.355	0.71	1.065	$e = (E - 0.71)/0.355 + 1$
F (mm): Inner front bumper	0.98	1.96	2.94	$f = (F - 1.96)/0.98 + 1$
G (mm): Front floor panel	0.355	0.71	1.065	$g = (G - 0.71)/0.355 + 1$
H (mm): Left and right outer CBN	0.415	0.83	1.245	$h = (H - 0.83)/0.415 + 1$
J (mm): Left and right front fenders	0.76	1.52	2.28	$j = (J - 1.52)/0.76 + 1$
K (mm): Left and right inner front rails	0.95	1.90	2.85	$k = (K - 1.9)/0.95 + 1$
Q (mm): Left and right outer front rails	0.76	1.52	2.28	$q = (L - 1.52)/0.76 + 1$
M (mm): Rear plate	0.355	0.71	1.065	$m = (G - 0.71)/0.355 + 1$
N (mm): Suspension frame	1.305	2.61	3.915	$n = (N - 2.61)/1.305 + 1$

TABLE 6.14.2 Fractional factorial design matrix of 3_{III}^{13-10} (L_{27}) for crashworthiness [Yildiz and Solanki 2011].

Run	a	b	c	d	e	f	g	h	j	k	q	m	n	Y_1	Y_2	Y_3
1	0	0	0	0	0	0	0	0	0	0	0	0	0	57	385	48.9
2	0	0	0	0	1	1	1	1	1	1	1	1	1	64	370	82.6
3	0	0	0	0	2	2	2	2	2	2	2	2	2	59	355	117
4	0	1	1	1	0	0	0	1	1	1	2	2	2	69	383	86.4
5	0	1	1	1	1	1	1	2	2	2	0	0	0	62	364	105.1
6	0	1	1	1	2	2	2	0	0	0	1	1	1	65	378	86.1
7	0	2	2	2	0	0	0	2	2	2	1	1	1	65	366	108.9
8	0	2	2	2	1	1	1	0	0	0	2	2	2	67	384	89.8
9	0	2	2	2	2	2	2	1	1	1	0	0	0	57	364	108.6
10	1	0	1	2	0	1	2	0	1	2	0	1	2	64	376	94.4
11	1	0	1	2	1	2	0	1	2	0	1	2	0	69	362	90.7
12	1	0	1	2	2	0	1	2	0	1	2	0	1	44	337	105.1
13	1	1	2	0	0	1	2	1	2	0	2	0	1	61	375	100.9
14	1	1	2	0	1	2	0	2	0	1	0	1	2	52	361	102
15	1	1	2	0	2	0	1	0	1	2	1	2	0	78	376	84.9
16	1	2	0	1	0	1	2	2	0	1	1	2	0	49	365	111.9
17	1	2	0	1	1	2	0	0	1	2	2	0	1	83	378	89.3
18	1	2	0	1	2	0	1	1	2	0	0	1	2	71	357	95.6
19	2	0	2	1	0	2	1	0	2	1	0	2	1	72	371	91.9
20	2	0	2	1	1	0	2	1	0	2	1	0	2	55	354	106.3
21	2	0	2	1	2	1	0	2	1	0	2	1	0	58	340	102.5
22	2	1	0	2	0	2	1	1	0	2	2	1	0	56	367	106.4
23	2	1	0	2	1	0	2	2	1	0	0	2	1	55	345	112.7
24	2	1	0	2	2	1	0	0	2	1	1	0	2	79	365	90
25	2	2	1	0	0	2	1	2	1	0	1	0	2	56	361	113.4
26	2	2	1	0	1	0	2	0	2	1	2	1	0	80	373	96.3
27	2	2	1	0	2	1	0	1	0	2	0	2	1	59	355	97.6
28	1	1	1	1	1	1	1	1	1	1	1	1	1	61	370	96.8

Notes:
Y_1 (mm): Indentation subjected to frontal impact; the smaller the better.
Y_2 (mm): Indentation subjected to side impact; the smaller the better.
Y_3 (kg): Mass of the parts listed in Table 6.14.1; the smaller the better.

The data analysis is done with Minitab (Stat → Regression → Regression → Fit Regression Model). The "most insignificant" factor or interaction that comes with the largest p-value is weeded out one by one until all p-values are less than 10% that is here defined as the parting significance level based on F distribution. The analysis of variances using the data of the first 27 runs in Table 6.14.2, yields the following predictive equations for the three different outputs as:

$$Y_1 = 63.41 + 2.500 \, b - 2.556 \, g - 8.056 \, h + 6.333 \, j + 1.556 \, q \tag{6.14.3}$$

(R = 94.0%, R_{adj} = 92.5%, and R_{pred} = 90.0%; p-value for constant = 0.0%)

$$Y_2 = 389.0 - 6.556 \, a + 2.944 \, b - 2.500 \, d - 6.778 \, e - 10.667 \, h \tag{6.14.4}$$

(R = 97.4%, R_{adj} = 96.8%, and R_{pred} = 95.7%; p-value for constant = 0.0%)

$$\text{and} \quad Y_3 = 48.733 + 4.65 \, a + 4.00 \, b + 2.30 \, c + 3.50 \, d + 1.35 \, e + 3.35 \, f + 6.55 \, g$$
$$+ 11.50 \, h + 2.35 \, j + 3.85 \, k + 2.05 \, q + 0.85 \, m + 2.20 \, n \tag{6.14.5}$$

(R = 100.00%, R_{adj} = 99.99%, and R_{pred} = 99.98%; p-value for constant = 0.0%)

The 28th run at the center point (1, 1, …, 1) is used for checking random error.

There are only eight factors that are statistically significant in either frontal collision or side impact in this study, while all 13 factors have direct impacts on the gross mass. The multipurpose optimization can be divided into different scenarios. Two example cases are demonstrated as follows:

1. Regardless of mass requirement: Consider the balance between frontal collision and side impact only. Each factor that is statistically insignificant in both frontal collision and side impact is set to have the minimum thickness, i.e., level 0, which is a reduced thickness as compared to the preliminary design level (level 1). Since Equations (6.14.3)–(6.14.5) are linear, the required particles in the multiobjective PSO are here generated using two extreme levels only, i.e., level 0 and level 2. Thus, 256 particles generated using the design matrix of factorial design 2^8 with factors A, B, D, E, G, H, J, and Q are given in Table 6.14.3. A chart for exploring Pareto frontier with front collision indentation Y_1 and side impact indentation Y_2 is plotted in Figure 6.14.2. It can be easily seen which one is the best design point (Particle 62: Y_{1p} = 42 mm and Y_{2p} = 336 mm, while the corresponding mass is Y_{3p} = 103.8 kg). The criterion space, i.e., projected mass (Y_{3p}) with all these 256 particles, is related to Y_{1p} and Y_{2p} as shown in Figure 6.14.3.

2. The mass is not to be more than 75 kg: A chart for exploring Pareto frontier (the red-dotted profile) with respect to front collision indentation Y_1 and side impact indentation Y_2 is plotted in Figure 6.14.4. The values of design factors corresponding to each particle on the Pareto frontier can be read out from the Minitab (or Excel) file that was used to generate the plot.

Material acquisition, part fabrication, and assembly costs can be incorporated into the objective function, if the cost function is of concern instead of mass.

TABLE 6.14.3 256 particles generated using factorial design 2^8 with statistically significant factors for front collision and side impact.

Run	a	b	d	e	g	h	j	q	Y_{1p}	Y_{2p}	Y_{3p}
1	0	0	0	0	0	0	0	0	63	389	48.7
2	2	0	0	0	0	0	0	0	63	376	58.0
3	0	2	0	0	0	0	0	0	68	395	56.7
4	2	2	0	0	0	0	0	0	68	382	66.0
5	0	0	2	0	0	0	0	0	63	384	55.7
6	2	0	2	0	0	0	0	0	63	371	65.0
7	0	2	2	0	0	0	0	0	68	390	63.7
8	2	2	2	0	0	0	0	0	68	377	73.0
...											
62	2	0	2	2	2	2	0	0	42	336	103.8
...											
249	0	0	0	2	2	2	2	2	58	354	96.3
250	2	0	0	2	2	2	2	2	58	341	105.6
251	0	2	0	2	2	2	2	2	63	360	104.3
252	2	2	0	2	2	2	2	2	63	347	113.6
253	0	0	2	2	2	2	2	2	58	349	103.3
254	2	0	2	2	2	2	2	2	58	336	112.6
255	0	2	2	2	2	2	2	2	63	355	111.3
256	2	2	2	2	2	2	2	2	63	342	120.6

Y_{3p} (kg): Predicted with the minimum thickness (level 0) for all insignificant factors.

FIGURE 6.14.2 Pareto frontier of indentations—front collision versus side impact.

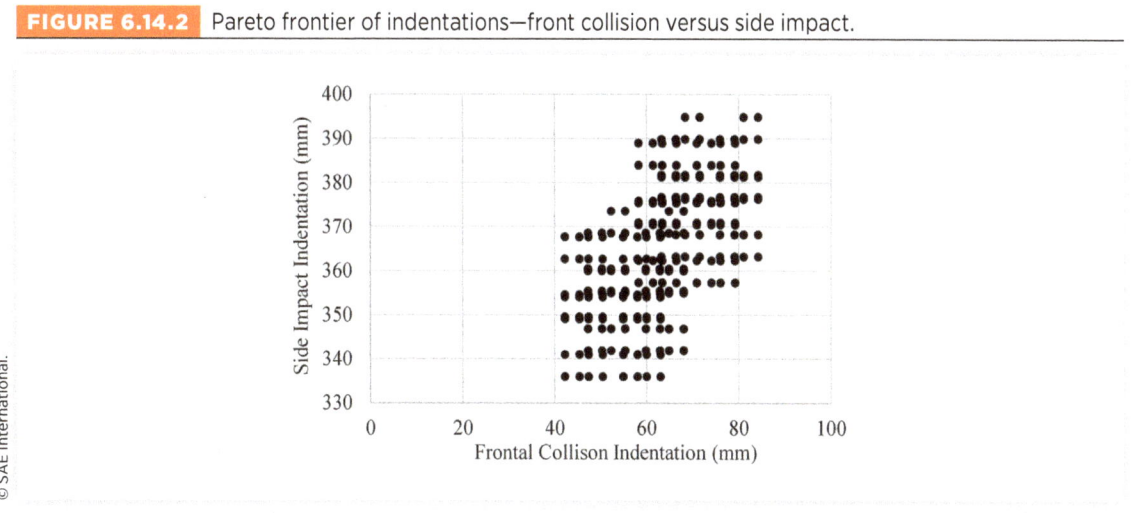

FIGURE 6.14.3 Criterion space: mass Y_{3p} versus indentations Y_{1p} and Y_{2p} with all 256 particles.

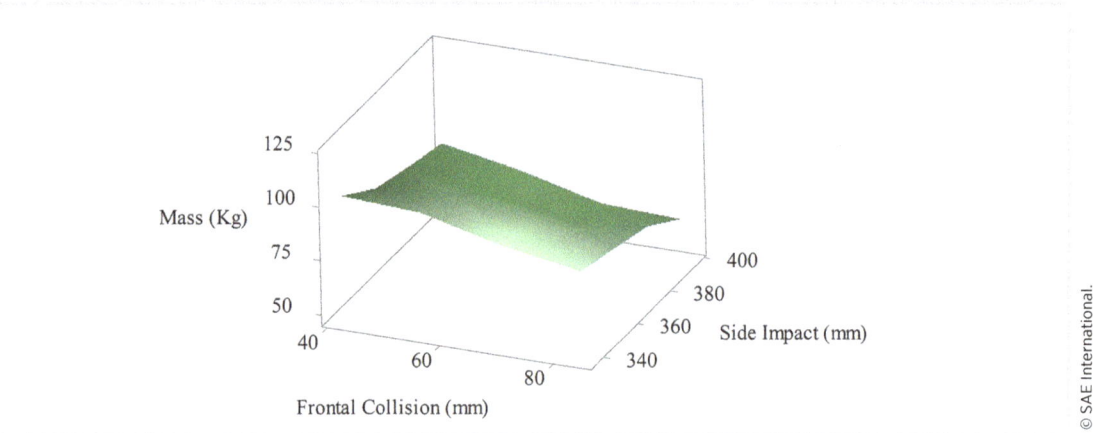

FIGURE 6.14.4 Pareto frontier: red-dotted profile given for mass $Y_{3p} \leq 75$ kg.

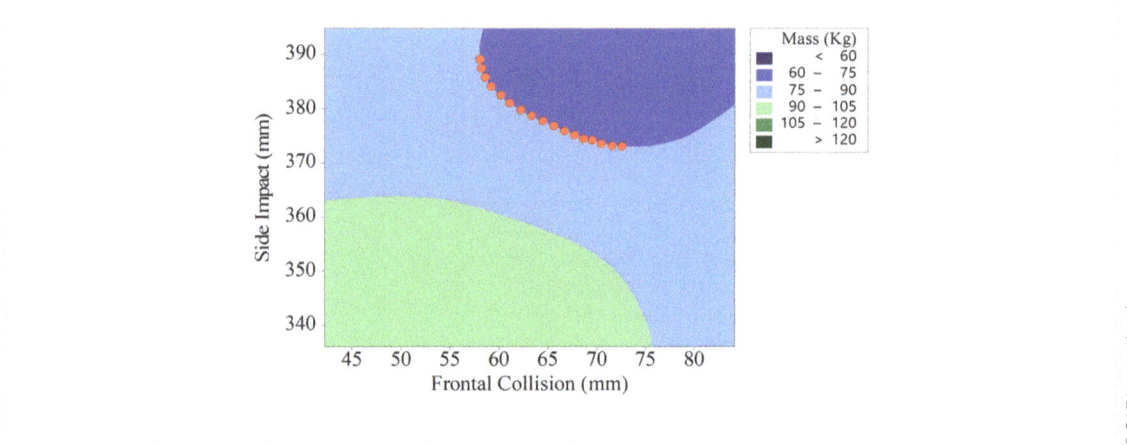

6.14.4. Design Optimization of S-Frames to Improve Crashworthiness

S-rails are S-shaped frames in front and rear of the automotive underbody with significant effects on absorbing collision energy during front and rear collisions of automobiles. Experimentally verified S-rail models are implemented through a nondominated sorting genetic algorithm with four-objective optimization, namely remaining energy, peak crushing force, and mass. Another concern is the product family penalty function (PFPF) that refers to the platform commonality of S-rails as applied within the product family [Hosseini 2019].

6.14.5. Pareto Frontier—Automotive Ride Comfort versus Road Holding Capability

The vibration and noise transmitted to passengers by tactile and visual paths are the major ride characteristics, by which a vehicle occupant is used to determine the ride quality of a vehicle [ISO 1997]. As being integrated from the roof to the ground, a simplified basic model of a vehicle ride system may consist of the following:

(a) Sprung mass: Vehicle components, including the body, engine, battery, seating system, and cross-members, being attached onto the suspensions and moving as an integral unit are regarded as the sprung mass, denoted by M_S.

(b) Suspension: The suspension is regarded as a combination of links, springs, and dampers (shock absorbers). As to the formulation and calculation for a closed-form solution to vehicle bounce, there is no rule defined for dividing the suspension. In practice, half the suspension mass, including control arms, struts, tie rods and springs, is added to the sprung mass; the other half to the unsprung mass.

(c) Unsprung mass: The axles and tire/wheel systems form the unsprung mass, which is regarded as a rigid body and transmits excitation forces from the ground to the sprung mass.

(d) Each tire is represented as a spring, because the amount of damping inherent to the viscoelastic tire is relatively small when compared with a shock absorber.

The vibration transferred to a vehicle occupant at the vehicle seat-occupant interface results in vehicle comfort, while the road holding capability is measured by the relative displacement between road and unsprung mass. A quarter car test rig is given in Figure 6.14.5.

FIGURE 6.14.5 Quarter car test rig [Mitra et al. 2016].

TABLE 6.14.4 Design factors and levels for examining ride comfort and road holding capability [Mitra et al. 2016].

Factor	Level (−1)	Level (+1)	Coded variable
A (rpm): Wheel speed	155	250	$a = (A − 202.5)/50$
B (kPa): Tire pressure	214.325	275.8	$b = (B − 258.5625)/17.2375$
C (degree): Camber angle	1	3	$c = (C − 120)/60$
D (kN/mm): Spring stiffness	18	26	$d = (A − 22)/4$
E (N·s/m): Damping coefficient	418	673	$e = (B − 545.5)/127.5$
G (mm): Toe	10	20	$f = (F − 15)/5$
G (kg): Sprung mass (added)	41	81	$c = (C − 61)/20$

TABLE 6.14.5 Factorial design 2_{IV}^{7-3} (I = ABDE, I = ABFG, I = −ACDG, and I = −ACEF) for examining vertical acceleration (ride comfort) and road holding capability.

Run	A	B	C	D	E	F	G	V (×2)		H (×2)	
1	155	241.3	3	18	418	10	41	0.48	0.45	−1.8	−1.5
2	155	241.3	3	26	673	20	81	0.67	0.65	1	1.2
3	155	241.3	1	26	673	10	41	0.8	0.72	−1.7	−1.5
4	155	241.3	1	18	418	20	81	1.24	1.3	−0.8	−1.1
5	250	241.3	3	18	673	10	81	0.54	0.62	−1	−0.7
6	250	241.3	3	26	418	20	41	1.2	1.06	−2.5	−1.8
7	250	241.3	1	26	418	10	81	1.7	1.5	−2.6	−2.1
8	250	241.3	1	18	673	20	41	1.15	1.15	0.6	0.6
9	155	275.8	3	26	418	10	81	1.12	1.17	−0.7	−0.8
10	155	275.8	1	18	673	10	81	0.55	0.57	−1.3	−1.2
11	155	275.8	1	26	418	20	41	0.9	0.82	−0.4	−0.3
12	155	275.8	3	18	673	20	41	0.54	0.44	−0.1	−0.2
13	250	275.8	1	18	418	10	41	0.97	0.91	1.4	2
14	250	275.8	1	26	673	20	81	1.6	1.6	0.2	0.6
15	250	275.8	3	26	673	10	41	0.61	0.69	1	0.7
16	250	275.8	3	18	418	20	81	1.75	1.75	−2.2	−2

where
V (m/s²): Vertical vibration, used for measuring ride comfort; two replicates
H (mm): Tire deflection against ground; +: off the ground and −: in contact; two replicates

The data analysis is done with Minitab (Stat ➜ DOE ➜ Factorial ➜ …). Based on the data of the 16 runs in Table 6.14.4, the DOE analysis yields the following predictive equations for ride comfort (vertical vibration in m/s²) and road holding capability (road contact in mm), respectively (Table 6.14.5).

$$V_p = 8.369 - 0.02148 \, A - 0.02904 \, B - 2.759 \, C + 0.0241 \, D - 0.000973 \, E - 0.05284 \, F$$

$$+ 0.00051 \, G + 0.0000779 \, A \, B + 0.0037 \, A \, C - 0.0000263 \, A \, D - 0.00000175 \, A \, E$$

$$+ 0.0003974 \, A \, F + 0.0000395 \, A \, G + 0.01072 \, B \, C - 0.0000168 \, A \, B \, C$$

$$(R = 99.5\%, \; R_{adj} = 99.0\%, \; R_{pred} = 98.0\%; \text{p-value for constant} = 0.0\%)$$

$$H_p = -7.15 - 0.0348\ A - 0.0549\ B + 1.82\ C + 0.3299\ D - 0.00479\ E + 0.4284\ F$$

$$+ 0.09141\ G + 0.000504\ A\ B + 0.0062\ A\ C - 0.001645\ A\ D + 0.000042\ A\ E$$

$$- 0.001974\ A\ F - 0.000513\ A\ G + 0.0000\ B\ C - 0.000061\ A\ B\ C$$

$$(R = 99.1\%,\ R_{adj} = 98.2\%,\ R_{pred} = 96.2\%;\ \text{p-value for constant} = 0.0\%)$$

According to the design matrix of factorial design 2^7 with factors A, B, C, D, E, F, and G, 128 particles are generated and plotted in Figure 6.14.6 that can be used for exploring Pareto frontier in the decision space with vertical acceleration (ride comfort) and road contact (road holding capability). The profile of Pareto frontier can be easily identified.

FIGURE 6.14.6   Pareto frontier—ride comfort (vertical acceleration) versus road holding capability (road contact).

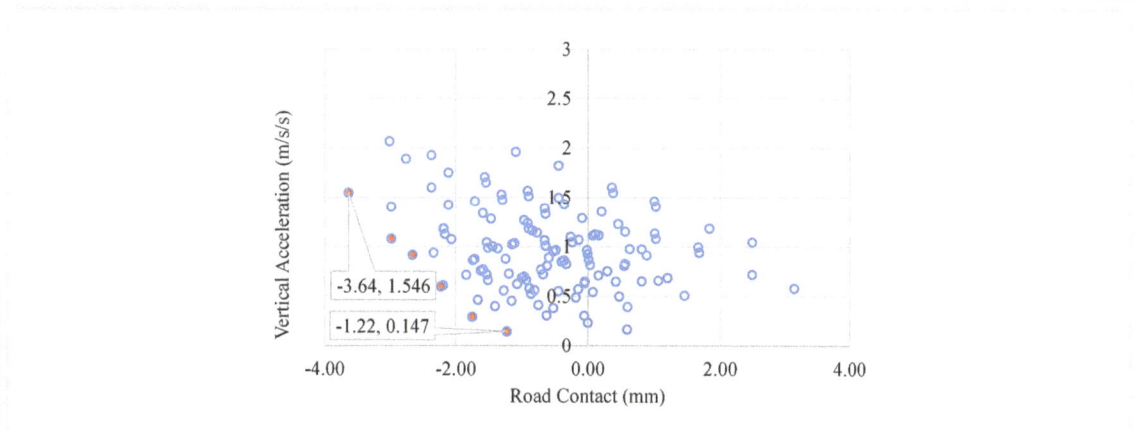

References

Abdi, M. et al., "Topology Optimization of Geometrically Nonlinear Structures Using an Evolutionary Optimization Method," *Engineering Optimization* 50, no. 11 (2018): 1850-1870.

Abdi, M. (2015), "Evolutionary Topology Optimization of Continuum Structures Using X-FEM and Isovalues of Structural Performance," PhD dissertation, University of Nottingham, UK.

Adan, I. and Resing, J. (2002), *Queueing Theory*, Department of Mathematics and Computing, Eindhoven University of Technology, Eindhoven, the Netherlands, 180 pages.

Anderson, M.J. and Whitcomb, P.J., *RSM Simplified, Optimizing Processes Using Response Surface Methods for Design of Experiments* (New York: Productivity Press, 2014a).

Anderson, M.J. and Whitcomb, P.J., "Practical Aspects for Designing Statistically Optimal Experiments," *Journal of Statistical Science and Application* 2 (2014b): 85-92.

Atkinson, A.C., Donev, A.N., and Tobias, R.D., *Optimum Experimental Designs, with SAS* (Oxford: Oxford University Press, 2007), ISBN:978-0-19-929660-6.

Augusto, J., Liu, J., McCullagh, P., Wang, H. et al., "Management of Uncertainty and Spatio-Temporal Aspects for Monitoring and Diagnosis in a Smart Home," *International Journal of Computational Intelligence Systems* 1, no. 4 (2008): 361-378.

Balta, B., Erk, O., Solak, H., and Durakbasa, N., "Pareto Optimization of Heavy-Duty Truck Rear Underrun Protection Design for Regulative Load Cases," *SAE Int. J. Commer. Veh.* 7, no. 2 (2014): 726-735, doi:https://doi.org/10.4271/2014-01-9027.

Baumgartner, U., Magele, C., and Renhart, W., "Pareto Optimality and Particle Swarm Optimization," *IEEE Transactions on Magnetics* 40, no. 2 (2004): 1172-1175.

Bayram, I.S., Devetsikiotis, M., and Jovanovic, R., "Optimal Design of Electric Vehicle Charging Stations for Commercial Premises," *Int'l Journal of Energy Research* 46, no. 8 (2022): 10040-10051.

Broekaart, D. (2015), "How Topology Optimization Adds Innovation for 3D-Printed Products," Simuleon FEA Blog, Posted since October 13, 2015, 12:55:32 PM; info.simuleon.com/blog/topic/how-topology-optimization.

Büyük, E., Zor, E., and Karaman, A., "Joint Modeling of Rayleigh Wave Dispersion and H/V Spectral Ratio Using Pareto-Based Multi-Objective Particle Swarm Optimization," *Turkish Journal of Earth Science* 29 (2020): 684-595.

Cauffriez, L. et al., "Robustness Study and Reliability Growth Based on Exploratory Design of Experiments and Statistical Analysis: A Case Study Using a Train Door Test Bench," *Int'l Journal of Advanced Manufacturing Technology* 66 (2013): 27-44.

Chen, S. et al., "Automotive Exterior Noise Optimization Using Grey Relational Analysis Coupled with Principal Component Analysis," *Fluctuation and Noise Letters* 12, no. 3 (2013): 1350017.

Chen, Q., Paulavicius, R., and Adjiman, C., "An Optimization Framework to Combine Operable Space Maximization with Design of Experiments," *AIChE Journal* 64, no. 11 (2018): 3944-3957.

Chiang, Y.J., "Robust Design of the Iosipescu Shear Tests Specimen for Composites," *Journal of Testing and Evaluation, ASTM* 24, no. 1 (1996): 1-11.

Clerc, M. and Kennedy, J., "The Particle Swarm-Explosion, Stability, and Convergence in a Multidimensional Complex Space," *IEEE Transactions on Evolutionary Computation* 6, no. 1 (2002): 58-73.

Coello, C.A.C. and Zacatenco, C.S.P. (2002), "Evolutionary multiobjective optimization: Past, present and future," Tutorial-CINVESTAV-IPN, Depto. de Ingeniería Eléctrica, Sección de Computación, p. 42.

Cook, R.D., Malkus, D.S., and Plesha, M.E., *Concepts and Applications of Finite Element Analysis* (New York: Wiley, 1998).

Cook, R.D. and Nachtsheim, C.J., "A Comparison of Algorithms for Constructing Exact D-Optimal Designs," *Technometrics* 22, no. 3 (1980): 315-324.

Costa, N.R. and Lourenco, J.A. (2015), "Exploring Pareto Frontiers in the Response Surface Methodology," in Yang, G.-C., Ao, S.-I., and Gelman, L. (eds), *Transactions on Engineering Technologies: World Congress on Engineering 2014*, Springer, Berlin/Heidelberg, pp. 399-412.

Cowper, G. and Symonds, P. (1958), "Strain Hardening and Strain Rate of Cantilever Beams," Applied Mathematics Report, Brown University.

Das, R.N. and Lin, D.K.J., "On D-Optimal Robust Designs for Lifetime Improvement Experiments," *Journal of Statistical Planning and Inference* 141 (2011): 3753-3759.

Deng, J., "Introduction to Grey System Theory," *The Journal of Grey System* 1 (1989): 1-24.

Deng, J.L., *The Foundation of Grey System* (Wuhan, China: Huazhong University of Science and Technology Press, 2002).

Diffey, S., Smith, A., Welsh, A., and Cullis, B., "A New REML (Parameter Expanded) EM Algorithm for Linear Mixed Models," *Australian & New Zealand Journal of Statistics* 59, no. 4 (2017): 433-448.

Donnelly, T. (2016), "Creating and Analyzing Definitive Screening Designs," Mastering JMP Webcast, April 1, 2016.

Draper, N.R. and Lin, D.K., "Small Response Surface Design," *Technometrics* 32, no. 2 (1990): 187-194.

Eberhart, R.C. and Kennedy, J. (1995), "A New Optimizer Using Particle Swarm Theory," in *Proceedings of the 6th International Symposium on Micromachine and Human Science*, Nagoya, Japan, March 13–16, 1995, pp. 39-43.

Emmerich, M. and Deutz, A., "A Tutorial on Multi-Objective Optimization: Fundamentals and Evolutionary Methods," *Natural Computing* 17 (2018): 585-609.

Freeman, L.J. (2010), "Statistical Methods for Reliability Data from Designed Experiments," PhD dissertation in Statistics, Virginia Polytechnic Institute and State University.

Fisher, R.A. (1966), *The Design of Experiments*, 8th Edition, 236 pages, Hafner Publishing Company, New York.

Gilmour, A., Thompson, R., and Cullis, B., "Average Information REML: An Efficient Algorithm for Variance Parameter Estimation in Linear Mixed Models," *Biometrics* 51, no. 4 (1995): 1440-1450.

Girish, B.M. et al., "Taguchi Grey Relational Analysis for Parametric Optimization of Severe Plastic Deformation Process," *SN Applied Sciences* 1 (2019): 937.

Goldfarb, H.B., Anderson-Cook, C.M., Borror, C.M., and Montgomery, D.C., "Fraction of Design Space Plots for Assessing Mixture and Mixture-Process Designs," *Journal of Quality Technology* 36, no. 2 (2004): 169-179.

Goos, P. and Jones, B. (July 2011), *Optimal Design of Experiments: A Case-Study Approach*, Wiley, New York, 459 pages; ISBN: 978-0470744611.

Goos, P. and Leemans, H., "Teaching Optimal Design of Experiments Using a Spreadsheet," *Journal of Statistics Education* 12, no. 3 (2004).

Harrant, M., Nirmaier, T., Grimm, C., and Pelz, G. (2014), "Configurable Load Emulation Using FPGA and Power Amplifiers for Automotive Power ICs," in: Haase J. (ed) *Models, Methods, and Tools for Complex Chip Design*; Lecture Notes in Electrical Engineering, Vol 265, Springer.

Hannane, F. et al., "Forecasting the PV Panel Operating Conditions Using the Design of Experiments Method," *Energy Procedia* 36 (2013): 479-487.

Hacioglu, G. and Sesli, E., "Improved RSS Based Distance Estimation for Autonomous Vehicles," *Wireless Personal Communications: An International Journal* 125, no. 1 (2022): 325-350.

Ho, H., Tsai, M.Y., Morton, J., and Farley, G.L., "Numerical Analysis of the Iosipescu Specimen for Composite Materials," *Composites Science and Technology* 46 (1993): 115-128.

Hosseini, S.M., "Optimal Design of the S-Rail Family for an Automotive Platform with Novel Modifications on the Product-Family Optimization Process," *Thin-Walled Structures* 138 (2019): 143-154.

Hsiao, J.C. et al., "Shape Design Optimization of a Robot Arm Using a Surrogate-Based Evolutionary Approach," *Applied Sciences* 10, no. 7 (2020): 2223.

Hu, X. et al., "Calculating Complete and Exact Pareto Front for Multi-Objective Optimization: A New Deterministic Approach for Discrete Problems," *IEEE Transactions on Cybernetics* 43, no. 3 (2013): 1088-1101.

Iosipescu, N., "New Accurate Procedure for Single Shear Testing of Metals," *Journal of Materials* 2, no. 3 (1967): 537-566.

ISO (1997) "Mechanical Vibration and Shock—Evaluation of Human Exposure to Whole-Body Vibration—Part 1: General Requirements," ISO 2631-1:1997.

JMP (2021), "Comparison of a D-Optimal and an I-Optimal Response Surface Design," JMP Statistical Discovery ➔ Examples of Custom Designs ➔ Response Surface Experiments ➔ Comparison of a D-Optimal and an I-Optimal Response Surface Design, November 10, 2021, https://www.jmp.com/support/help/en/16.2/index.shtml#page/jmp/comparison-of-a-doptimal-and-an-ioptimal-response-surface-design.shtml.

Johnson, G.R. and Cook, W.H. (1983), "A Constitutive Model and Data for Metals Subjected to Large Strains, High Strain Rates and High Temperatures," in *Proceedings of the 7th International Symposium on Ballistics*, The Hague, the Netherlands, 1983.

Jones, B. and Goos, P., "I-Optimal versus D-Optimal Split-Plot Response Surface Designs," *Journal of Quality Technology* 44, no. 2 (2012): 85-101.

Jones, B. and Nachtsheim, C., "Definitive Screening Designs with Added Two-Level Categorical Factors," *Journal of Quality Technology* 45, no. 2 (2013): 121-129.

Jones, B. and Nachtsheim, C.J., "A Class of Three Level Designs for Definitive Screening in the Presence of Second-Order Effects," *Journal of Quality Technology* 43 (2011): 1-14.

Kiefer, J. and Wolfowitz, J., "Optimum Designs in Regression Problems," *Annals of Mathematical Statistics* 30 (1959): 271-294.

Kovacs, I. et al., "Blocking in Design of Experiments," *ACTA Technica Napocensis* 54, no. 3 (2013): 42-46.

Kuhfeld, W.F., Tobias, R.D., and Garratt, M., "Efficient Experimental Design with Marketing Research Applications," *Journal of Marketing Research* 31, no. 4 (1994): 545-557.

Kumar, S. et al., "Application of GRA Method for Multi-Objective Optimization of Roller Burnishing Process Parameters Using a Carbide Tool on High Carbon Steel (AISI-1040)," *Grey Systems: Theory and Application* 9, no. 4 (2019): 449-463.

Laumanns, M. and Laumanns, N., "Evolutionary Multi-Objective Design in Automotive Development," *Applied Intelligence* 23 (2005): 55-70.

Li, G., Yan, L., and Qu, B., "Multi-Objective Particle Swarm Optimization Based on Gaussian Sampling," *IEEE Access* 8 (2020): 209717-209737.

Lin, D.K.J., "A New Class of Supersaturated Designs," *Technometrics* 35 (1993): 28-31.

Liu, J. and Wang, X., "Effect of Drying Temperature and Relative Humidity on Contraction Stress in Wood," *BioResources* 11, no. 3 (2016): 6625-6638.

Malhotra, G. et al., "Time-Varying Decision Boundaries: Insights from Optimality Analysis," *Psychonomic Bulletin & Review* 25, no. 3 (2018): 971-996.

Marengo, F. and Todeschini, R., "A New Algorithm for Optimal, Distance-Based Experimental Design," *Chemometrics and Intelligent Laboratory Systems* 16, no. 1 (1992): 37-44.

Mascitelli, R., *Mastering Lean Product Development: A Practical, Event-Driven Process for Maximizing Speed, Profits and Quality* (Northridge, CA: Technology Perspectives, 2011), ISBN:978-0966269741.

Meyer, R.K. and Nachtsheim, C.J., "The Coordinate-Exchange Algorithm for Constructing Exact Optimal Experimental Designs," *Technometrics* 37, no. 1 (1995): 60-69.

Miao, Q., Huang, B., and Jia, B., "Estimating Distances via Received Signal Strength and Connectivity in Wireless Sensor Networks," *Wireless Networks* 26 (2020): 971-982.

Minitab Statistical Software (n.d.), https://www.minitab.com/en-us/.

Mitra, A.C. et al., "Optimization of Automotive Suspension System by Design of Experiments: A Nonderivative Method," *Advances in Acoustics and Vibration* 2016 (2016): 1-10.

Monden, Y., *Toyota Production System: An Integrated Approach to Just-in-Time*, 4th ed. (Boca Raton, FL: CRC Press, 2011), ISBN:978-1439820971.

Nedjah, N., Mourelle, L., and Lopes, H. (Eds), *Evolutionary Multi-Objective System Design: Theory and Applications* (Boca Raton, FL: Chapman and Hall/CRC, 2018).

Nelson, B.L. (2013), "Experiment Design and Analysis," *Foundations and Methods of Stochastic Simulation* Springer, Boston, MA; ISBN: 978-1-4614-6160-9.

Nesterov, Y., "Efficiency of Coordinate Descent Methods on Huge-Scale Optimization Problems," *SIAM Journal on Optimization* 22, no. 2 (2012): 341-362.

Nguyen, N. and Stylianous, S., "Constructing Definitive Screening Designs Using Cyclic Generators," *Journal of Statistical Theory and Practice* 7 (2013): 713-724.

Oyejola, B.A. and Nwanya, J.C., "Selecting the Right Central Composite Design," *Int'l Journal of Statistics and Applications* 5, no. 1 (2015): 21-30.

Ozol-Godfrey, A. et al., "Fraction of Design Space Plots for Examining Model Robustness," *Journal of Quality Technology* 37, no. 3 (2005): 223-235.

Ozol-Godfrey, A. (2004), "Understanding Scaled Prediction Variance Using Graphical Methods for Model Robustness, Measurement Error and Generalized Linear Models for Response Surface Designs," PhD dissertation, Virginia Polytechnic Institute and State University, Blacksburg, VA.

Parsopoulos, K.E. and Vrahatis, M.N. (2008), "Multi-Objective Particles Swarm Optimization Approaches," in Bui, L.T. et al. (eds), *Multi-Objective Optimization in Computational Intelligence: Theory and Practice*, IGI Global, Hershey, PA, pp. 20-42; ISBN: 978-1599044989.

Patterson, H. and Thompson, R., "Recovery of Inter-Block Information When Block Sizes are Unequal," *Biometrika* 58 (1971): 545-554.

Peng, T. and Zhou, B., "Scheduling Multiple Servers to Facilitate Just-in-Time Part-Supply in Automobile Assembly Lines," *Assembly Automation* 38, no. 3 (2018): 347-360.

Piepel, G., Szychowski, J., and Loeppky, J., "Augmenting Scheffe Linear Mixture Models with Squared and/or Cross Product Terms," *Journal of Quality Technology* 34, no. 3 (2002): 297-314.

Phoa, F.K.H. and Lin, D.K.J., "A Systematic Approach for the Construction of Definitive Screening Designs," *Statistica Sinica* 25 (2015): 853-861.

Polhemus, N. (2018), "Definitive Screening Designs," The Statgraphics Blog, March 26, 2018.

Reyes-Sierra, M. and Coello, C.A., "Multi-Objective Particle Swarm Optimizers: A Survey of the State-of-the-Art," *International Journal of Computational Intelligence Research* 12, no. 3 (2006): 287-308.

Rodriguez, M., Jones, B., Borror, C., and Montgomery, D., "Generating and Assessing Exact G-Optimal Designs," *Journal of Quality Technology* 42, no. 1 (2010): 3-20.

SAE International, Ground Vehicle Standard (2021), "Taxonomy and Definitions for Terms Related to Driving Automation Systems for On-Road Motor Vehicles," SAE Standard J3016, April 2021.

Schaffer, J.D. (1984), "Multiple Objective Optimization with Vector Evaluated Genetic Algorithms," PhD dissertation, Vanderbilt University, Nashville, TN.

Škrlec, A. and Klemenc, J., "Estimating the Strain-Rate-Dependent Parameters of the Cowper-Symonds and Johnson-Cook Material Models using Taguchi Arrays," *Journal of Mechanical Engineering* 62, no. 4 (2016): 220-230.

Škrlec, A. and Klemenc, J., "Estimating the Strain-Rate-Dependent Parameters of the Johnson-Cook Material Model Using Optimization Algorithms Combined with a Response Surface," *Mathematics* 8 (2020): 1105.

Srinivas, N. and Deb, K., "Multi-Objective Optimization Using Nondominated Sorting in Genetic Algorithms," *IEEE Transactions on Evolutionary Computation* 2, no. 3 (1994): 221-248.

Stouwen, A. and Goos, P., "A Note on the Output of a Coordinate-Exchange Algorithm for Optimal Experimental Design," *Chemometrics and Intelligent Laboratory Systems* 192, no. 15 (2019): 103819.

Suresh, K. and Takalloozadeh, M., "Stress-Constrained Topology Optimization: A Topological Level-Set Approach," *Structural and Multidisciplinary Optimization* 48, no. 2 (2013): 295-309.

Suresh, K., "A 199-Line Matlab Code for Pareto-Optimal Tracing in Topology Optimization," *Structural and Multidisciplinary Optimization* 42, no. 5 (2010): 665-679.

Suresh, T. et al. (2021), "Comparative Machining Characteristics Studies on SS 304 Using Coated and Uncoated Brass Wire through Wire EDM," La Metallurgia Italiana, Maggio, pp. 32-42.

Thomsen, J., "Topology Optimization of Structures Composed of One or Two Materials," *Journal of Structural Optimization* 5 (1992): 108-115.

Vořechovský, M., Mašek, J., and Eliáš, J., "Distance-Based Optimal Sampling in a Hypercube: Analogies to N-Body Systems," *Advances in Engineering Software* 137 (2019): 102709.

Weese, M.L., Ramsey, P.J., and Montgomery, D., "Analysis of Definitive Screening Designs: Screening vs Prediction," *Applied Stochastic Models in Business and Industry* 34, no. 2 (2018): 244.

Xiao, L., Lin, D.K.J., and Fengshan, B., "Constructing Definitive Screening Designs Using Conference Matrices," *Journal of Quality Technology* 44, no. 1 (2012): 2-8.

Xie, S., Xu, L., Fang, K., and Wu, S., "Discussion on Strain Rate Effects in Numerical Simulation of Vehicle Crash," SAE Technical Paper 2008-01-0504 (2008), doi:https://doi.org/10.4271/2008-01-0504.

Xu, J. et al., "A Distance-Based Maximum Likelihood Estimation Method for Sensor Localization in Wireless Sensor Networks," *International Journal of Distributed Sensor Networks* 2016 (2016): 2080536.

Yao, J., Pan, J., Han, Y., and Wang, L. (2010), "Application of Particle Swarm Optimization with Stochastic Inertia Weight and Adaptive Mutation in Target Localization," in *Proceedings of International Conference on Computer Application and System Modeling*, October 2010, pp. 251-254.

Yildiz, A.R. and Solanki, K.N., "Multi-Objective Optimization of Vehicle Crashworthiness Using a New Particle Swarm-Based Approach," *The International Journal of Advanced Manufacturing Technology* 59 (2011): 367-376.

Zahran, A.R., Anderson-Cook, C.M., and Myers, R.H., "Fraction of Design Space to Assess Prediction Capability of Response Surface Designs," *Journal of Quality Technology* 35 (2003): 377-386.

Zheng, Q., Tain, S., and Zhang, Q., "Optimal Torque Split Strategy of Dual-Motor Electric Vehicle Using Adaptive Nonlinear Particle Swarm Optimization," *Mathematical Problems in Engineering* 2020 (2020): 1-21.

Zitzler, E. and Thiele, L., "Multi-Objective Evolutionary Algorithms: A Comparative Case Study and the Strength Pareto Approach," *IEEE Transactions of Evolutionary Computation* 3, no. 4 (1999): 257-271.

Problems

P6.1: A film coating process is studied to explore the impacts on the temperature and relative humidity at the output, i.e., $T_{air,out}$ (°C) and Relative humidity (%), by three input factors:

$$A: T_{air,input} \in [20 \ °C, 85 \ °C],$$
$$B: M_{coating \ speed} \in [10 \ g/min, 80 \ g/min]$$
and $$C: Q_{air \ flow} \in [150 \ ft^3/min, 450 \ ft^3/min]$$

Two test matrices are given as follows [Chen, Q. et al. 2018]:

Design 1:				Design 2:			
Run	A	B	C	Run	A	B	C
1	20	10	450	1	20	10	150
2	85	10	150	2	85	10	150
3	85	10	450	3	85	10	450
4	85	80	450	4	85	80	158

Both designs are highly orthogonal. Which one has the higher D-efficiency.

P6.2: Continued from P6.1. Five experimental treatments are applied instead of four. The test results are given as follows:

Run	A	B	C	Y_1	Y_2
1	20	10.5	150	16.3	33.7
2	20	10	450	18.8	20.1
3	23.7	80	450	14.2	71.0
4	85	10	450	83.4	16.2
5	85	80	157.6	50.2	87.4

Please:

(1) Estimate the D-efficiency of the design matrix given above

(2) Assess the impacts of these three factors on Y_1, i.e., output temperature $T_{air,out}$ (°C) and

(3) Assess the impacts of these three factors on Y_2, i.e., relative humidity (%)

P6.3: Two design matrices constructed with two different distinct optimal designs are given in Example 6.4.2. Which one is better in terms of SPV for fitting the response surface model $Y = \theta_0 + \theta_a \ a + \theta_b \ b + \theta_{ab} \ a \ b + \theta_{aa} \ a^2 + \theta_{bb} \ b^2$?

P6.4: Do power evaluation of the main effects of factorial design 2^3.

P6.5: An experimental design was set up by [Hannane et al. 2013] to explore the performance of a photovoltaic (PV) panel with two factors, i.e., source distance (factor A) and irradiance temperature (Factor B). Responses include electric voltage V (volt), electric current I (ampere) and maximum power P_{max} (Watt). The design matrix and test results (V, I, P_{max}) are given as follows:

Run	a	b	A	B	V (V)	I (A)	P_{max} (W)
1	−1	−0.1429	1.1	40	18.3	2.5	32.6
2	−1	−0.7143	1.1	30	18.3	2.5	32.3
3	−1	−0.9429	1.1	26	18.7	2.5	32.2
4	−1	−0.0857	1.1	44	18.1	2.5	31.3
5	−1	1	1.1	60	17.3	2.5	31.2
6	0.0588	−0.7143	2.0	30	17.7	1.5	19.0
7	0.0588	−1	2.0	25	18.1	1.5	19.2
8	0.0588	−0.1429	2.0	40	17.7	1.5	19.2
9	0.0588	−1	2.0	60	17.1	1.5	18.4
10	1	−0.6571	2.8	31	17.1	1.0	12.0
11	1	−0.2571	2.8	38	16.9	1.0	12.5

where coded variables a = (A − 1.95)/0.85 and b = (B − 42.5)/17.5

1. Do power evaluation of the main effects.
2. Find out the predictive equations, including linear terms, quadratic terms, and 2-factor interactions, for all the three responses.
3. Find out the correlations between the predicted values and the 11 experimental data.

P6.6: A study based on DOE aims to explore the effect of temperature and humidity on the shrinkage and contraction stress of elm wood during drying [Liu and Wang 2016]. In the final stabilized state, results from nine experimental treatments are shown in Table P6.6. Please identify the factors that have significant influences on the shrinkage and moisture content. Present the Pareto frontier using Excel if both shrinkage and moisture content are the smaller the better.

TABLE P6.6 Factorial design for exploring shrinkage and moisture content of elm wood.

Run	A	B	C	D	S	M_c
1	50	30	Tan	25	0.48	7.08%
2	50	40	Radial	50	1.96	8.13%
3	50	50	Tan	50	1.58	9.39%
4	70	30	Radial	50	3.11	5.90%
5	70	40	Tan	25	0.62	7.32%
6	70	50	Tan	50	1.69	8.36%
7	90	30	Tan	50	1.53	5.30%
8	90	40	Tan	50	1.07	6.25%
9	90	50	Radial	25	0.92	7.56%

where
A (°C): Temperature—3 levels
B (%): Moisture—3 levels
C: Direction—2 levels, in either tangential (Tan) or radial (Rad) direction
D (mm): Thickness—2 levels
S (mm): Shrinkage (Response 1)
M_c (%): Moisture content (Response 2)

P6.7: Burnishing is a cold-work operation, by which a smooth surface finish is produced (polished) on a metallic part by rubbing actions that soothe scratch marks and other small defects. For example, in a carbide roller burnishing process, the material displacement in the outer case of steel takes place in form of plastic deformation when the burnishing tool transverses along the rotating roller that is mounted on a lathe. Process parameters and their design levels used for improving cold roller burnishing quality of AISI 1040 steel are listed as follows [Kumar et al. 2019].

Factor	Level 1	Level 2	Level 3
A: Machined surface condition	Wet	Dry	—
B (rpm): Roller speed	371	557	835
C (mm/min): Feed rate	0.06	0.11	0.17
D (mm): Roller penetration depth	1	1.5	2
E: Number of passes	1	2	3

As given in Table P6.7, design matrix $2 \times 3^{4-2}$ is employed to explore the effects of the above factors on the hardness (H_{RB}) and surface roughness (R_a) of a steel part [Kumar et al. 2019]. The calculated gray relational grades listed in the last column by [Kumar et al. 2019]. Does the Pareto frontier plot based on DOE lead to the same conclusion?

TABLE P6.7 Design matrix $2 \times 3^{4-2}$ used for improving cold burnishing quality by rollers [Kumar et al. 2019].

Run	A	B	C	D	E	H_{RB}	R_a	Grade
1	Wet	371	0.06	1	1	89.66	0.6	0.469119
2	Wet	371	0.11	1.5	2	91.66	0.35	0.607312
3	Wet	371	0.17	2	3	93.33	0.19	0.748712
4	Wet	557	0.06	1	2	93.40	0.13	0.798517
5	Wet	557	0.11	1.5	3	93.33	0.24	0.715664
6	Wet	557	0.17	2	1	89.66	0.39	0.548453
7	Wet	835	0.06	1.5	1	88.66	0.2	0.614157
8	Wet	835	0.11	2	2	87	0.57	0.449918
9	Wet	835	0.17	1	3	92	0.71	0.503396
10	Dry	371	0.06	2	3	94.33	1.28	0.503957
11	Dry	371	0.11	1	1	96.66	0.78	0.734694
12	Dry	371	0.17	1.5	2	92.33	0.54	0.55552
13	Dry	557	0.06	1.5	3	96	0.44	0.764749
14	Dry	557	0.11	2	1	95	0.94	0.579692
15	Dry	557	0.17	1	2	90	0.56	0.496252
16	Dry	835	0.06	2	2	91.33	0.41	0.57395
17	Dry	835	0.11	1	3	90	0.6	0.485302
18	Dry	835	0.17	1.5	1	91	0.4	0.547899

P6.8: A study is conducted to find out the influences of five operating parameters on the surface roughness and material hardness (Vickers) subjected to wire cutting with EDM [Suresh T. et al. 2021]. The workpiece is AISI 304 stainless material, while the wire is made of brass with or without coating.

Factor	Level (1)	Level (2)	Level (3)
A: Wire material—Brass	Zn-coated	Uncoated	—
B (μs): Pulse on time	105	110	115
C (μs): Pulse off time	25	35	45
D (Amp): Peak current	1	3	5
E (Volt): Gap voltage	20	30	40

As listed in Table P6.8, factorial design $2 \times 3^{4-2}$ is employed to deploy the design matrix and two sequences of objective functions (Y_{qr}, r = 1 and 2) are

(1) Surface roughness (Y_{q1}, q = 1, 2, 3, ..., Q): the smaller the better
(2) Hardness, Vickers (Y_{q2}, q = 1, 2, 3, ..., Q): the higher the better

Please do the Pareto frontier plot based on DOE. Compare the result with what obtained by gray relational coefficients (Z_{qr}) of these 18 run conditions published by [Suresh T. et al. 2021].

TABLE P6.8 Factorial design $2\times3^{4-2}$ for surface roughness and material hardness [Suresh T. et al. 2021].

q	A	B	C	D	E	Y_{q1}	Y_{q2}
1	Zn-coated	105	25	1	20	1.782	199
2	Zn-coated	105	35	3	30	2.091	194
3	Zn-coated	105	45	5	40	2.042	193
4	Zn-coated	110	25	1	30	2.789	179
5	Zn-coated	110	35	3	40	2.757	188
6	Zn-coated	110	45	5	20	3.108	202
7	Zn-coated	115	25	3	20	3.432	189
8	Zn-coated	115	35	5	30	2.681	186
9	Zn-coated	115	45	1	40	2.836	198
10	Uncoated	105	25	5	40	2.235	188
11	Uncoated	105	35	1	20	2.205	193
12	Uncoated	105	45	3	30	2.303	179
13	Uncoated	110	25	3	40	2.668	201
14	Uncoated	110	35	5	20	2.786	191
15	Uncoated	110	45	1	30	2.508	176
16	Uncoated	115	25	5	30	2.890	190
17	Uncoated	115	35	1	40	2.823	188
18	Uncoated	115	45	3	20	3.016	203

P6.9: A gray relational analysis example given by [Girish et al. 2019] is to be rephased here. The study is about refining the microstructure of Al 6061 (an aluminum alloy) to the nanometer levels for improving its mechanical properties. The process resembles the burnishing process by cold work. Design parameters and their design levels are listed as follows:

Factor	Level (−1)	Level (0)	Level (+1)
A (mm/min): Displacement rate	1	1.5	2
B: Number of passes	1	3	5
C (mm): Thickness of workpiece	3	4	5

As shown in Table P6.9, factorial design 3^3 is employed to deploy the design matrix and three objective functions (Y_{qr}, r = 1, 2, 3; q = 1, 2, ..., 27):

(1) Y_{q1}: Micro-hardness (the higher the better)
(2) Y_{q2}: (MPa): Tensile strength (the higher the better)
(3) Y_{q3}: (μm): Grain size (the smaller the better)

Please calculate gray relational coefficients (Z_{qr}) of these 27 treatments. Are they the same as the data given in Table P6.4, from Girish et al. 2019, which was published by the original authors. Please also calculate gray relational grades and rank the influences of these 27 treatments on the representative objective function (performance characteristic).

TABLE P6.9 Factorial design 3^3, test data Y_{qr}, coded response Z_{qr}, and gray relational value ξ_{qr} [Girish et al. 2019].

Run/q	A	B	C	Y_{q1}	Y_{q2}	Y_{q3}	Z_{q1}	Z_{q2}	Z_{q3}	ξ_{q1}	ξ_{q2}	ξ_{q3}
1	1	1	3	44.48	94.82	7.7	0.206	0.405	0.061	0.386	0.457	0.348
2	1	1	4	43.27	97.31	7.9	0.090	0.459	0.020	0.355	0.480	0.338
3	1	1	5	43.02	76.38	8.0	0.066	0.000	0.000	0.349	0.333	0.333
4	1	3	3	46.17	106.0	6.5	0.367	0.649	0.306	0.441	0.588	0.419
5	1	3	4	45.50	109.8	6.4	0.303	0.734	0.327	0.418	0.653	0.426
6	1	3	5	45.70	89.29	5.6	0.322	0.283	0.490	0.425	0.411	0.495
7	1	5	3	49.20	113.0	3.8	0.657	0.804	0.857	0.593	0.718	0.778
8	1	5	4	48.65	120.0	3.1	0.605	0.957	1.000	0.559	0.922	1.000
9	1	5	5	47.98	96.00	4.0	0.541	0.431	0.816	0.521	0.468	0.731
10	1.5	1	3	46.01	97.70	8.0	0.352	0.468	0.000	0.436	0.484	0.333
11	1.5	1	4	44.38	96.92	7.2	0.196	0.451	0.163	0.383	0.477	0.374
12	1.5	1	5	42.79	79.07	6.7	0.044	0.059	0.265	0.343	0.347	0.405
13	1.5	3	3	47.62	105.0	6.7	0.506	0.627	0.265	0.503	0.573	0.405
14	1.5	3	4	46.19	109.6	5.8	0.369	0.729	0.449	0.442	0.648	0.476
15	1.5	3	5	45.45	91.10	5.0	0.299	0.323	0.612	0.416	0.425	0.563
16	1.5	5	3	52.78	114.7	4.2	1.000	0.841	0.776	1.000	0.759	0.690
17	1.5	5	4	52.73	119.0	3.5	0.995	0.936	0.918	0.991	0.886	0.860
18	1.5	5	5	50.08	96.25	4.7	0.742	0.436	0.673	0.659	0.470	0.605
19	2	1	3	46.33	95.92	8.0	0.383	0.429	0.000	0.448	0.467	0.333
20	2	1	4	44.82	96.09	7.6	0.238	0.433	0.082	0.396	0.468	0.353
21	2	1	5	42.33	78.69	6.4	0.000	0.051	0.327	0.333	0.345	0.426
22	2	3	3	47.91	104.5	6.3	0.534	0.617	0.347	0.518	0.567	0.434
23	2	3	4	46.02	112.1	5.6	0.353	0.784	0.490	0.436	0.698	0.495
24	2	3	5	44.18	88.44	5.1	0.177	0.265	0.592	0.378	0.405	0.551
25	2	5	3	52.13	115.1	4.2	0.938	0.851	0.776	0.889	0.770	0.690
26	2	5	4	52.23	121.9	3.8	0.947	1.000	0.857	0.905	1.000	0.778
27	2	5	5	49.63	97.10	4.6	0.699	0.455	0.694	0.624	0.478	0.620

Bibliography: Books

Akao, Y. (Editor; 1990), *Quality Function Deployment: Integrating Customer Requirements into Product Design*, Productivity Press, Cambridge, MA, 392 pages; ISBN: 9781563273131.

Allen, P. (2020), *Design of Experiments for 21st Century Engineers*, Lulu.com, 214 pages; ISBN: 978-0244584504.

Altshuller, G. (1984), *Creativity as an Exact Science*, Gordon & Breach, New York.

American Supplier Institute (1993), *Quality Function Deployment*, American Supplier Institute, Dearborn, MI.

Anderson, V. L. and McLean, R. A. (1974), *Design of Experiments: A Realistic Approach*, Marcel Dekker, New York.

Anderson, M. J. and Whitcomb, P. J. (2015), *DOE Simplified: Practical Tools for Effective Experimentation*, 3rd Edition, Productivity Press, London, UK, 268 pages; ISBN: 978-0429258022.

Anderson, M. J. and Whitcomb, P. J. (2016), *RSM Simplified: Optimizing Processes Using Response Surface Methods for Design of Experiments*, Productivity Press, New York, 311 pages; ISBN: 978-1315382326.

Antony, J. (2014), *Design of Experiments for Engineers and Scientists*, 2nd Edition, Elsevier, San Diego, CA; ISBN: 978-0-08-099417-8.

Armold, S. F. (1990), *Mathematical Statistics*, Prentice-Hall International, Englewood Cliffs, NJ; ISBN: 978-0135630990.

Atkinson, A. C., Donev, A. N., and Tobias, R. D. (2007), *Optimum Experimental Designs, with SAS*, Oxford University Press, Oxford, UK; ISBN: 978-0-19-929660-6.

Automotive Industry Action Group (2010), *Measurement System Analysis Manual*, 2nd Edition, Southfield, MI.

Bailey, R. A. (2008), *Design of Comparative Experiments*, Cambridge University Press, Cambridge, UK.

Beckwith, T. G. (1990), *Mechanical Measurements*, Addison-Wesley Publishing Company, Reading, MA.

Bertsche, B. (2008), *Reliability in Automotive and Mechanical Engineering*, Springer, Berlin, 511 pages; ISBN: 978-3540681892.

Bhote, K. (1996), *Going beyond Customer Satisfaction to Customer Loyalty*, American Management Association, New York, 140 pages; ISBN: 978-0814423622.

Bhote, K. R. and Bhote, A. K. (1999), *World Class Quality Using Design of Experiments to Make it Happen*, 2nd Edition, American Management Association, New York, 487 pages; ISBN: 978-0814426425.

Bicheno, J. (2004), *New Lean Toolbox: Towards Fast Flexible Flow*, PICSIE Books, Buckingham, UK.

Bjorke, O. (1989), *Computer-Aided Tolerancing*, 2nd Edition, ASME Press, New York.

Blischke, W. R., Karim, M. R., and Prabhakar Murthy, D. N. (2011), *Warranty Data Collection and Analysis*, Springer, London, UK; ISBN: 978-0-85729-646-7.

Boothroyd, G., Dewhurst, P., and Knight, W. (2010), *Product Design for Manufacture and Assembly*, 3rd Edition, CRC Press, Hoboken, NJ.

Box, G. E. P. and Draper, N. R. (1998), *Evolutionary Operation: A Statistical Method for Process Improvement*, First Printing Edition, John Wiley & Sons, New York, 237 pages; ISBN: 978-0471255512.

Box, G. E. P. and Draper, N. R. (2007), *Response Surfaces, Mixtures, and Ridge Analyses*, 2nd Edition, John Wiley & Sons, Hoboken, NJ, 857 pages; ISBN: 978-0470053577.

Box, G. E. P., Hunter, J. S., and Hunter, W. G. (2005), *Statistics for Experimenters: Design, Innovation, and Discovery*, 2nd Edition, John Wiley & Sons, London, UK, 672 pages; ISBN: 978-0-471-71813-0.

Brook, R. J. and Arnold, G. C. (1985), *Applied Regression Analysis and Experimental Design*, Taylor & Francis Group/CRC Press, Boca Raton, FL, 256 pages; ISBN: 0-8247-7252-0.

Brue, G. and Launsby, R. G. (2003), *Design for Six Sigma*, McGraw-Hill, New York.

Bryson, A. E. and Ho, Y. C. (1975), *Applied Optimal Control*, John Wiley & Sons, New York.

Chiang, Y. J. (2019), *Mechanics and Design for Product Life Prediction*, Chongqing University Press; ISBN: 978-7-5689-1917-6.

Chiang, Y. J. (2022), *Automotive Engineering Materials—Thermomechanical Properties*, Chongqing University Press; ISBN: 978-7-5689-3293-6.

Cochran, W. G. and Cox, G. M. (1957), *Experimental Design*, John Wiley & Sons, New York.

Cogorno, G. R. (2020), *Geometric Dimensioning and Tolerancing for Mechanical Design*, 3rd Edition, McGraw-Hill, New York; ISBN: 978-1260453782.

Coles, S. (2001), *An Introduction to Statistical Modeling of Extreme Values*, Springer-Verlag, London, UK.

Condra, L. (2019), *Reliability Improvement with Design of Experiment*, 2nd Edition, CRC Press, Boca Raton, FL; ISBN: 978-0824705275.

Cooper, W., Lawrence, M., and Joe, Z., (2011), *Handbook on Data Envelopment Analysis*, Springer, New York; ISBN: 978-1-4419-6150-1.

Cox, N. D. (2006), *How to Perform Statistical Tolerance Analysis*, ASQC, Milwaukee, WI; ISBN: 978-0-87389-010-6.

Czitrom, V. and Spagon, P. D. (1997), *Statistical Case Studies for Industrial Process Improvement*, ASA-SIAM Series on Statistics and Applied Probability, SIAM, Philadelphia, PA.

Das, R. N. (2014), *Robust Response Surfaces, Regression, and Positive Data Analyses*, Chapman and Hall/CRC, Boca Raton, FL, 336 pages; ISBN: 978-1466506770.

Davenport, W. B. and Root, F.S. (1980), *An Introduction to the Theory of Random Signals and Noise*, McGraw-Hill, New York.

Davies, O. L. (Editor; 1971), *The Design and Analysis of Industrial Experiments*, McMillan Co., New York.

Davim, P. J. (Editor; 2016), *Design of Experiments in Production Engineering*, Springer, Cham; ISBN: 978-3319238371.

Dean, A., Voss, D., and Draguljic, D. (2017), *Design and Analysis of Experiments*, 2nd Edition, Springer, Cham; ISBN: 978-3319522487.

Deming, W. E. (1986), *Out of the Crisis*, Massachusetts Institute of Technology, Cambridge, MA.

Diamond, W. J. (1981), *Fractional Experimental Designs*, Wadsworth, Belmont, CA.

Dodson, B. (2006), *The Weibull Analysis Handbook*, 2nd Edition, American Society for Quality, Quality Press, Milwaukee, WI.

Doebelin, E. O. (1990), *Measurement Systems, Applications and Design*, 4th Edition, McGraw-Hill, New York.

Drake, P. Jr. (1999), *Dimensioning and Tolerancing Handbook*, McGraw-Hill, New York; ISBN: 0-07018131-4.

Eriksson, L. et al. (2008), *Design of Experiments Principles and Applications*, 3rd Edition, Umetrics Academy, Umeå, Sweden; ISBN: 91-973730-0-1.

Fang, K., Li, R., and Sudjianto, A. (2006), *Design and Modeling for Computer Experiments*, Chapman & Hall/CRC, London, UK; ISBN: 978-1584885467.

Fienberg, S. E. (1987), *The Analysis of Cross-Classified Categorical Data*, The MIT Press, Cambridge, MA.

Fisher, R. A. (1966), *The Design of Experiments*, 8th Edition, Hafner Publishing Company, New York, 236 pages.

Fleiss, J. L. (1981), *Statistical Methods for Rates and Proportions*, 2nd Edition, John Wiley & Sons, New York.

Foster, T. (2004), *Managing Quality: An Integrative Approach*, 2nd Edition, Prentice Hall, Upper Saddle River, NJ, 544 pages; ISBN: 978-0138759643.

Fowlkes, W. Y. and Creveling, C. M. (1995), *Engineering Methods for Robust Product Design: Using Taguchi Methods in Technology and Product Development*, Addison-Wesley Publishing Company, Reading, MA; ISBN: 978-0133007039.

Funkenbusch, P. D. (2004), *Practical Guide to Designed Experiments: A Unified Modular Approach*, CRC Press, Boca Raton, FL; ISBN: 978-0824753887.

Good, P. I. and Hardin, J. W. (2009), *Common Errors in Statistics (And How to Avoid Them)*, 3rd Edition, John Wiley & Sons, Hoboken, NJ.

Goos, P. and Jones, B. (2011), *Optimal Design of Experiments: A Case-Study Approach*, Wiley, New York, 459 pages; ISBN: 978-0470744611.

Grant, E. L. and Leavenworth, R. S. (2004), *Statistical Quality Control*, Indian Edition, McGraw-Hill, India; ISBN 13: 9780070435551.

Grunditz, E. A. (2016), "Design and Assessment of Battery Electric Vehicle Powertrain, with Respect to Performance, Energy Consumption and Electric Motor Thermal Capability," PhD thesis, Dept. of Energy and Environment, Chalmers University of Technology, Gothenburg, Sweden.

Gunter, B. and Coleman, D. (2014), *A DOE Handbook: A Simple Approach to Basic Statistical Design of Experiments*, Createspace Independent Publishing Platform, 118 pages; ISBN: 978-1497511903.

Gwet, K. L. (2014), *Handbook of Inter-Rater Reliability*, 4th Edition, Advanced Analytics, LLC, Gaithersburg; ISBN: 978-0970806284.

Harrington, M. (2011), *The Design of Experiments in Neuroscience*, 3rd Edition, SAGE, 261 pages; ISBN: 978-1412974325.

Hicks, C. R. and Turner K. V. (1999), *Fundamental Concepts in the Design of Experiments*, 5th Edition, Oxford University Press, New York, 565 pages; ISBN: 978-0195122732.

Hines, W. W. and Montgomery, D. G. (1990), *Probability and Statistics in Engineering and Management Science*, John Wiley & Sons, New York.

Hinkelmann, K. and Kempthorne, O. (2008), *Design and Analysis of Experiments Set*, Wiley, Hoboken, NJ, 1411 pages; ISBN: 978-0-470-38551-7.

Hite, J. A., Schmidt, J. W., and Bennett, G. H. (1975), *Analysis of Queuing Systems*, Elsevier, New York.

Huhn, S. and Drechsler, R. (2021), *Design for Testability, Debug and Reliability: Next Generation Measures Using Formal Techniques*, Springer, Cham; ISBN: 978-3030692087.

Huitema, B. E. (2011), *The Analysis of Covariance and Alternatives: Statistical Methods for Experiments, Quasi-Experiments, and Single-Case Studies*, 2nd Edition, Wiley, Hoboken, NJ, 688 pages.

International Bureau of Weights and Measures (2006), *The International System of Units*, 8th Edition; ISBN: 92-822-2213-6.

Joens, B. and Montgomery, D. C. (2019), *Design of Experiments: A Modern Approach*, 1st Edition, Wiley, Hoboken, NJ, 272 pages; ISBN: 978-1-119-61119-6.

Juran, J. M. and Gryna, F. M. (1993), *Quality Planning and Analysis from Product Development through Use*, 3rd Edition, McGraw-Hill, New York, p. 256.

Kalpakjian, S. (Editor; 1992), *Manufacturing, Engineering and Technology*, 2nd Edition, Addison Wesley, Reading, MA.

Kapur, K. C. and Lamberson, L. R. (1991), *Reliability in Engineering Design*, Wiley, Hoboken, NJ, 608 pages; ISBN: 978-0-471-51191-5.

Keyte, B. and Locher, D. (2004), *The Complete Lean Enterprise: Value Stream Mapping for Administrative and Office Processes*, Productivity Press, New York.

Khuri, A. I. and Cornell, J. A. (2019), *Response Surfaces Design and Analyses*, Taylor & Francis Group, Boca Raton, FL, 536 pages; ISBN: 978-0367401252.

King, J. P. and Jewett, W. S. (2010), *Robustness Development and Reliability Growth: Value-Adding Strategies for New Products and Processes*, Prentice Hall, Upper Saddle River, NJ; ISBN: 978-0-13-222551-9.

Kuehl, R. O. (2000), *Design of Experiments: Statistical Principles of Research Design and Analysis*, 2nd Edition, Duxbury Press, North Scituate, MA; ISBN: 978-0534368340.

Kverneland, K. O. (1996), *Metric Standards for Worldwide Manufacturing Engineering*, ASME, New York.

Lawless, J. F. (2003), *Statistical Models and Methods for Lifetime Data*, 3rd Edition, John Wiley & Sons, New York.

Lawson, J. (2014), *Design and Analysis of Experiments with R*, Chapman & Hall/CRC Press, Boca Raton, FL; ISBN: 978-1439868133.

Liker, J. Z (2000), *The Toyota Way: 14 Management Principles from the World's Greatest Manufacturer*, 2nd Edition, McGraw-Hill, New York; ISBN: 0071392319.

Liston, C. and Sheth, N. J. (1973), *Statistical Design and Analysis of Engineering Experiments*, McGraw-Hill, New York.

Long, J. S. (1997), *Regression Models for Categorical and Limited Dependent Variables*, SAGE Publications, Thousand Oaks, CA.

Lorenzen, T. J. and Anderson, V. L. (1993), *Design of Experiments: A No-Name Approach*, Marcel Dekker, Inc., New York.

Mabie, H. H. and Reinholtz, C. F. (1987), *Mechanisms and Dynamics of Machinery*, 4th Edition, John Wiley & Sons, New York.

Mantz, H. F. and Waller, R. A. (1977), *Bayesian Reliability Analysis*, John Wiley & Sons, New York.

Mascitelli, R. (2011), *Mastering Lean Product Development: A Practical, Event-Driven Process for Maximizing Speed*, Profits and Quality, Northridge, CA; ISBN: 978-0966269741.

Mason, R. L., Gunst, R. F., and Hess, J. L. (2003), *Statistical Design & Analysis of Experiments with Applications to Engineering and Science*, 2nd Edition, John Wiley & Sons, Inc., New York, 760 pages; ISBN: 978-0-471-37216-5.

Mathews, P. G. (2004), *Design of Experiments with MINITAB: Homework Problems*, ASQ Quality Press, Milwaukee, WI.

Meeker, W. Q., Escobar, L. A., and Pascual, F. G. (1998), *Statistical Methods for Reliability Data*, 2nd Edition, John Wiley & Sons, New York, 704 pages; ISBN: 978-1-118-11545-9.

Military and Government Specs & Standards (Naval Publications and Form Center) (NPFC) (1980), MIL-STD-785, Reliability Program for Systems and Equipment Development and Production, DOD, Washington, DC.

Milliken, G. A. and Johnson, D. E. (1984), *Analysis of Messy Data*, Vol. 1: Designed Experiments, Van Nostrand Reinhold, New York.

Modarres, M., Amiri, M., and Jackson, C. (2017), *Probabilistic Physics of Failure Approach to Reliability: Modeling, Accelerated Testing, Prognosis and Reliability Assessment*, Wiley-Scrivener, 288 pages; ISBN: 978-1119388630.

Moen, R. et al. (2012), *Quality Improvement through Planned Experimentation*, McGraw-Hill, New York; ISBN: 978-0071759663.

Montgomery, D. C. (2019), *Design and Analysis of Experiments*, 10th Edition, John Wiley & Sons, Inc., Hoboken, NJ, 688 pages; ISBN: 978-1-119-49244-3.

Morris, M. D. (2017), *Design of Experiments: An Introduction Based on Linear Models*, CRC Press, Boca Raton, FL; ISBN: 978-1138628021.

Murthy, D. P., Xie, M., and Jiang, R. (2004), *Weibull Models*, John Wiley & Sons, Inc., Hoboken, NJ.

NASA (2010), NASA System Safety Handbook, Volume 1, System Safety Framework and Concepts for Implementation, Version 1, SP-2010-580, Washington, DC.

Nelson, W. B. (2004), *Accelerated Testing- Statistical Models, Test Plans, and Data Analysis*, John Wiley & Sons, New York; ISBN: 978-0471697367.

Nelson, B. L. (2013), *Foundations and Methods of Stochastic Simulation*, Springer, Boston, MA; on-line ISBN: 978-1-4614-6160-9.

Nevins, J. L. and Whitney, D. E. (1990), *Concurrent Design of Products and Processes*, McGraw-Hill, New York.

NIST, *Engineering Statistics Handbook*, National Institute of Standards and Technology, Washington, DC.

NIST/SEMATECH (2012), *e-Handbook of Statistical Methods*, Springer, Berlin, http://www.itl.nist.gov/div898/handbook.

O'Connor, P. and Kleyner, A. (2011), *Practical Reliability Engineering*. John Wiley & Sons, New York.

Oehlert, G. W. (2010), *A First Course in Design and Analysis of Experiments*, University of Minnesota Digital Conservancy, Minneapolis, MN, 683 pages; ISBN: 9780716735106; Retrieved from the University of Minnesota Digital Conservancy, https://hdl.handle.net/11299/168002.

Ohring, M. and Kasprzak, L. (2014), *Reliability and Failure of Electronic Materials and Devices*, 2nd Edition, Elsevier, New York, 734 pages; ISBN: 978-0120885749.

Onyiah, L. C. (2008), *Design and Analysis of Experiments: Classical and Regression Approaches with SAS*, Chapman and Hall/CRC, London, UK; ISBN (eBook): 9780429140273.

Peace, G. S. (1993), *Taguchi Methods: A Hands-On Approach*, Addison-Wesley Publishing Company, Reading, MA.

Petroski, H. (1996), *Invention by Design How Engineers Get from Thought to Thing*, Harvard University Press, Cambridge, MA.

Petrov, V. (2019), *Laws of System Evolution*, TRIZ, 57 pages; ISBN: 978-1696068833.

Phadke, M. S. (1989), *Quality Engineering Using Robust Design*, Prentice Hall, Piscataway, NJ.

Proust, M. (2010), *Design of Experiments Guide: JMP*, A Business Unit of SAS, Cary, NC.

Pyzdek, T. and Keller, P. (2019), *The Six Sigma Handbook*, 5th Edition, McGraw-Hill, New York, 720 pages; ISBN: 978-1260121827.

Rhinehart, R. R. (2016), *Nonlinear Regression Modeling for Engineering Applications: Modeling, Model Validation, and Enabling Design of Experiments*, Wiley, Chichester, UK, 400 pages; ISBN: 978-1-118-59796-5.

Rigdon, S. E., Pan, R., Montgomery, D. C., and Borror, C. M. (2022), *Experiments for Reliability Achievement*, Wiley, Boca Raton, FL, 416 pages; ISBN: 978-1119237693.

Rodrigues, M. and Iemma, A. (2014), *Experimental Design and Process Optimization*, CRC Press, 336 pages; ISBN: 978-0429161865.

Ross, S. M. (1987), *Introduction to Probability and Statistics for Engineers and Scientists*, John Wiley & Sons, New York.

Ross, P. J. (1988), *Taguchi's Techniques for Quality Engineering*, McGraw-Hill, New York.

Rössler, A. (2014), *Design of Experiment for Coatings*, Vincentz Network, Hanover, Germany; ISBN: 978-3-86630-885-5.

Rother, M. (2009), *Toyota Kata*, McGraw-Hill, New York.

Rother, M. and Shook, J. (2009), *Learning to See*, Lean Enterprise Institute, Boston, MA.

Roy, R. K. (2001), *Design of Experiments Using the Taguchi Approach: 16 Steps to Product and Process Improvement*, Wiley, New York, 560 pages; ISBN: 978-0-471-36101-5.

Russell, K. G. (2021), *Design of Experiments for Generalized Linear Models*, Chapman and Hall/CRC, Boca Raton, FL, 240 pages; ISBN: 978-1032094052.

Sahai, H. and Ageel, M. I. (2000), *The Analysis of Variance*, Birkhauser, Boston, MA.

Samuels, M. L., Witmer, J. A., and Schaffner, A. A. (2016), *Statistics for the Life Sciences*, 5th Edition, Pearson Education, Harlow, UK; ISBN: 978-1-292-10181-1.

Santner, T. J., Williamns, B. J., and Notz, W. I. (2018), *The Design and Analysis of Computer Experiments*, 2nd Edition, Springer, New York; ISBN: 978-1493988457.

Schleich, B. (2017), *Skin Model Shapes: A New Paradigm for the Tolerance Analysis and the Geometrical Variations Modelling in Mechanical Engineering*, VDI Verlag, Düsseldorf.

Schmidt, S. R. and Launsby, R. G. (1994), *Understanding Industrial Designed Experiments*, 4th Edition, Air Academy Press, Colorado Springs, CO, 768 pages.

Shigley, J. E. and Mischke, C. R. (1989), *Fundamentals of Machine Component Design*, 5th Edition, McGraw-Hill, New York; ISBN: 978-1118012895.

Shingo, S. (1985), *A Revolution in Manufacturing: The SMED System*, Productivity Press, Stamford, CT; ISBN: 978-0915299034.

Silva, V. (2018), *Statistical Approaches with Emphasis on Design of Experiments Applied to Chemical Processes*, Intechopen, London, UK, 180 pages; ISBN: 978-953-51-3878-5.

Taguchi, G. (1987), *System of Experimental Design*, UNIPUB, Kraus International Publications, New York.

Taguchi, G. and Konishi, S. (1987), *Taguchi Methods, Orthogonal Arrays and Linear Graphs: Tools for Quality Engineering*, American Supplier Institute, Dearborn, MI.

Taguchi, G., Elsayed, E, and Hsiang, T. (1989), *Quality Engineering in Production Systems*, McGraw-Hill Book Company, New York.

Taguchi, G., Chowdhury, S., and Taguchi, S. (2000), *Robust Engineering*, McGraw Hill, New York.

Taguchi, G., Chowdhur, S., and Wu, Y. (2005), *Taguchi's Quality Engineering Handbook*, John Wiley & Sons, Inc., Hoboken, NJ.

Tamhane, A. C. (2009), *Statistical Analysis of Designed Experiments Theory and Application*, Wiley, New York, 720 pages; ISBN: 978-0-471-75043-7.

Taylor, W. (1991), *Optimization and Variation Reduction in Quality*, McGraw Hill, New York; ISBN: 0-07-063255-3.

Thomke, S. H. (2003), *Experimentation Matters: Unlocking the Potential of New Technologies for Innovation*, Harvard Business School Press, Boston, MA.

Toutenburg, H. and Shalabh (2009), *Statistical Analysis of Designed Experiments*, 3rd Edition, Springer, New York, 633 pages; ISBN: 978-1441911476.

Vuchkov, I. N. and Boyadjieva, N. L. (2001), *Quality Improvement with Design of Experiments: A Response Surface Approach*, Springer, Dordrecht, 508 pages; ISBN: 978-1402003929.

Watson, G. H. (1993), *Strategic Benchmarking: How to Rate Your Company's Performance against the World's Best*, ASQC Quality Press, Milwaukee, WI.

Weber, D. C. and Skillings, J. H. (1999), *A First Course in the Design of Experiments: A Linear Models Approach*, CRC Press, New York.

Winer, B. J. (1971), *Statistical Principles in Experimental Design*, 2nd Edition, McGraw-Hill, New York.

Wu, C. F. J. and Hamada, M. S. (2009), *Experiments: Planning, Analysis and Parameter Design Optimization*, 2nd Edition, Wiley, New York; ISBN: 978-0471699460.

Yates, F. (1937), *The Design and Analysis of Factorial Experiments*, Imperial Bureau of Social Science, Harpenden, England.

Index

About the Authors

Young J. Chiang, PhD The author earned his BS degree from the Department of Power Mechanical Engineering, Tsing Hua University, Taiwan, in 1976. When studying his PhD degree (Department of Engineering Mechanics, University of Wisconsin-Madison), the author was honored to take two graduate courses from professor George Box in the early 1980s as part of his minor program (Department of Statistics, University of Wisconsin-Madison). He obtained his PhD degree in 1983. Ever since, Dr. Chiang has had opportunities to apply design of experiments in combination with finite element methods (computer simulation) and experimental mechanics (experimental life tests) to improve product performance and solve complex reliability problems in the automotive industry. He has 30 years of engineering and management experiences with product R&D, CAE, design, reliability, quality, manufacturing, and program management in the US automotive industry, including Uniroyal (Michelin), Cummins, Magna, Edscha, Danaher, CTS, Coda, and Chrysler. Dr. Chiang was a member of ASME and SAE. After retiring from the US automotive industry, he worked as the vice director of Weichai Power Research Center (Shandong, China) and founded Weichai Institute of Reliability Research (>100 research engineers). He was also named the vice director of China Key State Laboratory of Engine Reliability and the executive commissioner on Reliability Committee, Chinese Mechanical Engineering Society.

Besides technical papers and reports, Dr. Chiang has put forth a great effort into a nonprofit business engaged in the development of potential automotive engineering textbooks for graduate students and practitioners in the automotive industry, ever since the early 1990s. He has published two books on how to improve product performance and grow product reliability based on physics of failure and material constituents: *Mechanics and Design for Product Life Prediction* (ISBN 978-7-5689-1917-6) and *Automotive Engineering Materials—Thermomechanical Properties* (ISBN 978-7-5689-3293-6) with Chongqing University Press (Chongqing, China). This publication (*Fundamentals of Design of Experiments for Automotive Engineering*) is the first volume of the book series on *Design of Experiments for Product Reliability Growth* brought forth by the author. The corresponding manuscript was even used as a textbook-to-be by the author, in the capacity of full professor, for teaching graduate students in the Department of Automotive Engineering, Chongqing University.

Young J. Chiang, PhD

Amy L. Chiang, MS After the author obtained her BS degree in electrical engineering from the University of Michigan, Ann Arbor, in 2012, she works for Viasat, Inc. (Carlsbad, CA) with much of her focus on product development and testing. While working for the company, she completed the MS degree of "Masters of Advanced Study Degree (MAS) in Wireless Embedded Systems," offered by the Departments of Electrical and Computer Engineering and Computer Science and Engineering at the University of California-San Diego.

The author is an active participant in businesses engaged in the development of new technologies and multidisciplined products. With insights into viable paradigms and advancements in practice, she assists product teams in developing wireless embedded software, program strategy, and methodology of statistical thinking.

Amy L. Chiang, MS

www.ingramcontent.com/pod-product-compliance
Lightning Source LLC
Chambersburg PA
CBHW040140200326
41458CB00025B/6327